www.ingramcontent.com/pod-product-compliance
Lightning Source LLC
Chambersburg PA
CBHW050451200326
41458CB00014B/5137

Meteorology: An Atmospheric Science

Meteorology: An Atmospheric Science

Edited by Dorothy Rambola

SYRAWOOD
PUBLISHING HOUSE

New York

Published by Syrawood Publishing House,
750 Third Avenue, 9th Floor,
New York, NY 10017, USA
www.syrawoodpublishinghouse.com

Meteorology: An Atmospheric Science
Edited by Dorothy Rambola

International Standard Book Number: 978-1-68286-758-7 (Hardback)

Cataloging-in-Publication Data

Meteorology : an atmospheric science / edited by Dorothy Rambola.
    p. cm.
Includes bibliographical references and index.
ISBN 978-1-68286-758-7
1. Meteorology. 2. Atmosphere. 3. Atmospheric physics. I. Rambola, Dorothy.
QC861.3 .M48 2019
551.5--dc23

TABLE OF CONTENTS

PREFACE

Meteorology is a branch of the atmospheric sciences that is concerned with weather forecasting. The different variables of the Earth's atmosphere such as temperature, air pressure, water vapor and mass flow are subject to change with time and their interaction with each other. These changes contribute to variations in the weather. The quantities of temperature, pressure, humidity and wind are measured by using the thermometer, barometer, hygrometer and the anemometer. Air quality sensors for analyzing carbon monoxide, methane, carbon dioxide, ozone, dust or smoke in the air are also widely used. Flood sensor, rain gauge, seismometer, lightning sensor, etc. are other instruments that are crucial for gathering meteorological data. The applications of meteorology are in agriculture, military, energy production, transport and construction. This book is a compilation of chapters that discuss the most vital concepts and emerging trends in the field of meteorology. It aims to shed light on some of the unexplored aspects and the recent researches in this field. It is a collective contribution of a renowned group of international experts. Those in search of information to further their knowledge will be greatly assisted by this book.

Various studies have approached the subject by analyzing it with a single perspective, but the present book provides diverse methodologies and techniques to address this field. This book contains theories and applications needed for understanding the subject from different perspectives. The aim is to keep the readers informed about the progresses in the field; therefore, the contributions were carefully examined to compile novel researches by specialists from across the globe.

Indeed, the job of the editor is the most crucial and challenging in compiling all chapters into a single book. In the end, I would extend my sincere thanks to the chapter authors for their profound work. I am also thankful for the support provided by my family and colleagues during the compilation of this book.

Editor

Informativeness of wind data in linear Madden–Julian oscillation prediction

Theodore L. Allen,[1]*[†] Brian E. Mapes[1] and Nicholas Cavanaugh[2]

[1]*Department of Meteorology and Physical Oceanography, Rosenstiel School of Marine and Atmospheric Science, University of Miami, FL, USA*
[2]*Climate and Ecosystem Sciences, Lawrence Berkeley National Laboratory, Berkeley, CA, USA*

*Correspondence to:
T. L. Allen, The International
Research Institute for Climate
and Society, Columbia University,
61 Route 9 W, Palisades, NY
10964–1000, USA. E-mail:
tallen@iri.columbia.edu

†Currently at The International
Research Institute for Climate
and Society, Columbia University.

Abstract

Linear inverse models (LIMs) are used to explore predictability and information content of the Madden–Julian Oscillation (MJO). Hindcast skill for outgoing longwave radiation (OLR) related to the MJO on intraseasonal timescales in the tropics has been examined for a variety of LIMs using OLR and optionally 200 and 850 hPa zonal wind information channels. The dependence of OLR hindcast skill on wind channels was evaluated by randomizing in time, averaging in space, or omitting data entirely. Results show positive prediction skill (relative to climatology) up to 3 weeks and wind information, mostly at the largest scales, adds 1–2 days of skill.

Keywords: linear inverse modeling; Madden–Julian Oscillation; sub-seasonal prediction

1. Introduction

The Madden–Julian Oscillation (MJO) is an intraseasonal zonally propagating atmospheric signal in tropical rainfall and related fields (Madden and Julian, 1971; Madden and Julian, 1972; Zhang, 2005; Wang, 2006; and Lau and Waliser, 2011). Besides impacting weather directly in the tropics it also has impacts in the extra-tropics through teleconnections and can impact short-term climate events (Martin and Schumacher, 2011; Zhang, 2013). The long timescale of the MJO suggests it could be a source of extended-range predictability.

Linear statistical models can have comparable MJO prediction skill to dynamical models (Newman *et al.*, 2009; Xavier *et al.*, 2014; Klingaman and Woolnough, 2014) and offer unique opportunities for decomposition that may reflect on the MJO's incompletely understood dynamics. Cavanaugh *et al.* (2014, hereafter C14) explored the skill of linear inverse models (LIMs) in hindcasting the MJO, and this article extends and complements that work methodologically and scientifically. Klingaman and Woolnough (2014) include LIM results from both C14 and the methods described here, as a baseline for evaluating numerical model hindcasts. Those hindcasts as well as C14 were scored in the time-longitude (latitudinally averaged) space of Wheeler and Hendon (2004)'s Realtime Multivariate MJO index (RMM) encompassing Outgoing Longwave Radiation (OLR) and zonal wind at 850 and 200 hPa (u850 and u200). The relevance of including wind field information in addition to OLR has been questioned, for both MJO definition and diagnosis (Kiladis *et al.*, 2014) and for aspects of forecasting such as initiation of a new MJO event (Straub, 2013), whose final paragraph notes that "the RMM index is dominated by its circulation components. However, the clouds and rain are of special interest for impacts, so we will score all hindcasts in terms of OLR anomaly (OLR')".

The goals of this article are: (1) to illustrate the workings of LIMs in a more intuitive physical channel space (time-longitude sections), rather than C14's truncated space of empirical orthogonal functions (EOFs); (2) to explore how close the resulting large number of channels brings us to the problem of statistical overfitting (the usual justification for such EOF truncation approaches); and (3) to estimate the value of wind information and small-scale information in statistical predictions of intraseasonal cloudiness signals, and consider the implications as a potential partial clue to MJO dynamics.

2. LIM summary and statistical forecasting issues

Linear inverse modeling (Penland and Sardeshmukh, 1995) is a generalization of the simple idea that anomalies in a stationary time series decay with time – exponentially, in the case of a postulated system obeying $dx/dt = -\mathbf{B}X + \text{noise}$. In a multi-channel LIM (where a 'channel' is meant in the sense of an information stream, i.e. a time series or a column in a dynamical state vector), anomalies can oscillate and propagate among channels as they decay, because the complex exponential function has those behaviors in addition to the simple decay of the real exponential function.

In fitting a LIM from data, one postulates that those data came from a linear stochastically forced system with the form:

$$\frac{dx}{dt} = \mathbf{B}\mathbf{X} + \text{noise} \qquad (1)$$

where the state vector $\mathbf{X}$ comprises m columns of anomaly values and $\mathbf{B}$ is an $m \times m$ matrix. All linearly predictable dynamical interactions among the system variables are represented in the linear operator $\mathbf{B}$, also known as the deterministic linear feedback matrix or the system sensitivity matrix (Shin et al., 2010). It can be shown for a system of form (Equation (1)) that if the noise term is white (uncorrelated in time, but not necessarily uncorrelated among channels) and Gaussian, then for any specific time lag τ_0, $\mathbf{B}$ is related to the time-lagged covariance matrix $\mathbf{C}(\tau_0)$ by:

$$\mathbf{B} = \frac{1}{\tau_0} \ln \left[\mathbf{C}\left(\tau_0\right) \mathbf{C}\left(0\right)^{-1} \right] \qquad (2)$$

This result is formally identical to how one would estimate a decay coefficient from lagged autocorrelation in a univariate ODE, but here $\mathbf{C}$ and $\mathbf{B}$ are matrices and the $\ln[\cdot]$ function is the matrix generalization of the ordinary logarithm. The optimum forecast (indicated by the caret) for such a system, optimal in the sense of minimizing squared error, is:

$$\hat{\mathbf{x}}(t + \tau) = \exp\left(\mathbf{B}\tau\right) \mathbf{x}(t) = \mathbf{G}_\tau \mathbf{x}(t) \qquad (3)$$

where $\mathbf{G}_\tau$ is known as the propagator matrix that evolves initial anomalies, $\mathbf{x}(t)$, forward by any desired lead time (τ).

When working from a finite, real-world data sample, we must view the $\mathbf{B}$ obtained from Equation (2) as an estimate, and view Equation (1) as a postulate of how the real world (which generated the data) acts. One can estimate $\mathbf{B}$ from Equation (2) for various training lags τ_0. The similarity of these various estimates for $\mathbf{B}$ has been viewed as a test (called the 'tau test') of the validity of the postulate that the form (Equation (1)) characterizes the real system adequately (Penland and Sardeshmukh, 1995). Referring again to the simpler univariate case: if the autocorrelation decay rate estimated at different lags is similar, then indeed the decay curve must be close to exponential, which bolsters the case that the data-generating system acts like the simple linear decay equation being postulated (or fitted). We found that Equation (3) gives similar hindcast skill using $\mathbf{B}$ matrices estimated from τ_0 values ranging from 1 to 4 days in Equation (2) (not shown), supporting the LIM approach. Only a little more skill is gained by using lagged regression (LR) rather than LIM (not shown). In LR, one estimates $\mathbf{C}$ separately for each forecast lead time τ, so that Equation (3) simply becomes $\hat{\mathbf{x}}(t + \tau) = \mathbf{C}(\tau) \mathbf{C}(0)^{-1} \mathbf{x}(t)$. LIM results offer similar scientific lessons to LR (not shown), but with more elegance and simplicity, and so will be the main focus of this article.

3. Data and experiments

This study utilizes time-longitude sections of daily data from 1979 to 2011, including interpolated outgoing long wave radiation (OLR) observed from satellites (Liebmann and Smith, 1996) along with zonal wind u at the 850 and 200 hPa levels derived from the NCEP-NCAR Reanalysis project (Kalnay et al., 1996). Each variable was averaged from 15°S to 15°N. A 25-year composite annual cycle (1979–2004) was removed from each variable to produce anomalies, and a 120-day mean prior to each day was subtracted to remove low frequency signals, following Wheeler and Hendon (2004), which means that usable data begins 120 days into the time series. Each channel in the training set thus consists of a time series of more than 7000 daily observations. These high-passed anomalies are denoted with a prime, for example OLR'. These data contain many kinds of variability, but the hindcast skill here mostly bears the hallmarks of the MJO (timescale and eastward propagation), so we have used that moniker in the text and title.

It is helpful to define a LIM baseline or control case: all three variables, in 15 degree longitude bins, using $\tau_0 = 2$ days. Each of the 24 longitude bins thus contains three channels consisting of a daily time series of anomalous MJO index variables (OLR', u850' and u200'). In summary, the baseline LIM has a total of 72 input channels, 3 for each longitude bin. For clean comparisons to this baseline, including wind information denial experiments, we choose to score the hindcasts based on the twenty-four 15-degree OLR bins only. When other channels (u850', u200') are used, their impact is evaluated only in terms of the OLR prediction skill. Likewise, when additional longitude fine structure is included, we score its effect only on 15 degree scale OLR.

In many LIM studies (including C14), principal component series truncation has been used to minimize channel numbers while maximizing the variance represented. However, this encoding of the information channels makes a LIM's workings somewhat opaque. Following the examples of Shin et al. (2010) and Hakim (2013), we leave our channels as spatial boxes, and furthermore they are ordered by adjacent longitudes, so LIM forecasts and their errors can simply be contoured in longitude-time space.

We train LIMs on data from 1979 to 1999 and verify on the independent set 2000–2011, thus eliminating any chance of artificial skill (DelSole and Shukla, 2009). Seasonal masks are also optionally applied to the training period to seek an optimal LIM construction for verification in that specific season.

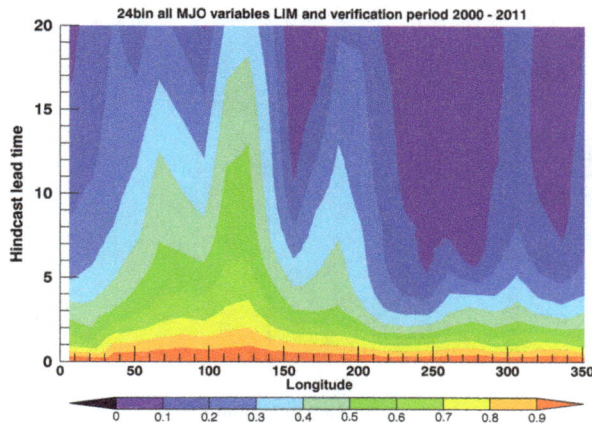

Figure 1. OLR anomaly hindcast correlation skill score for each of the 24 longitude bins from the baseline LIM.

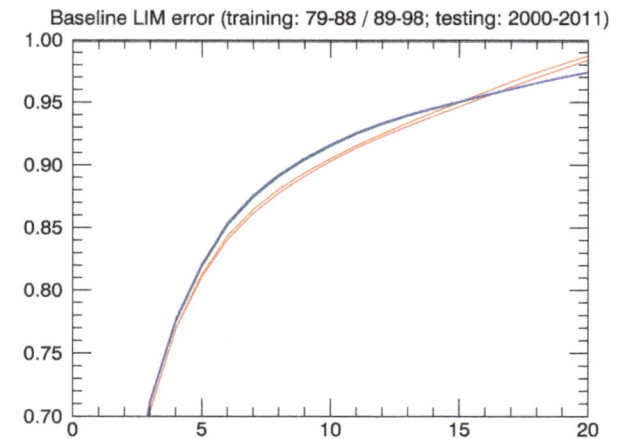

Figure 2. Squared error of the baseline (red) and 144 channel (blue) LIMs trained on two consecutive non-overlapping 10-year epochs between 1979 and 1998. Two curves are shown in each color; their (very close) spacing indicates the level of accuracy for further deductions involving hindcast skill differences in subsequent figures.

4. Results

4.1. Longitude dependence of OLR' predictability

LIM skill can be displayed as a function of longitude (Figure 1). Figure 1 illustrates the correlation coefficient between predicted OLR anomalies from the baseline LIM and observed OLR anomalies for the 2000–2011 period. Regional differences in skill are evident. The highest correlation at all lead times is found in the region of the maritime continent between 110°E and 130°E. Here, the correlation remains above +0.6 for 6 days and remains above +0.5 for 13 days. By contrast, the east Pacific region (longitude 230 in Figure 1) has the lowest one week hindcast correlation skill ($r < +0.1$). Summarizing Figure 1, three hindcast skill hot-spot peaks are identified within the central Indian Ocean, the maritime continent and central Pacific, with areas of low predictive skill from the east Pacific to the Pacific coast of Central America. These results are consistent with the notion that the MJO is the basis of long-range prediction skill, as presupposed in our title, even though the prediction is really for OLR' including all phenomena.

To have a single scalar skill score, we define the verification error score (to be minimized) in future experiments as a global sum of squared OLR' hindcast errors. The no-skill asymptote of this score is the global climatological variance, which is the skill of a forecast of zero anomaly every day (climatology used as a forecast).

4.2. Impacts of using a large number of channels

Is a 72-channel LIM too large? That is, will statistical overfitting of so many coefficients from finite training data samples lead to poor skill when tested on independent data? The skill of our baseline results, comparable to results from C14's reduced EOF space (Klingaman and Woolnough, 2014), suggests that the answer is no. To further address this question, we push the numbers much further by doubling the number of

longitude bins from 24 to 48, making each longitude bin 7.5 rather than 15 degrees wide. This doubles the number of channels from 72 to 144, quadrupling the number of coefficients in the **G** and **B** matrix estimates. Furthermore, we reduce the training set into two independent and consecutive training periods to estimate the effects of sampling error in our final results graphical space. This experiment reduced the number of data points used per coefficient estimated from 50 to 25.* The change from 15 degree to 7.5 degree longitude resolution has a negligible effect on hindcast error (black curves are only slightly above the red curves in Figure 2), thus providing little evidence of overfitting at 7.5 degree longitude bin resolution. However, overfitting becomes steeply worse once longitude resolution increases to 2.5 degrees (8 data values per coefficient, not shown).

4.3. Estimated value of wind information: illustrating randomization method

Excluding u850 and u200 from the LIM provides physical insight regarding the impact of wind anomalies on the prediction of OLR anomalies associated with the MJO (Figure 3). The skill score omitting or randomizing wind channels during training and hindcasting (blue and green curves) is compared to the baseline LIM (red curves repeated from Figure 2). Figure 3 indicates that the wind information in the baseline LIM does indeed contribute unique and valuable information content to the LIM for OLR' prediction out to 15 days, a result clearly significant with respect to sampling noise (the

---

*Of course, the statistically important information measure is independent degrees of freedom per coefficient estimated, not data points per coefficient. But daily timescale OLR is not excessively autocorrelated, and the rules of thumb for relating degrees of freedom to data points using autocorrelation are imprecise and debatable, so we use data points here for simplicity.

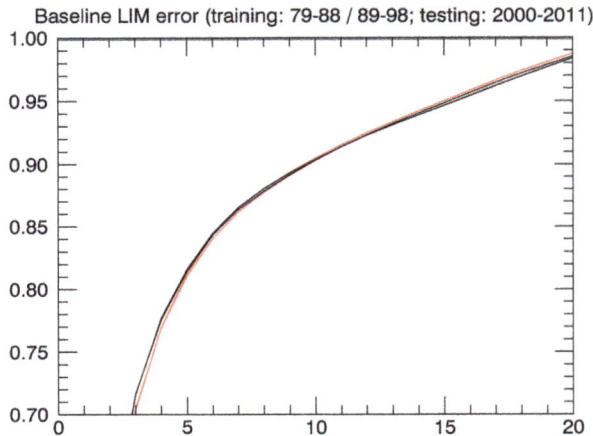

Baseline LIM error (training: 79-88 / 89-98; testing: 2000-2011)

Figure 3. As in Figure 2 but for an OLR only LIM (blue), the baseline LIM with scrambled winds (green), and the baseline LIM (red). Zoomed in insert provided as well.

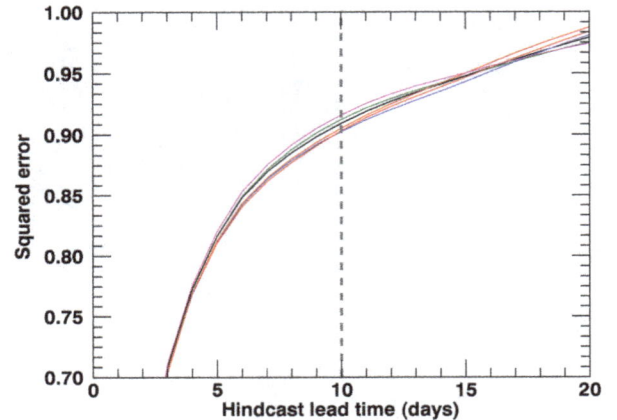

Figure 4. Squared error for five LIMs calculated using 15 degree longitude bins and trained on the following data channels. Listed in order from highest error to lowest error at a 10-day hindcast lead time are (1) OLR only (magenta), (2) OLR and wave numbers 1 and 2 for u850 and u200 (green), (3) OLR and zonal mean wind (black), (4) OLR, zonal mean winds, and wave numbers 1 and 2 (blue), and (5) the baseline LIM with all wind information (double red curves, for two non-overlapping half-length training periods). The vertical dashed grey line provides a visual reference at the 10-day lead time. Zoomed in insert provided as well.

gap between the red curves). Furthermore, we can conclude that omitting vs. randomizing the wind channels has a very similar effect, indicating again that our LIMs are far from the danger of overfitting, with the modest number of channels and large amounts of training and verification data used here.

4.4. Decomposition of the value of wind information

We also want to know what aspects of the wind field contain the important information content for predicting OLR: the zonal mean wind, the first two zonal wavenumbers (indicative of large-scale convergence and divergence) or other aspects? To address this question, we construct a 30-channel LIM built from an anomalous state vector with OLR' for each 15 degree longitude bin (24 channels), the zonal mean from u850' and u200' (two channels), and the first two wavenumbers of u850' and u200' (four channels). We apply the channel randomization technique introduced in the previous section to test the impact of various wind channels on the prediction of OLR'.

To partition the wind channels' information content, Figure 4 shows results from four 30-channel LIMs: (1) OLR data only (randomized zonal mean and the first two wavenumbers of u850 and u200), (2) OLR and the first two wavenumbers (randomized zonal means), (3) OLR and the zonal means (randomized wavenumbers 1 and 2), and (4) OLR with all six wind components. For reference, the 'baseline' LIM with all wind information at 7.5 degree scale is repeated in red, with two lines trained from independent halves of the training epoch as an indicator of the statistical significance of difference due to finite-sample effects.

The OLR' only LIM exhibits the worst skill (greatest error, magenta curve in Figure 4 up to 2 weeks lead time) while the 'baseline' is the best (red curves). The cases in between essentially map the amount of useful information content in various aspects of u850' and u200'. Results may be summarized as follows:

- Value of all wind information ('baseline' vs. OLR only): 1.5−2 days of additional skill between 5 and 10 day hindcast lead time (red curves on Figure 4; as in Figure 3)
- Zonal means plus wavenumbers 1 + 2: about the same as the 'baseline' LIM (blue curve on Figure 4)
- Zonal means alone: about 1/2 of total wind signal (black curve on Figure 4)
- Wavenumbers 1 + 2 alone: about 1/3 of total wind signal (green curve in Figure 4)

The contributions of 'about 1/2' and 'about 1/3' need not sum to unity, because the channels are not orthogonal, merely linearly independent. In particular, wavenumbers 1 + 2 may contain sample-specific noise as well as robustly useful signal, and overfitting of that noise in the training period could yield lower skill (merely 1/3 instead of 1/2 of the total value of all wind information) in the evaluation period.

4.5. Seasonality of predictability

The MJO is seasonally strongest in the northern autumn and winter seasons, with summer intraseasonal variability sometimes given a different name such as MISO or BSISO (Kikuchi et al., 2012; Sharmila et al., 2013). Our findings so far suggest that our dataset is plentiful enough to give robust results even if subdivided. Might OLR' hindcast skill be increased if the training set and verification are confined to certain seasons, rather than pooling all data? To test this notion, the baseline LIM from Figure 2 was subdivided into an all season and boreal winter (DJF) datasets for both training and evaluation. OLR' hindcast skill is much better for DJF training and scoring, compared to the 72 channel all-season LIM (black vs red in Figure 5). If a

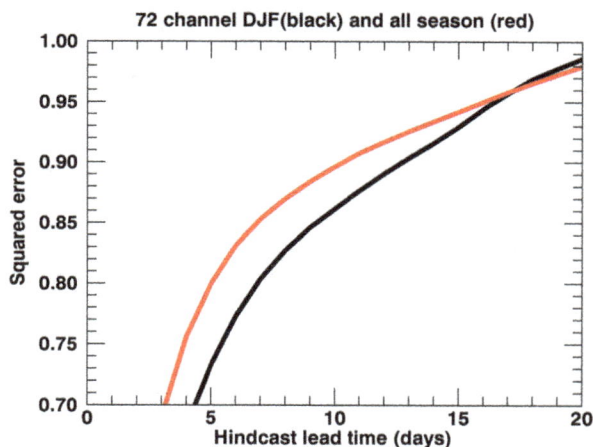

Figure 5. As in Figure 2 but for a LIM trained with DJF seasonal (black) and all season (red) LIM.

normalized squared error of 0.85 is used as a no-skill threshold, that is achieved after 7 days for the all season LIM and 9 days for the DJF LIM. Alternatively, if differences are measured vertically on the graph, for a 1-week lead time prediction, the control LIM has a 7% increase in OLR anomaly hindcast error compared to the DJF trained LIM from a 1-week hindcast lead time. All squared errors asymptote toward 1 after a 14-day hindcast lead time. In summary, it is preferable to train the LIM on less data, but on the proper season (DJF), rather than a using a larger set of training data including data from all seasons.

5. Conclusions

Hindcast skill for 15°N–15°S averaged OLR' in time-longitude space has been examined for a variety of LIM models using daily OLR data and optionally 200 and 850 hPa zonal wind from 1979 to 2011. Results show some positive prediction skill (relative to climatology) up to 3 weeks, consistent with Cavanaugh *et al.* (2014) who used a LIM built with a reduced channel space of EOFs from maps (not just latitude belt averages). Klingaman and Woolnough (2014) shows that the present approach, with many more channels but simpler spatial interpretation, is just as skillful, or even more so for the Year of Tropical Convection intercomparison case presented there.

The dependence of OLR' anomaly prediction skill on information in other channels of the dynamical state vector was evaluated. LIM predictive skill is robust to the number of input channels up to 144, giving similar skill whether excess channels are omitted or randomized in their time ordering. Using 15-degree longitude bins for the 3-variable LIM results in a 72×72 lagged covariance matrix consisting of 5184 coefficients fitted to two training periods between 1979 and 1999 (3652 days $\times$ 72 channels), or about 50 data values per coefficient. Even with double the number of channels (7.5 degree bins; 1/2 as many data-per-coefficient), no

skill loss (evidence of overfitting) was evident. Skill loss and possible overfitting is finally evident with 2.5 degree longitude bin channels or in this case at about 8 data values per coefficient.

Wind data (u850' and u200') adds 1.5–2 days of additional skill. Wavenumbers 1 and 2 contribute less than that in isolation (perhaps because they also contribute 'noise' distractions: sample-dependent patterns that are not repeatable in the verification period). All higher wavenumbers contribute negligibly (the blue line is statistically indistinguishable from the red lines in Figure 4).

In general, winds and OLR are correlated predictors, so their information content is mixed and they are not cleanly separable despite the labels which sound like they are two independent physical quantities. The results here do not necessarily shed light on fundamental MJO dynamics. Combined-variable indices always have debatable relative normalizations for the different variables (Liu *et al.*, 2016). From this study's point of view, predicting anomalous clouds and rain (OLR), the wind information may help the system avoid being misinterpreted by happenstance occurrences of MJO-shaped equatorial cloud patterns that are not actually part of a predictable wave in the real physical memory variables (inertia or water vapor or perhaps SST). Predictability can be limited as much by the strength of distractions and noise as it is by the dynamics of the predictable subsystem, so results about predictability may be results about such noise, not about dynamics. LIM OLR' hindcast prediction error is better during the DJF season and in Indo-Pacific longitudes, both indicative of the region and season where the MJO contributes the most to OLR anomalies and where tropical OLR variance is greatest.

Acknowledgements

The authors gratefully acknowledge financial support from NSF grant 0731520, NASA CYGNSS grant NNX13AQ50G, ONR grant N000141310704, DOE grant DE-SC0006806, NOAA grant NA13OAR4310156, and Government of India EarthMM/SERP/Univ_Miami_USA/2013/INT-1/002. The authors are also grateful for the two anonymous reviewer's constructive comments.

References

Cavanaugh N, Allen T, Subramanian A, Mapes B, Miller AJ. 2014. The skill of tropical linear inverse models in hindcasting the Madden-Julian Oscillation. *Climate Dynamics* **44**: 897–906, doi: 10.1007/s00382-014-2181-x.

DelSole T, Shukla J. 2009. Artificial skill due to predictor screening. *Journal of Climate* **22**(2): 331–345, doi: 10.1175/2008JCLI2414.1.

Hakim GJ. 2013. The variability and predictability of axisymmetric hurricanes in statistical equilibrium. *Journal of the Atmospheric Sciences* **70**(4): 993–1005, doi: 10.1175/JAS-D-12-0188.1.

Kalnay E, Kanamitsu M, Kistler R. Collins W, Deaven D, Gandin L, Iredell M, Saha S, White G, Woollen J, Zhu Y, Leetmaa A, Reynolds R, Chelliah M, Ebisuzaki W, Higgins W, Janowiak J, Mo KC, Ropelewski C, Wang J, Jenne R, Joseph D. 1996. The NCEP/NCAR 40 year reanalysis project. *The Bulletin of the American Meteorological Society* **77**: 437–471.

Kikuchi K, Wang B, Kajikawa Y. 2012. Bimodal representation of the tropical intraseasonal oscillation. *Climate Dynamics* **38**(9–10): 1989–2000, doi: 10.1007/s00382-011-1159-1.

Kiladis GN, Dias J, Straub KH, Wheeler MC, Tulich SN, Kikuchi K, Weickmann KM, Ventrice MJ. 2014. A comparison of OLR and circulation-based indices for tracking the MJO. *Monthly Weather Review* **142**(5): 1697–1715, doi: 10.1175/MWR-D-13-00301.1.

Klingaman NP, Woolnough SJ. 2014. The role of air-sea coupling in the simulation of the Madden-Julian Oscillation in the Hadley Centre Model: air-sea coupling and the MJO. *Quarterly Journal of the Royal Meteorological Society* **140**(684): 2272–2286, doi: 10.1002/qj.2295.

Lau KH, Waliser DE. 2011. Intraseasonal Variability in the Atmosphere-Ocean Climate System. Praxis, 646 pp.

Liebmann B, Smith CA. 1996. Description of a Complete (Interpolated) Outgoing Longwave Radiation Dataset. *Bulletin of the American Meteorological Society* **77**: 1275–1277.

Liu P, Zhang Q, Zhang C, Zhu Y, Khairoutdinov M, Kim H-M, Schumacher C, Zhang M. 2016. A revised real-time multivariate MJO index. *Monthly Weather Review* **144**: 627–642, doi: 10.1175/MWR-D-15-0237.1.

Madden RA, Julian PR. 1971. Detection of a 40–50 day oscillation in the zonal wind in the tropical Pacific. *Journal of the Atmospheric Sciences* **28**: 702–708, doi: 10.1175/1520-0469(1971)028<0702:DOADOI>2.0.CO;2.

Madden RA, Julian PR. 1972. Description of global-scale circulation cells in the tropics with a 40–50 day period. *Journal of the Atmospheric Sciences* **29**: 1109–1123, doi: 10.1175/1520-0469(1972)029,1109:DOGSCC.2.0.CO;2.

Martin ER, Schumacher C. 2011. Modulation of Caribbean precipitation by the Madden–Julian Oscillation. *Journal of Climate* **24**(3): 813–824, doi: 10.1175/2010JCLI3773.1.

Newman M, Sardeshmukh PD, Penland C. 2009. How important is air–sea coupling in ENSO and MJO evolution? *Journal of Climate* **22**(11): 2958–2977, doi: 10.1175/2008JCLI2659.1.

Penland C, Sardeshmukh P. 1995. The optimal growth of tropical sea surface temperature anomalies. *Journal of Climate* **8**: 1999–2024.

Sharmila S, Pillai PA, Joseph S, Roxy M, Krishna RPM, Chattopadhyay R, Abhilash S, Sahai AK, Goswami BN. 2013. Role of ocean–atmosphere interaction on northward propagation of Indian Summer Monsoon Intra-Seasonal Oscillations (MISO). *Climate Dynamics* **41**(5–6): 1651–1669, doi: 10.1007/s00382-013-1854-1.

Shin S-I, Sardeshmukh PD, Pegion K. 2010. Realism of local and remote feedbacks on tropical sea surface temperatures in climate models. *Journal of Geophysical Research* **115**(D21), doi: 10.1029/2010JD013927.

Straub KH. 2013. MJO Initiation in the Real-Time Multivariate MJO Index. *Journal of Climate* **26**(4): 1130–1151, doi: 10.1175/JCLI-D-12-00074.1.

Wang B. (ed.) 2006. *The Asian Monsoon*. Springer, 787 pp.

Wheeler M, Hendon HH. 2004. An all-season real-time multivariate MJO index: Development of an index for monitoring and prediction. *Monthly Weather Review* **132**: 1917–1932.

Xavier P, Rahmat R, Cheong WK, Wallace E. 2014. Influence of Madden-Julian Oscillation on Southeast Asia rainfall extremes: observations and predictability. *Geophysical Research Letters* **41**(12): 4406–4412, doi: 10.1002/2014GL060241.

Zhang C. 2005. Madden-Julian Oscillation. *Reviews of Geophysics* **43**: RG2003, doi: 10.1029/2004RG000158.

Zhang C. 2013. Madden–Julian Oscillation: bridging weather and climate. *Bulletin of the American Meteorological Society* **94**(12): 1849–1870, doi: 10.1175/BAMS-D-12-00026.1.

Teleconnections and variability in observed rainfall over Saudi Arabia during 1978–2010

H. Athar*

Department of Meteorology, COMSATS Institute of Information Technology, Islamabad, Pakistan

*Correspondence to:
H. Athar, Department of
Meteorology, COMSATS Institute
of Information Technology, Park
Road, Chak Shahzad, Islamabad
44000, Pakistan.
E-mail:
athar.hussain@comsats.edu.pk

Abstract

The analyses of monthly and annual variability, and the seasonal teleconnections with the Indian Ocean Dipole (IOD), El-Niño Southern Oscillation (ENSO), and the North Atlantic Oscillation (NAO) circulation indices, for rainfall from 26 stations in Saudi Arabia (SA), for 33-year period (1978–2010), are presented. High interannual variability with non-monsoonal annual cycle characterizes the rainfall climate of SA in recent 33-year period. Eight of 26 stations display statistically significant simultaneous teleconnections with IOD and ENSO in the September–October–November (SON) season. These eight stations are situated north of 21.50°N (nSA) and are located in northern and in southeastern regions of nSA.

Keywords: Saudi Arabia; observed monthly and annual rainfall variability; seasonal rainfall teleconnections; IOD and ENSO

1. Introduction

In an arid climate and data sparse country such as Saudi Arabia, SA (see Figure 1), rainfall plays a crucial role in several important socio-economic aspects, including the water resources management and the agriculture (see, for instance, Kotwicki and Al-Sulaimani, 2009). SA alone covers almost 80% of the Arabian Peninsula area (Vincent, 2008). Understanding variability characteristics of the rainfall in such a country would allow policy makers to gain an accurately informed view of the current climatic conditions (IPCC, 2013). Analysis of observed rainfall variability is also crucial for any climate change projection study (Kotwicki and Al-Sulaimani, 2009). The Intergovernmental Panel on Climate Change (IPCC) AR5 indicates that the trends in rainfall over the SA and Arabian Peninsula region show decreasing rainfall for some of the stations, and increasing rainfall in others, during the past 30 years. However, this region has a considerable amount of incomplete or missing data. It is important, therefore, to analyze the station data more thoroughly in order to assess these trends.

The rainfall occurrence in SA may be divided into wet and dry seasons (see, for instance, Almazroui et al., 2012a, 2012b, and references cited therein). The wet season is from November to April, whereas the dry season is from June to September. During the wet/dry season, rainfall occurs predominantly in northern SA/southern SA (nSA/sSA).The nSA rainfall is only 4% of the sSA rainfall during the dry season (see, for instance, Almazroui et al., 2013). The sSA stations are influenced by the Indian monsoon during the dry season, whereas the nSA stations are impacted by relatively weak Mediterranean migratory air mass systems (Walters and Sjoberg, 1988). The rainfall mechanisms in the sSA include the modulations of the Indian monsoon system moist and hot air intrusions from the southwest as well as the northward movement of the inter tropical convergence zone (Charabi, 2009, and references cited therein). In fact, for several locations, the maximum rainfall occurs during these months, such as the stations at Khamis Mushait, Abha, and Al-Baha. Also, in the south-western region of SA, the orographic uplift as well as the moisture transport from the Red Sea contributes toward rainfall occurrence (Subyani, 2004). Essentially, all of the dry season rain falls in the south-western parts of the country and is dominated by the Indian monsoon south westerly circulations (Subyani, 2004). During the months of September, October, and November, the monsoon system retreats, and the Mediterranean systems are in seasonal transition (from June–July–August or JJA to December–January–February or DJF), thus both systems influence the rainfall occurrence in SA (Walters and Sjoberg, 1988).

The station-based rainfall characteristics in SA have been considered in several previous studies (Almazroui et al., 2012a, 2012b). AlSarmi and Washington (2011) studied the total annual station-based and regional rainfall trends (in addition to temperature) for six stations (Tabuk, Riyadh, Jeddah, Khamis Mushait, Gizan, and Dammam) over SA for a 24-year period (1985–2008). A negative trend of −20.9 mm decade^{-1} for Tabuk for the above period was noticed, which is statistically significant relative to the 24-year base period.

Several earlier studies have pointed out the impacts of the large-scale climatic indices for rainfall in neighboring regions including Middle East (Chakraborty et al., 2006; AlSarmi and Washington, 2011), Oman (Charabi, 2009), Eastern Mediterranean including Iran (Kahya, 2011), and United Arab Emirates (Kumar and Ouarda,

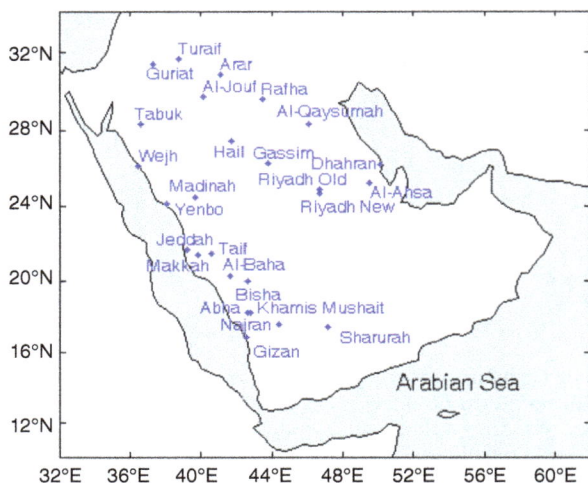

Figure 1. The geographical locations of all the 26 stations in SA from which the rainfall data are used.

2014). The purpose of this study is thus to present the monthly and annual variability characteristics of the observed rainfall over SA, in particular for the recent decade of 2001–2010, not discussed in any of the earlier studies. Furthermore, the teleconnections between the seasonal rainfall over SA at station level (for all 26 stations) and the large-scale climatic indices (Indian Ocean Dipole abbreviated as IOD, El-Niño Southern Oscillation abbreviated as ENSO, and North Atlantic Oscillation abbreviated as NAO) are also studied, for the first time, for a 33-year period (1978–2010).

2. Data and methodology

The monthly rainfall datasets were obtained from the Presidency of Meteorology and Environment (PME), Jeddah, SA, for the period 1978–2010. Table S1, Supporting Information, provides a list of the stations used in this study. The obtained rainfall datasets were subject to quality control, including but not limited to, the identification of negative and duplicate values, following Athar (2014). Figure 1 displays the geographical locations of all the 26 stations included in this analysis and indicates that there are no stations in the south-eastern region of SA, which harbors world's largest continuous sand desert, called Ruba Al'khali (Edgell, 2006).

Commonly used statistical techniques including the ordinary least squares (OLS) method for trend analysis is employed, to describe the salient variability features of the observed rainfall datasets, derived from the stations (Wilks, 2011). The OLS method consists in minimizing the sum of squares of the errors between the data points and the linearly regressed fit.

The following four seasons are considered: DJF, March–April–May (MAM), JJA, and September–October–November (SON). Additionally, the analysis is performed for the wet (NDJFMA) and dry (JJAS) seasons also. These latter seasons span over a period of 6 and 4 months, respectively (with two transitional

months), in contrast to the conventional symmetric 3-month seasons described above.

The teleconnections with the IOD, ENSO, and NAO are computed for each of the 26 stations, on a seasonal basis, for the duration 1978–2010. The monthly Indian ocean anomaly index datasets were obtained from the Japan Agency for Marine-Earth Science and Technology site (http://www.jamstec.go.jp/frsgc/research/d1/iod/). The intensity of the IOD is represented by the anomalous sea surface temperature (SST) gradient between the western equatorial Indian Ocean (50°–70°E and 10°S–10°N) and the south-eastern equatorial Indian Ocean (90°–110°E and 10°S–0°N). This gradient is named as the Dipole Mode Index (DMI). When the DMI is positive, the phenomenon is referred to as a positive IOD, and when it is negative, it is referred to as a negative IOD. The SST DMI monthly dataset derived from HadISST is used for the period 1978–2010. The monthly Niño 3.4 anomaly datasets were obtained from the climate prediction center (CPC) database of National Centers for Environmental Prediction (NCEP), based on the 1981–2010 climatology period (http://www.cpc.ncep.noaa.gov/data/indices/sstoi.indices); for more details, see Reynolds and Smith (1994) and Hurrell and Trenberth (1999). The NAO monthly anomaly index datasets were obtained from the CPC at the National Oceanic and Atmospheric Administration (NOAA) website (http://www.cpc.noaa.gov/products/precip/CWlink/zpna/znao.shtml). Among all the seasons, the magnitude of Niño 3.4 anomaly index is the largest in absolute terms in the DJF season. The monthly anomaly index data are averaged to obtain values for all the six seasons mentioned above.

3. Results and discussion

3.1. Monthly rainfall variability

The maximum rainfall (46.7 mm) was recorded by the station at Abha during March (when averaged over the data availability period). The next to maximum rainfall (44.9 mm) was also recorded by the same station (Abha) but during April (figures not shown). This station is situated in the coastal south-western region of SA at an elevation of 2100 m. A monthly rainfall variability pattern is noticeable in terms of the amount of rainfall received during the months of JJAS over SA. During this season, nSA stations receive minimal amounts of rainfall, whereas sSA stations receive maximal amounts of rainfall. When averaged over the data availability period for all the 26 stations, the total rainfall for the dry season for n(s)SA stations is 2.9 mm (69.4 mm).

Figure 2 displays the monthly rainfall over SA, based on the 26 stations. The maximal (minimal) values occur in the month of April (September), amounting to 418 mm (38.4 mm), when summed over all stations and averaged over the duration of the available dataset period for each station. Overall, the monthly total mean

Total monthly Saudi Arabia rainfall

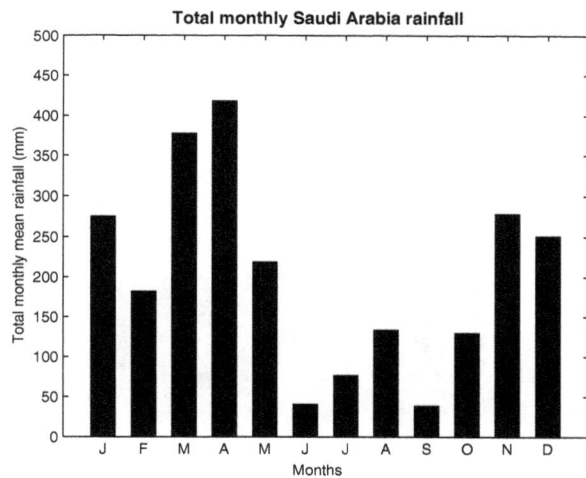

Figure 2. The total monthly mean rainfall (mm) over SA for all the 26 stations.

rainfall pattern is indicative of the non-monsoonal dominance of rainfall in SA.

3.2. Annual rainfall variability

Figure 3 displays the annual variability for the selected stations in SA, relative to the mean rainfall for each station. The displayed annual rainfall is based on the total monthly rainfall for each station. Figure 3 indicates that during the period 1978–1993, several stations received more than the climatological mean for the total annual rainfall. On the other hand, during the latter half of the past decade (2001–2010), the stations received less than the climatological mean for the total annual rainfall, except for the following stations: Wejh, Madina, Yenbo, Jeddah, and Khamis Mushait. This may relate to the occurrence of the several extreme rainfall events that occurred in the region of these five stations in recent years (see, for instance, Almazroui *et al.*, 2012a). The rainfall in SA occurs mostly in the form of short but high-intensity events, and this is reflected in large annual variability for each station in nSA as well as in sSA, when quantified in terms of standard deviation (not shown).

In terms of rainfall range analysis, the station at Abha has the maximum range (479.8 mm), whereas the minimum rainfall range is for Tabuk (79.2 mm). The maximal and minimal mean values for the total annual rainfall are for Abha and Wejh, amounting to 226.7 and 29.5 mm, respectively. For the entire rainfall dataset, station-wise, the maximum rainfall (567.3 mm) was recorded by Abha station in 1997. The minimum rainfall (0 mm) was recorded by Jeddah station (in 1984 and 1986) and by Najran station (in 1981, 1982, and 1984).

3.3. Annual rainfall trend analysis

The following nine stations indicate a positive trend using the OLS method (see Section 2 for details), in the total annual rainfall amount: Al-Jouf, Wejh, Riyadh

Old, Yenbo, Jeddah, Makkah, Najran, Sharurah, and Gizan. The first five stations are in nSA, whereas the remaining four are in the sSA (see Figure 1). Gizan (Al-Jouf) station displays the maximum (minimum) positive trend of $1.9\,\mathrm{mm\,year^{-1}}$ ($0.03\,\mathrm{mm\,year^{-1}}$). However, the maximum value for the coefficient of determination, R^2, is 0.06 for the station at Najran. All other stations, except for the nine above, display a negative trend. Abha (Taif) station displays the maximum (minimum) negative trend of $-4.2\,\mathrm{mm\,year^{-1}}$ ($-0.1\,\mathrm{mm\,year^{-1}}$). The maximum value for R^2 is 0.27, for the station at Arar. For the nSA stations, the cumulative trend for total annual rainfall is $-13.5\,\mathrm{mm\,year^{-1}}$, whereas for the sSA stations, the cumulative trend for the total annual rainfall is $-4.2\,\mathrm{mm\,year^{-1}}$. The cumulative trend for all the SA stations vis-à-vis total annual rainfall is $-17.7\,\mathrm{mm\,year^{-1}}$, based on the OLS method. Overall, it may be concluded that the negative trend is more prevalent than the positive one, and that the maximal and minimal variations in the total annual rainfall occur in sSA. The decreasing trend in the rainfall, especially for the southern stations (<21.50°N), was alluded to in the IPCC 2013 report.

Our results for the trend analysis are in general agreement with those of Nasrallah and Balling (1996), who pointed out a statistically insignificant annual negative trend for Dhahran for the duration 1940–1989. Our results are also in agreement with the recent results of AlSarmi and Washington (2011). In support of these, for the duration 1985–2008, we find a decreasing annual trend for Tabuk of $-2.3\,\mathrm{mm\,year^{-1}}$. However, after the inclusion of the extended data used in this study, we obtain a negative trend of $-0.7\,\mathrm{mm\,year^{-1}}$. This indicates relatively large annual variability in the total annual rainfall trend for Tabuk after the inclusion of the extended data set. This suggests that longer periods of data should be used in climate change assessment studies.

3.4. Seasonal teleconnections

In this sub-section, the results for the statistically significant, with confidence level (CL) $\geq 90\%$, Pearson pair-wise linear correlation coefficient (CC) teleconnections between the station rainfall amounts and IOD, ENSO, and NAO are presented on a seasonal basis, for the entire period of the study duration.

The upper (middle) panel of Figure 4 indicates that a total of 40 (29) statistically significant correlations exist between the IOD (ENSO) and the station rainfall amounts over the whole of SA, during all seasons. The results for NAO index are displayed in the lower panel of Figure 4. A total of 21 correlations exist with CL $\geq 90\%$. The IOD impact thus seems to be the dominant one. The relatively close proximity and the strength of the IOD oscillation pattern may be a contributing factor. There are both positive and negative correlations with the stations' seasonal rainfall amounts over both the nSA and the sSA, for all three indices. In nSA, the rainfall amounts seem to be better correlated with the

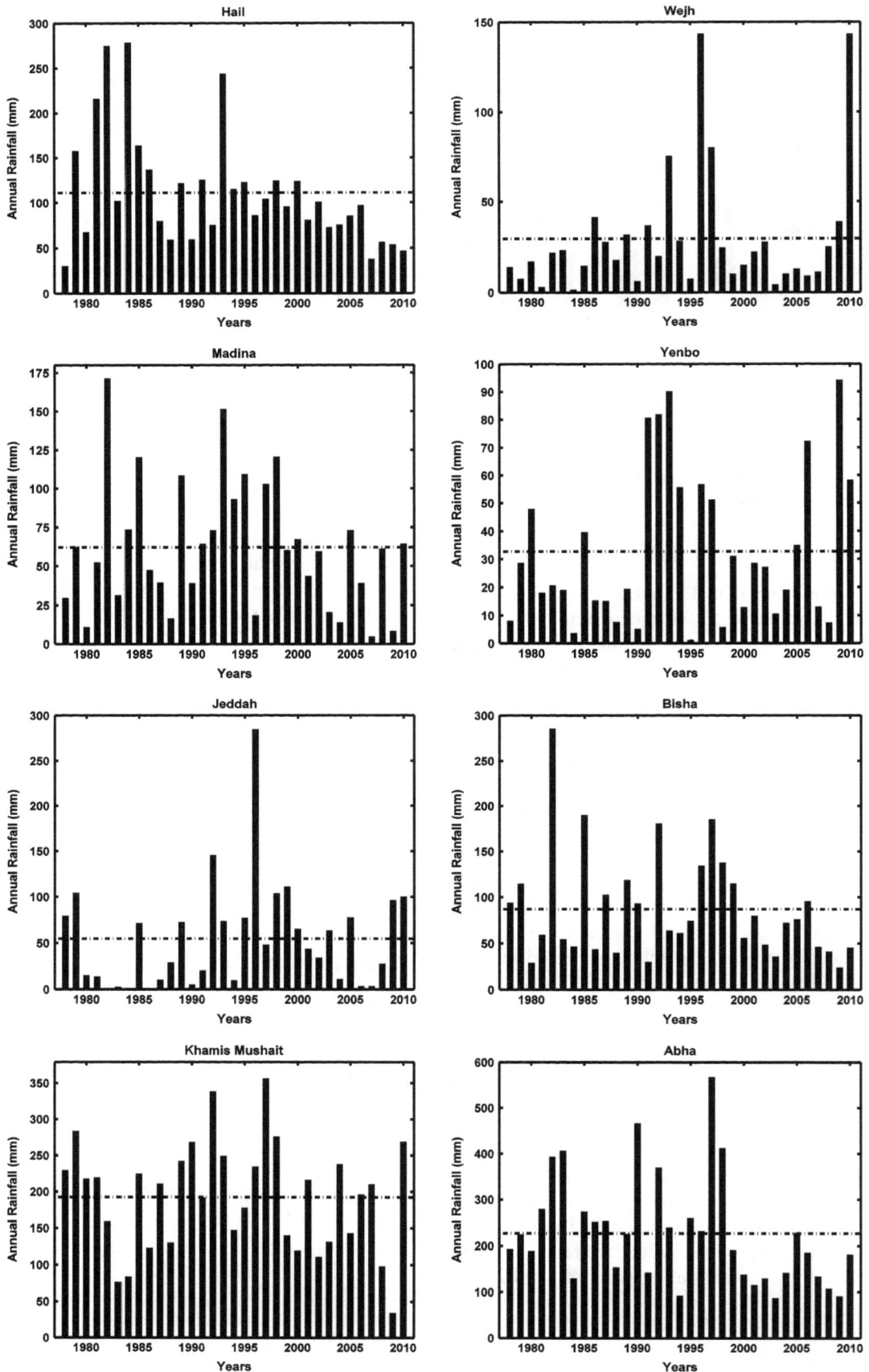

Figure 3. The annual rainfall variability for the selected stations. The horizontal dashed dotted line displays the mean annual rainfall for each station.

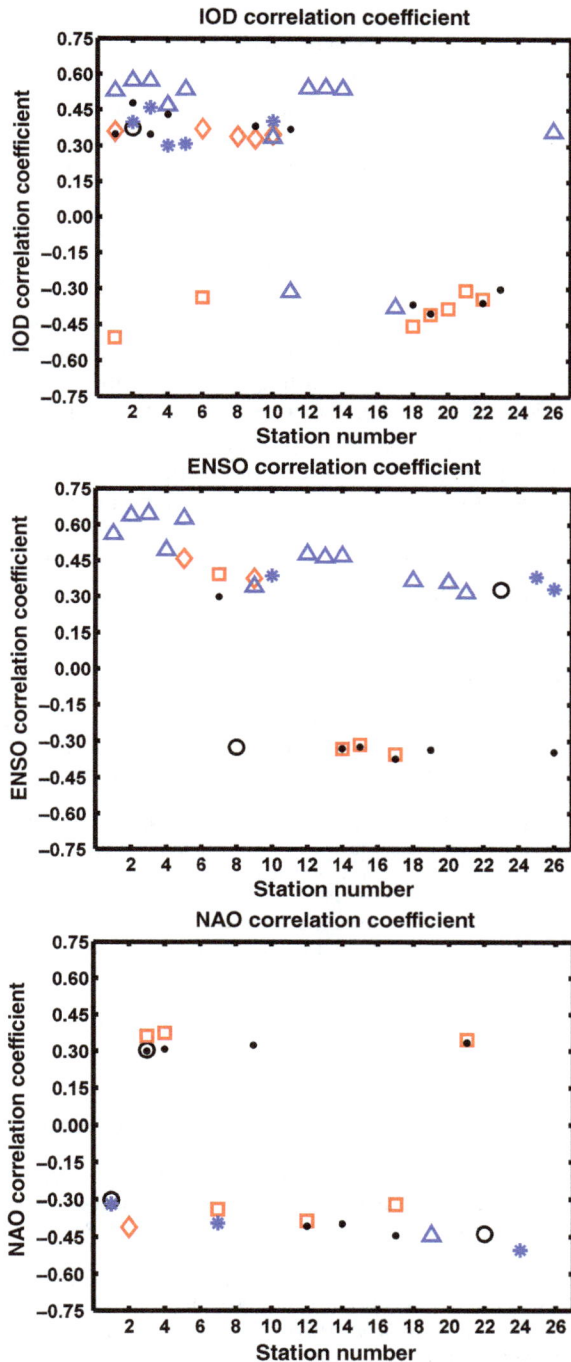

Figure 4. The correlation coefficients of the IOD (upper panel), the ENSO (middle panel), and the NAO (lower panel) indices with the station based seasonal rainfall for CL ≥ 90%. In each panel, the horizontal axis displays the station numbers. The black open circle (o) indicates the DJF season, diamond (◇) MAM season, square (□) JJA season, triangle (△) SON season, asterisk (*) the wet season, and the black filled circle (•) the dry season.

IOD and ENSO indices. For the JJA season, the dominant *negative* CC contribution is from the IOD and is absent for the ENSO and NAO indices for station numbers 18, 19, 20, 21, and 22; all these stations are *situated* in the sSA (see Table S1). These findings may have implications for the seasonal rainfall predictions.

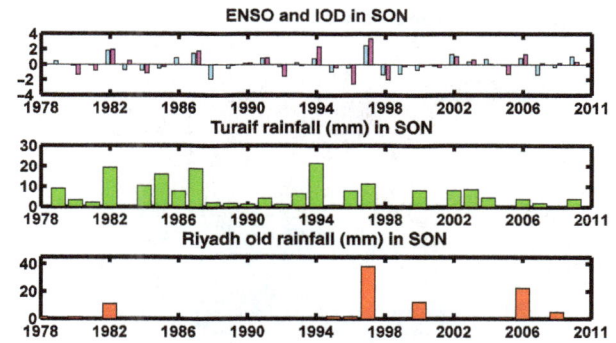

Figure 5. Upper panel: the interannual variability of the ENSO (cyan) and IOD (magenta) indices, in the SON season. Middle and lower panel: the interannual variability of the rainfall (mm) for two selected stations (Turaif and Riyadh Old), in the SON season, for the 33-year study period.

The following stations have the highest number of statistically significant correlations: station 1 and station 3 each has 7 correlations, whereas station 2 and station 4 each has 6 correlations. The highest positive CC (0.65) is for the station at Arar during the SON season and is with the ENSO index. The highest negative CC (−0.50) is for the station at Turaif (Najran) during the JJA (wet) season and is with the IOD (NAO) index. For all three circulation indices, the negative but statistically significant CCs are more apparent for the stations in sSA. Only the station at Yenbo has no statistically significant correlation in any season with any of the three indices.

Although the SON months are considered as a transitional season (see Section 1) with respect to the movement of the Indian monsoonal and Mediterranean weather systems, the IOD and the ENSO circulation indices have the highest and equal number of statistically significant CCs (both positive and negative) with the station rainfall amounts during *this* season. Thus, a possible physical mechanism for this seasonal correlation is briefly discussed next.

Figure 5 displays a comparison of time series of the IOD and ENSO indices with the rainfall amounts for two selected stations in the SON season. We note that the IOD and ENSO when in phase (for instance, positive IOD and El-Niño) leads to higher rainfall (Chakraborty *et al.*, 2006; Charabi, 2009). Figure 6 displays an example of equatorward shift in the upper level subtropical jet stream when the IOD is positive and the ENSO is in El-Niño phase, relative to when they are not, in the adjacent year.

The in phase occurrence of higher equatorial SSTs in the Indian and Pacific oceans lead to equatorward shift in the upper air subtropical jet stream and a vorticity source for the Rossby wave formation; this leads to local vorticity generation at lower levels (see, for instance, Kumar and Ouarda, 2014, and references cited therein), resulting in higher moisture flux uplift from the adjacent sea areas (see, for instance, Chakraborty *et al.*, 2006). However, even though the number of statistically significant correlations is the highest for stations in the nSA, in SON, when IOD and ENSO are in phase, not all

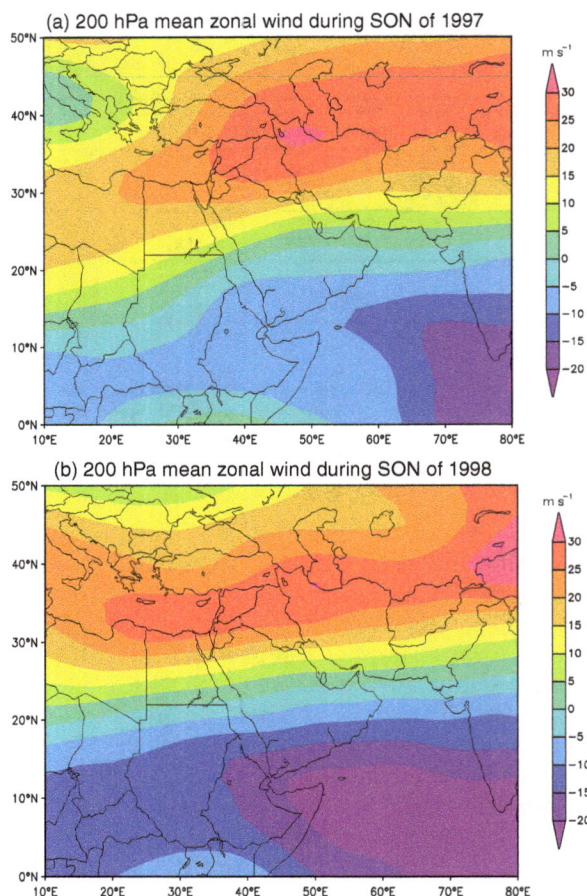

Figure 6. (a) The horizontal distribution of 200 hPa mean zonal wind for the SON season during 1997 over and in the vicinity of Saudi Arabia, (b) same as (a) except for 1998, based on NCEP/National Center for Atmospheric Research (NCAR) reanalysis data.

the CCs are positive (see Figure 4). Thus, local impacts (such as topography based) also seem to play some role in establishing these correlations. A further detailed discussion on this topic is beyond the scope of this paper.

4. Conclusions

The variability features of the rainfall over SA for the 33-year period (1978–2010) are presented using commonly used statistics from a total of 26 stations on seasonal and annual bases. The rainfall characteristics were stratified according to the conventional four seasons (DJF, MAM, JJA, and SON), as well as according to the dry (JJAS) and the wet seasons (NDJFMA). The 17 stations in nSA (latitude > 21.50°N) received minimal amounts of rainfall (2.9 mm) during the dry season, as compared to 69.4 mm for the nine stations in sSA.

Although the trends in total annual rainfall are statistically insignificant, several stations display statistically significant correlations between seasonal rainfall variations and teleconnection indexes such as the IOD, ENSO, and the NAO, at a CL ≥ 90%. The following eight (two) nSA (sSA) stations have statistically

significant seasonal CCs with all the three circulation indices: Turaif, Guriat, Arar, Al-Jouf, Gassim, Al-Ahsa, Riyadh Old, and Jeddah (Makkah and Bisha). The equatorward shift of the upper level subtropical jet stream in SON season with in phase IOD and ENSO (positive IOD and El-Niño) results in relatively higher supply of moisture flux into SA from adjacent sea areas. Also, local topographic effects may be the relevant factors contributing to these statistically significant CCs.

As noticed in several previous studies, the overall rainfall occurrence trend is decreasing (when measured over different time periods), and our findings are consistent with previous findings, in particular, for the duration 1978–2010. On a decadal basis, during the 2001–2010 decade, years in which rainfall was above the climatological mean, extreme rainfall events were dominant. Overall, the temporal evolution of the rainfall occurrence in Saudi Arabia is characterized by the relatively large variability on the annual basis. Similar large variability is noticeable on the seasonal basis as well.

References

Almazroui M, Islam MN, Athar H, Jones PD, Rahman MA. 2012a. Recent climate change in the Arabian Peninsula: annual rainfall and temperature analysis of Saudi Arabia for 1978–2009. *International Journal of Climatology* **32**: 953–966.

Almazroui M, Islam MN, Jones PD, Athar H, Rahman MA. 2012b. Recent climate change in the Arabian Peninsula: seasonal rainfall and temperature climatology of Saudi Arabia for 1979–2009. *Atmospheric Research* **111**: 29–45.

Almazroui M, Abid MA, Athar H, Islam MN, Ehsan MA. 2013. Interannual variability of rainfall over the Arabian Peninsula using the IPCC AR4 Global Climate Models. *International Journal of Climatology* **33**: 2328–2340.

AlSarmi S, Washington R. 2011. Recent observed climate change over the Arabian Peninsula. *Journal of Geophysical Research* **116**: D11109.

Athar H. 2014. Trends in observed extreme climate indices in Saudi Arabia during 1979–2008. *International Journal of Climatology* **34**: 1561–1574.

Chakraborty A, Behera SK, Mujumdar M, Ohba R, Yamagata T. 2006. Diagnosis of tropospheric moisture over Saudi Arabia and influences of IOD and ENSO. *Monthly Weather Review* **134**: 598–617.

Charabi Y. 2009. Arabian summer monsoon variability: teleconexion to ENSO and IOD. *Atmospheric Research* **91**: 105–117.

Edgell HS. 2006. *Arabian Deserts: Nature, Origin and Evolution.* Springer: Dordrecht, The Netherlands.

Hurrell JW, Trenberth KE. 1999. Global sea surface temperature analyses: multiple problems and their implications for climate analysis, modeling, and reanalysis. *Bulletin of American Meteorological Society* **80**: 2661–2678.

IPCC (Intergovernmental Panel on Climate Change). 2013. Climate change 2013: the physical science basis. In *Contribution of Working Group I to the Fifth Assessment Report of the Intergovernmental Panel on Climate Change*, Stocker TF, Qin D, Plattner G-K, Tignor M, Allen SK, Boschung J, Nauels A, Xia Y, Bex V, Midgley PM (eds). Cambridge University Press: Cambridge, UK; 1535 pp.

Kahya E. 2011. The impacts of NAO on the hydrology of the eastern Mediterranean. *Advances in Global Change Research* **46**: 57–71.

Kotwicki V, Al-Sulaimani Z. 2009. Climates of the Arabian Peninsula – past, present, future. *International Journal of Climate Change Strategies and Management* **1**: 297–310.

Kumar KN, Ouarda TBMJ. 2014. Precipitation variability over UAE and global SST teleconnections. *Journal of Geophysical Research* **119**(17): 10313–10322.

Nasrallah HA, Balling RC Jr. 1996. Analysis of recent climatic changes in the Arabian Peninsula region. *Theoretical and Applied Climatology* **53**: 245–252.

Reynolds RW, Smith TM. 1994. Improved global sea surface temperature analyses using optimum interpolation. *Journal of Climate* **7**: 929–948.

Subyani AM. 2004. Geostatistical study of annual and seasonal mean rainfall patterns in southwest Saudi Arabia. *Hydrological Science Journal* **49**: 803–817.

Vincent P. 2008. *Saudi Arabia: an Environmental Overview.* Taylor and Francis: London.

Walters KR Sr, Sjoberg WF. 1988. The Persian gulf region: a climatological study. Technical Report No. USAFETAC/TN–88/002 (AD–A222 654), USAF Environmental Technical Applications Center, Scott Air Force Base, IL.

Wilks DS. 2011. *Statistical Methods in the Atmospheric Sciences*, 3rd ed. Elsevier Publishers: New York, NY.

Bias in closed-form gamma parameter estimates for disdrometer data

Dan Brawn*

Department of Mathematical Sciences, University of Essex, Colchester, UK

*Correspondence to:
 D. Brawn, Department of
 Mathematical Sciences,
 University of Essex, Colchester,
 UK.
 E-mail: dbrawn@essex.ac.uk*

Abstract

Bias in a recently introduced closed-form estimator for gamma parameters is described via simulations over a wide range of feasible parameters. The simulated data are lower truncated and grouped with a Joss-Waldvogel disdrometer bin regime. A crude model for the bias in a ratio of adjacent moments is sufficient to eliminate most of the gamma parameter bias. This provides confidence in the parameter estimates, at least for near gamma distributed data and larger sample sizes.

Keywords: bias; gamma; closed-form; disdrometer; estimation; moments

1. Introduction

The gamma distribution has been a popular choice for modelling the distribution of raindrop sizes for decades (see e.g. Ulbrich, 1983; Willis, 1984). Brawn and Upton (2007) introduced a new simple closed-form procedure (BU) for estimating gamma distribution parameters with grouped or binned data. The motivation for that approach was to find a simple closed formulation to fit a gamma distribution to disdrometer data and that the new method has much reduced bias compared with the historically used untruncated moment methods. Instruments known as disdrometers attempt to continuously monitor the number and size of raindrops, but in so doing they typically lower truncate (lose smaller drops) and also group each drop size into bins rather than record each drop's size as a continuous variable. Probably the best known instrument of this type is the Joss-Waldvogel disdrometer (JWD), which is an impact disdrometer. The output from such instruments is a set of drop counts over a given period collated into a range of bins, producing an observed drop size distribution. Brawn and Upton (2008) discuss the practical details of disdrometers further and apply BU to various data sets. Recently, Johnson *et al.* (2014) explored the performance of BU against a full implementation of the maximum likelihood method and given sophisticated nonlinear optimization software, their method seems ideal for unbiased estimates. The authors of that paper highlighted the bias in BU estimates when used in simulations with known population parameters, even as sample sizes increase towards infinity. It is that bias which this article seeks to more fully describe, track down and correct.

2. Bias in BU gamma parameter estimates

Let f_j be the observed frequency of drops in bin j. Also let D_j (in mm) be the midpoint of bin j, standing for the typical drop diameter of all drops falling within that bin. Finally let w_j be the width of bin j (in mm). Assume, as is common in atmospheric science, that the drop size distribution may be modelled using a gamma density function $N(D)$ over a continuous drop diameter (in mm) D, where

$$N(D) = N_0 D^\mu e^{-\lambda D} \tag{1}$$

with the units of $N(D)$, N_0, and λ being, respectively, $mm^{-1}\, m^{-3}$, $mm^{-\mu-1}\, m^{-3}$, and mm^{-1}. The BU procedure requires the successive calculation of

$$\widehat{R}_k = \sum_j f_j D_j^k / \sum_j f_j D_j^{k-1}$$

$$b_j = \ln(D_j)\widehat{R}_k - D_j \quad c_j = \ln(f_j) + k\ln(D_j) - \ln(w_j)$$

where $k = 3.5$ is chosen. Using these quantities, the BU estimates of λ, μ, and N_0 are given by

$$\widehat{\lambda} = \frac{\left(\sum_j f_j\right)\left(\sum_j f_j b_j c_j\right) - \left(\sum_j f_j c_j\right)\left(\sum_j f_j b_j\right)}{\left(\sum_j f_j\right)\left(\sum_j f_j b_j^2\right) - \left(\sum_j f_j b_j\right)^2}$$

$$\widehat{\mu} = \widehat{\lambda}\widehat{R}_k - k \tag{2}$$

$$\ln\left(\widehat{N}_0\right) = \frac{\left(\sum_j f_j c_j\right)\left(\sum_j f_j b_j^2\right) - \left(\sum_j f_j b_j c_j\right)\left(\sum_j f_j b_j\right)}{\left(\sum_j f_j\right)\left(\sum_j f_j b_j^2\right) - \left(\sum_j f_j b_j\right)^2}$$

A derivation of this estimator is given in the study by Brawn and Upton (2007, 2008) based on weighted regression and briefly again described by Johnson *et al.* (2014).

The procedure BU, originally described by Brawn and Upton (2007), gives estimates of the values of λ, N_0, and μ with reduced bias compared to traditional untruncated-moment-based estimates but still retains a degree of bias for certain combinations of μ and λ. Johnson *et al.* (2014) for example (see Figure 1 of their paper), cite the case $\mu = 5.0$ and $\lambda = 13.48$ as showing clear and persistent bias in BU estimates even as total drop counts (sample sizes) increase towards infinity, yielding an underestimated $\widehat{\mu} = 4.3$ and $\widehat{\lambda} = 11.9$ for a truncated sample size of 1000.

For the study at hand, a series of gamma distribution simulations using the methods detailed by Brawn and Upton (2007) are run with 1000 simulations of 10 000 drops each. The very high drop numbers are intended to determine the bias. The range of μ is $0 \leq \mu \leq 10$ in steps of 0.5 with $\lambda > 0$ also in steps of 0.5 within a feasible range. Two feasibility constraints are imposed to exclude less practical drop size distributions. The first condition is that the JWD upper truncation (5.6 mm) applied to a continuous gamma function (for a given μ and λ) must not create more than 0.1 % error in the moment ratio compared to the same moment ratio for an untruncated gamma function. Explicitly, this may be written as

$$\left| \frac{\int_0^{5.6} D^{\mu+k} e^{-\lambda D} \, dD}{\int_0^{5.6} D^{\mu+k-1} e^{-\lambda D} \, dD} - \frac{(\mu+k)}{\lambda} \right| < 0.001 \left| \frac{(\mu+k)}{\lambda} \right|$$

the calculation is performed using standard incomplete gamma function routines. The second condition is that the variance (which is given by $(\mu+1)/\lambda^2$) of the untruncated continuous drop size distribution must be greater than 0.03. The first condition is to ensure that JWD upper truncation plays no part in the bias model developed in this study. Experiments suggest that upper truncation should be modelled separately and is excluded here by suitable choices for μ and λ. The second constraint insures that a reasonable number of bins are inhabited before lower truncation.

The lower truncation (loss of observations for drops less than 0.313 mm for a JWD) creates a small but influential distortion in $\widehat{R}_k$ which feeds into bias in the BU estimates. Smith *et al.* (2009) note that a moment ratio can be relatively well estimated from the data and that in general this fact underpins the success of BU estimates. However, in many particular cases, such as the previously highlighted $\mu = 5.0$ and $\lambda = 13.48$, the value of $\widehat{R}_k$ deviates significantly from the corresponding moment ratio for the population continuous untruncated gamma density (given by $(\mu+k)/\lambda$). At least as far as BU gamma parameter estimates are concerned.

Figure 1 exhibits the results from the simulation experiments and first applications of BU, no adjustment to the estimated moment ratio is applied in Figure 1. Points are the mean estimated parameters over 1000 simulations with very high total drop counts per simulation (10 000 drops). The 414 empty circles are centred

at the population parameters, chosen by the two criteria given above. Clearly corresponding to those circles are lines of dots showing the position of the mean estimated parameter values. The particular case $\mu = 5.0$ and $\lambda = 13.48$ falls within the plot in Figure 1. For a given μ the bias clearly increases as λ increases as might be expected given the proximity of the lower truncation point for these distributions. Recall that the mode of a gamma distribution is given by μ/λ.

3. Modelling the bias within the ratio of moments

Let the BU estimates for μ and λ be $\widehat{\mu}$, and $\widehat{\lambda}$ respectively. $\widehat{R}_k$ is an estimate of the population value $(\mu+k)/\lambda$. Two sources of bias in the BU estimates are lower truncation, in cases where there is significant lower truncation and the assumption that all drop counts are located at the mid-bin points. It is notable that if both these sources of bias are removed from $\widehat{R}_k$, prior to its use in BU, then the resulting BU gamma parameter estimates are also largely unbiased.

The first step of the analysis is to calculate, from the raw drop counts, a modified form for the estimated ratio of moments:

$$R_k^{\text{corr}} = \widehat{R}_k - 0.0125 \qquad (3)$$

The empirically derived (see details below) constant 0.0125 models bias created by the mid-bin approximations.

The second step is to apply BU using R_k^{corr} as the new estimated ratio of moments, to obtain $\widehat{\mu}$ and $\widehat{\lambda}$. Specifically, calculate in turn

$$b_j = \ln\left(D_j\right) R_k^{\text{corr}} - D_j \; c_j = \ln\left(f_j\right) + k \ln\left(D_j\right) - \ln\left(w_j\right)$$

where $k = 3.5$; using these quantities, the bias modified BU estimates of λ, μ are given by the expressions in Equation (2) but using R_k^{corr} in place of $\widehat{R}_k$

$$\widehat{\lambda} = \frac{\left(\sum_j f_j\right)\left(\sum_j f_j b_j c_j\right) - \left(\sum_j f_j c_j\right)\left(\sum_j f_j b_j\right)}{\left(\sum_j f_j\right)\left(\sum_j f_j b_j^2\right) - \left(\sum_j f_j b_j\right)^2}$$

$$\widehat{\mu} = \widehat{\lambda} R_k^{\text{corr}} - k \qquad (4)$$

The estimator for N_0 is given above.

A third step is implemented only in those cases where the estimated mode is less than 0.45 mm $\left[\left(\widehat{\mu}/\widehat{\lambda}\right) < 0.45\right]$. This is a further modification of the estimated ratio of moments to recalculate a value for R_k^{corr}

$$R_k^{\text{corr}} = \widehat{R}_k - 0.0125 - \left(-0.11\left(\widehat{\mu}/\widehat{\lambda}\right) + 0.0495\right) \quad (5)$$

Note that when $\widehat{\mu}/\widehat{\lambda} = 0.45$, Equations (3) and (5) are the same.

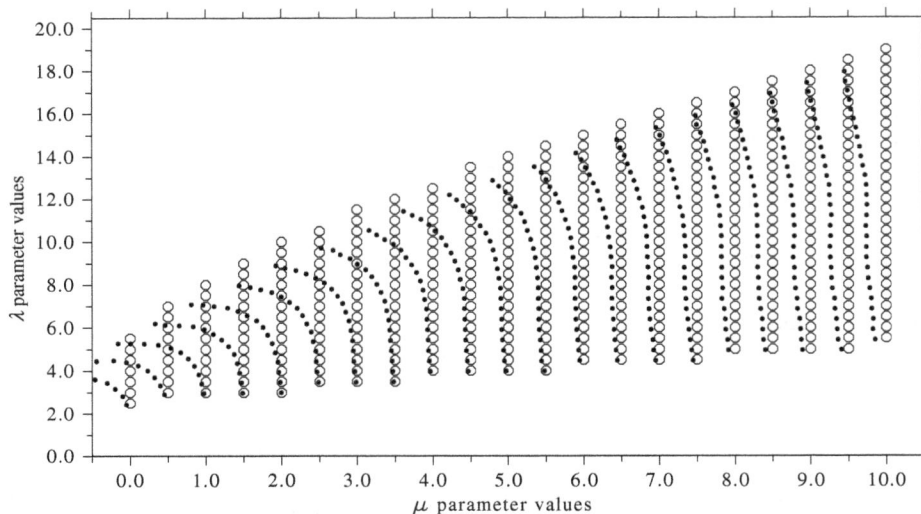

Figure 1. Actual and mean BU estimated gamma parameter values over 1000 simulations, each of 10 000 drops. Open circles for the selected population parameters (see main text) and dots for corresponding estimated parameters.

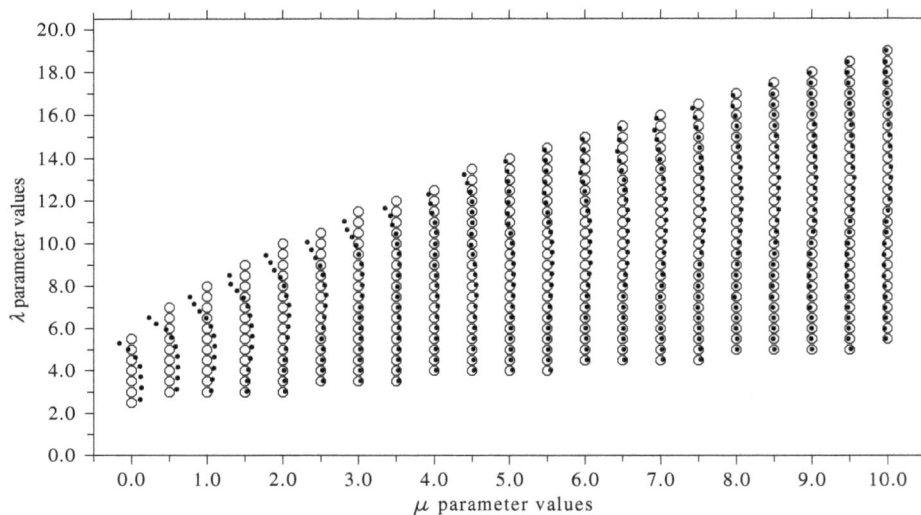

Figure 2. Same as Figure 1 but with the estimated parameters calculated using the bias corrected moment ratio within BU (see main text).

Following on from the third step, the final fourth step is to use the revised R_k^{corr} from Equation (5) to calculate new values, in turn, for b_j, c_j, $\widehat{\lambda}$, $\widehat{\mu}$, and $\widehat{N}_0$.

The third and fourth steps crudely model the effect of the lower cut-off on the moment ratio estimate. For cases where $\left(\widehat{\mu}/\widehat{\lambda}\right) > 0.45$, a simple general correction like Equation (3) to $\widehat{R}_k$ suffices with a single application of BU. That is to say that only the first and second steps enumerated above are required when $\left(\widehat{\mu}/\widehat{\lambda}\right) > 0.45$.

Figure 2 presents the same population values μ and λ as for Figure 1 but after the implementation of the bias correction steps enumerated above. The pattern of estimated points in Figure 2 is similar to the results obtained by providing BU with population values for the moment ratio. The latter observation reflects the success of the model, in general, for matching bias in the moment ratio. It is notable that

the bias in $\widehat{R}_k$ explains most of the bias within BU estimates of μ and λ for very large sample sizes. This is reminiscent of the serious bias inherent within all untrucated moment methods. The near exponential distributions such as $\mu = 1$, $\lambda = 8.0$ are relatively poorly modelled here.

This article attempts to model bias with very high total drop counts to avoid the undesirable feature of mean estimates not reasonably converging to population values as sample size tends towards infinity. However, the correction is useful over all sample sizes. To illustrate the behaviour at sample sizes ranging from very low (total recorded drop count of only 25) to an effectively infinite total drop count, two population parameter sets are presented in four tables. Firstly, again taking the case of $\mu = 5.0$ and $\lambda = 13.48$ (with modal drop size below 0.45 mm), results for the μ parameter are presented in Table 1.

Table 1. Mean estimated μ for population $\mu = 5.0$, $\lambda = 13.48$.

Total drop count	25	100	400	1600	6400	25 600	102 400
Original BU	7.43	4.88	4.39	4.32	4.32	4.33	4.34
Bias corrected BU	7.56	5.39	4.98	4.94	4.95	4.97	4.97

Table 2. Mean estimated μ for population $\mu = 10.0$, $\lambda = 10.0$.

Total drop count	25	100	400	1600	6400	25 600	102 400
Original BU	8.94	9.28	9.58	9.70	9.73	9.75	9.73
Bias corrected BU	9.14	9.49	9.80	9.92	9.96	9.97	9.96

Table 3. Mean estimated λ for population $\mu = 5.0$, $\lambda = 13.48$.

Total drop count	25	100	400	1600	6400	25 600	102 400
Original BU	17.95	13.01	12.13	11.94	11.94	11.96	11.97
Bias corrected BU	18.63	14.28	13.48	13.33	13.34	13.37	13.37

Secondly, taking the case of $\mu = 10.0$ and $\lambda = 10.0$ (with modal drop size well above 0.45 mm), results for the μ parameter are as given in Table 2.

Both Tables 1 and 2 show a similar pattern. Bias at very low sample sizes reduces with increasing sample size, reaching a plateau value. BU with the bias correction proposed in this article has the effect of shifting the limiting or plateau estimate of the parameter much closer to the true population value. At about 100 drops, for Table 1 and without considering the large variances of such estimates, it seems that the uncorrected estimate is closer to 5.0. However, the general behaviour of the corrected estimates is to be preferred and it is suggested that applying the correction is advantageous at all sample sizes. Tables 3 and 4 present corresponding results for the λ parameter, exhibiting a similar pattern with increasing sample size.

Experiments suggest that changes in the variances of $\widehat{\mu}$ and $\widehat{\lambda}$, compared to those from unmodified BU, are small and do not significantly alter reported results. For brevity , they are not reconsidered here. Equations (3) and (5) would need to be altered for other bin regimes with different disdrometers. The method of determining the constants for the two linear functions given by Equations (3) and (5) is now described. p_j by integration over bin j is defined as

$$p_j = \int_{l_j}^{u_j} D^\mu e^{-\lambda D} \, dD \qquad (6)$$

where l_j and u_j are the lower and upper limits of bin j. Then a measure of bias in the moment ratio is computed by

$$BR_k(\mu, \lambda) = \frac{\sum_j D_j^k p_j}{\sum_j D_j^{k-1} p_j} - \frac{(\mu + k)}{\lambda} \qquad (7)$$

Table 4. Mean estimated λ for population $\mu = 10.0$, $\lambda = 10.0$.

Total drop count	25	100	400	1600	6400	25 600	102 400
Original BU	9.43	9.46	9.63	9.71	9.73	9.74	9.72
Bias corrected BU	9.68	9.71	9.88	9.96	9.98	10.00	9.98

Summation here is over all available bins. $BR_k(\mu, \lambda)$ quantifies the overall effect of truncation and the mid-bin approximation for a given μ and λ. The model constants are derived by considering a plot of $BR_k(\mu, \lambda)$ against the population modal drop size, μ/λ. This plot is roughly constant for modal drop sizes above approximately 0.45 mm. For modes well below 0.45 mm, such a plot shows not only a good deal of scatter but also a clear negative correlation. The constant 0.0125 was first fitted by trial and error and testing via simulated data with higher modes. Linear regression and testing using simulated data with lower modes yield the remaining constants. This process may not be optimal. Similar or perhaps improved models for other bin regimes could be derived in this way.

4. Discussion

A simple correction to the estimated moment ratio before its use in the BU closed-form gamma parameter estimator is effective in reducing bias in gamma parameter estimates. This observation points to a crucial role for unbiased estimates of the single moment ratio and that alternative models for bias in this moment ratio may be valuable. The proposed model matches the JWD disdrometer case but similar procedures may be developed for bin regimes which differ from the JWD.

References

Brawn D, Upton GJG. 2007. Closed-form parameter estimates for a truncated gamma distribution. *Environmetrics* **18**: 633–645.

Brawn D, Upton GJG. 2008. Estimation of an atmospheric gamma drop size distribution using disdrometer data. *Atmospheric Research* **87**: 66–79.

Johnson RW, Kliche DV, Smith PL. 2014. Maximum likelihood estimation of gamma parameters for coarsely binned and truncated raindrop size data. *Quarterly Journal of the Royal Meteorological Society* **140**(681): 1245–1256.

Smith PL, Kliche DV, Johnson RW. 2009. The bias and error in moment estimators for parameters of drop-size distribution functions: Sampling from gamma distributions. *Journal of Applied Meteorology and Climatology* **48**: 2118–2126.

Ulbrich CW. 1983. Natural variations in the analytical form of raindrop size distributions. *Journal of Climate and Applied Meteorology* **22**: 1764–1775.

Willis PT. 1984. Functional fits to some observed drop size distribution and parameterization of rain. *Journal of the Atmospheric Sciences* **41**: 1648–1661.

4

Recent changes on land use/land cover over Indian region and its impact on the weather prediction using Unified model

Unnikrishnan C.K.,[1,*] Biswadip Gharai,[2] Saji Mohandas,[1] Ashu Mamgain,[1] E. N. Rajagopal,[1] Gopal R. Iyengar[1] and P. V. N. Rao[2]

[1] ESSO, MoES, National Centre for Medium Range Weather Forecasting, Noida, India
[2] Atmospheric and Climate Sciences Group, Earth & Climate Science Area, National Remote Sensing Centre, ISRO, Hyderabad, India

*Correspondence to:
U. C.K., ESSO, MoES,
National Centre for Medium
Range Weather Forecasting,
A50, Noida 201309, India.
E-mail:
unnikrishnan@ncmrwf.gov.in

Abstract

This study compares the changes of land use/land cover (Lu/Lc) or the surface type in last decades over India. Recent surface-type fractions show few major regional changes over India. There is a decrease in vegetation fraction, increase in urban and bare soil fractions over India. The Unified Model coupled with Joint UK Land Environment Simulator land surface model was used to investigate the recent Lu/Lc impact on weather prediction. Preliminary results show improvement in weather prediction by the incorporation of the recent Lu/Lc data. This highlights the need to incorporate more realistic Lu/Lc in the dynamical models for better weather prediction.

Keywords: land use land cover; land surface model; weather prediction; India; unified model

1. Introduction

Land surface acts as the lower boundary for the weather prediction models. The land surface forces and modifies the atmosphere above by transferring surface fluxes (latent heat flux, sensible heat flux, momentum and CO_2). The energy, water and carbon balance at surface are characterized by the regional features like topography, land use/land cover (Lu/Lc), soil type, etc. The regional heterogeneity of land surface directly impacts the surface fluxes to atmosphere and its evolution. The surface heterogeneity of land surface is accounted by different types of Lu/Lc data in land surface models. The Lu/Lc plays a key important role in the modulation of regional and local weather. Recent climate studies also suggest that the Lu/Lc changes can have local and remote (teleconnection) impact in dynamical model prediction (Devaraju et al., 2015).

Importance of Lu/Lc change on precipitation was investigated and documented by Pielke et al. (2007). Pielke et al. (2011) had suggested that the intensive Lu/Lc change over regions like India has more direct impact on regional climate. Mahmood et al. (2010) had stressed the importance of global monitoring of Lu/Lc change for both observational and modelling studies. Studies suggest that there are impacts on diurnal changes and mean surface warming as a result of the Lu/Lc change (Kalnay and Cai, 2003). The study of Feddema et al. (2005) suggested that the changes in land cover may influence Hadley and monsoon circulations. Most of these studies had focused on impact on long-term climate.

Studies on impact of land surface processes are limited over Indian region. Unnikrishnan et al. (2013) showed that the weekly satellite-observed vegetation fraction improves land surface parameter prediction over Indian region through better surface flux estimation. Similarly, Kumar et al. (2013) observed that updating vegetation fraction improves regional climate model predictions. Recent study of Xu et al. (2015) noted that Lu/Lc change shows enhanced 2-m air temperature variability in India. It is worth to investigate the impact of recent changes in Lu/Lc over Indian region on weather prediction.

Indian Space Research Organisation (ISRO) has developed meso-scale models compatible Lu/Lc data over Indian region derived from Advanced Wide Field Sensor (AWiFS) (Biswadip, 2014). The National Centre for Medium Range Weather Forecasting (NCMRWF) Unified Model (NCUM) uses by default the International Geosphere and Biosphere Programme (IGBP) Lu/Lc dataset which is based on National Oceanic and Atmospheric Administration's (NOAA) Advanced Very High Resolution Radiometer (AVHRR) data during 1992–1993 period. This paper investigates the recent changes in the nine surface-type fractions and its impact using NCUM coupled with Joint UK Land Environment Simulator (JULES) land surface model by using both IGBP and ISRO Lu/Lc datasets. Two separate prediction experiments (one wet and one dry) are performed to investigate the impact of ISRO Lu/Lc data on weather prediction.

The details of data and model are provided in Section 2. Section 3 compares the surface-type fraction

Table 1. Look-up table for converting 18 class Lu/Lc to 9 class JULES surface type fraction.

IGBP Class (n)		Fraction of surface types (Fm)								
		Broad leaf tree	Needle leaf tree	C3 grass	C4 grass	Shrubs	Urban	Water	Bare soil	Ice
1	EN forest	0	70	20	0	0	0	0	10	0
2	EB forest	85	0	0	10	0	0	0	5	0
3	DN forest	0	65	25	0	0	0	0	10	0
4	DB forest	60	0	5	10	5	0	0	20	0
5	Mixed forest	35	35	20	0	0	0	0	10	0
6	Closed shrubs	0	0	25	0	60	0	0	15	0
7	Open shrubs	0	0	5	10	35	0	0	50	0
8	Woody savannah	50	0	15	0	25	0	0	10	0
9	Savannah	20	0	0	75	0	0	0	5	0
10	Grassland	0	0	70	15	5	0	0	10	0
11	Permanent wetland	0	0	80	0	0	0	20	0	0
12	Cropland	0	0	75	5	0	0	0	20	0
13	Urban	0	0	0	0	0	100	0	0	0
14	Crop/natural mosaic	5	5	55	15	10	0	0	10	0
15	Snow and ice	0	0	0	0	0	0	0	0	100
16	Barren	0	0	0	0	0	0	0	100	0
17	Water body	0	0	0	0	0	0	100	0	0

from both IGBP and ISRO. The impact of Lu/Lc on weather prediction is presented in Section 4. Section 5 summaries and conclude the results.

2. Data and model

The NCUM adapted from Met office, UK is used at NCMRWF (Rajagopal *et al.*, 2012) for daily weather prediction. This is a grid point model with approximately 25 km horizontal resolution at mid latitude regions and it has 70 vertical levels. It also uses 4D variational data assimilation for creating model initial conditions. The surface parameters like soil moisture, snow depth, sea ice and SST are assimilated using a surface analysis scheme (SURF). JULES land surface model (Best *et al.*, 2011; Clark *et al.*, 2011) is coupled to the Unified Model. JULES has four vertical levels for soil moisture and temperature prediction. It is a tiled land surface model with sub-grid heterogeneity and computes surface temperatures and fluxes separately for each surface type in a grid-box. It can represent a grid box with nine major Lu/Lc types (surface type fractions) namely broad leaf trees, needle leaf trees, temperate grass, tropical grass, shrubs, urban, inland water, bare soil and land ice. JULES exchanges surface fluxes (latent heat flux, sensible heat flux, and CO_2) and momentum to the atmospheric model at each time step. At the same time atmospheric component of Unified Model forces the evolution of JULES land surface model by precipitation, surface short-wave and long-wave radiation, surface wind speed, pressure and moisture.

The NCUM uses by default the AVHRR-based Lu/Lc data from IGBP with 18 class Lu/Lc data (Loveland and Belward, 1997) to derive nine surface types for JULES land surface scheme. The dataset was derived from AVHRR data covering the period between April

1992 and March 1993 and the data have a resolution of 30 arc-second (~1 km) globally.

Recently ISRO IRS P6 satellite-derived Lu/Lc data over Indian region have become available (Biswadip, 2014). AWiFS sensor on board IRS P6 satellite during 2012–2013 period was used to derive these IGBP 18 surface types with a resolution of 30 s. Over Indian region, data in global IGBP data are replaced by ISRO Lu/Lc data. This global data was further processed using Central Ancillary Program (CAP) utility. CAP is a collection of UM utilities to make necessary ancillary input files for Unified Model-like topography, surface-type fraction, etc. Documentation of CAP is available at https://puma.nerc.ac.uk/trac/UM_TOOLS/wiki/ANCIL/CAPbuild# Introduction. CAP utility is used for converting 18 classes of Lu/Lc to 9 classes of JULES surface-type fractions. The aggregation method is used for the conversion to nine surface-type fraction of the target model grid boxes in CAP utility. The grid box surface types fraction (F_m) is calculated as below:

$$F_m = \Sigma \left(fm * \alpha_{mn} \right)$$

where F_m is the fraction of nine surface types ($m = 1–9$). fm is the fraction of each 18 IGBP class m and α_{mn} is the fraction of each nine surface-type m in each IGBP class n. The look up table used for α_{mn} in CAP is shown in Table 1.

The comparison of surface-type fractions in both datasets are discussed in the next section.

3. Surface-type fraction comparison

The recent period surface-type fraction from ISRO Lu/Lc data shows changes in type fractions compared to the IGBP data. The spatial pattern and area average of all surface types (broad leaf trees, needle leaf

Table 2. Area average fraction of surface types over Central India (70–85°E and 17–28°N) and throughout India.

	Central India		All India	
	IGBP	**ISRO**	**IGBP**	**ISRO**
Broad leaf tree	0.0784	0.1008	0.0981	0.1160
Needle leaf tree	0.0077	0.0218	0.0178	0.0253
C3 grass	0.5458	0.5038	0.4274	0.4021
C4 grass	0.0631	0.0598	0.0745	0.0738
Shrubs	0.0761	0.0343	0.1021	0.0793
Urban	0.00073	0.00479	0.0008	0.0034
Water	0.00914	0.0184	0.0149	0.0200
Bare soil	0.2187	0.2560	0.2640	0.2796
Ice	0	0	0	0

trees, temperate grass, tropical grass, shrubs, urban, inland water, bare soil and land ice) are compared in this section. Table 2 shows the changes Lu/Lc type in both datasets. Figure 1 shows the spatial variations of all nine surface-type fractions. The Figure 1(a) shows the surface-type fractions from IGBP and Figure 1(b) shows same from ISRO data. There is no land ice over the region. The average fractions of surface types are calculated over Central India region (17–28°N and 70–85°E) and all India is shown in Table 2. We can see from Figures 1 and Table 2 that there is a decrease in total vegetation type fractions (5.05%) over central India. Another major change is seen in area average bare soil fraction, which has increased in recent period (+3.72%), and area average urban fraction (+0.93%) over Central India. Similar change is also observed in all India. The increase in total average urban fraction, bare soil fraction and reduction in vegetation fraction are results of the anthropogenic activities during last two decades.

4. Impact on weather prediction

The two weather events are selected based on India Meteorological Department (IMD) weather daily/weekly report. The heavy rainfall event over Jammu and Kashmir on 3 September 2014 led to floods over the region. This event was selected for the wet case study. IMD-NCMRWF satellite-merged rainfall (Mitra et al., 2009) was used for the comparison of model rainfall forecasts. This daily rainfall data are available at 0.5° resolution from IMD (www.imdpune.gov.in). The 26 March 2014 was selected as the dry case, on that day there were above normal temperatures (3–4 K anomalies as per IMD weather report) over Western Ghats in Kerala. We have evaluated only the first 24-h model forecasts in this study.

The above normal maximum 2-m temperature was reported over Western Ghats, Kerala, India on 26 March 2014. The NCUM forecasts could reproduce the spatial pattern of above normal maximum 2-m temperatures over Western Ghats. Figure 2 shows the comparison of 2-m temperature model prediction using both Lu/Lc datasets. Even though the model is not able to predict

the actual maximum 2-m temperature reported by IMD (40.3 °C), it is seen that there is an increase of air temperature up to 1–2 K with the use of ISRO Lu/Lc data. The bias of model is also reduced by 1–2 K over the case study region. Lu/Lc types have different albedo, roughness length and surface conductance in the land surface model. The Lu/Lc type can directly impact surface fluxes in the model. The albedo and surface fluxes directly impact the surface energy budget and temperature forecast in the model. The change in Lu/Lc contributes to the change in temperature bias in the experiments.

A heavy rainfall was observed on 3 September 2014 over the Jammu and Kashmir region. Figure 3 shows comparison of model simulated rainfall with observations. It is seen from the figure that the use of the new Lu/Lc dataset has resulted in improved prediction of regional rainfall pattern. The rainfall biases (observation-model) from experiments are shown in Figure 4. The rainfall bias is reduced with ISRO Lu/Lc experiment. The observed average rainfall over the region (74.5–78°E and 33–36.5°N) was 19.2 mm, the model predictions with IGBP Lu/Lc gave an average of 8.9 mm while the ISRO Lu/Lc gave 11.1 mm. There is an improvement of rainfall by 2 mm day^{-1} (~20% of model rainfall) over the region. Lu/Lc types can impact the surface fluxes and lower boundary layer stability, any change in the surface evaporation and lower stability is reflected in the rainfall prediction in the model. This contributes to the change in the rainfall prediction in these experiments.

5. Summary and conclusion

The comparison of surface-type fraction from old IGBP data and recent ISRO data shows major regional changes. The major changes observed in the recent period are reduced total vegetation fractions and an increase in urban and bare soil fractions. These changes are the result of both anthropogenic activity and natural interannual variability of monsoon. The ISRO Lu/Lc data over Indian region were incorporated into NCUM and tested for two cases.

The preliminary results of both wet and dry weather case study of prediction show improvement in forecast by incorporating the ISRO Lu/Lc. Above normal temperature was improved by 1–2 K. This also raises the question whether the maximum 2-m air temperature over Indian region is increasing due to recent the Lu/Lc changes? This question should be addressed with a set of ensemble multi-model predictions along with strong observational evidences. This result also matches with the observation of Sertel et al. (2010), who reported that the incorporating of recent land cover dataset produced more accurate temperature simulation in a regional model over Turkey.

The prediction of Jammu and Kashmir rainfall event by incorporation of ISRO Lu/Lc data showed increased rainfall of 2 mm day^{-1} (around 20% of model rainfall).

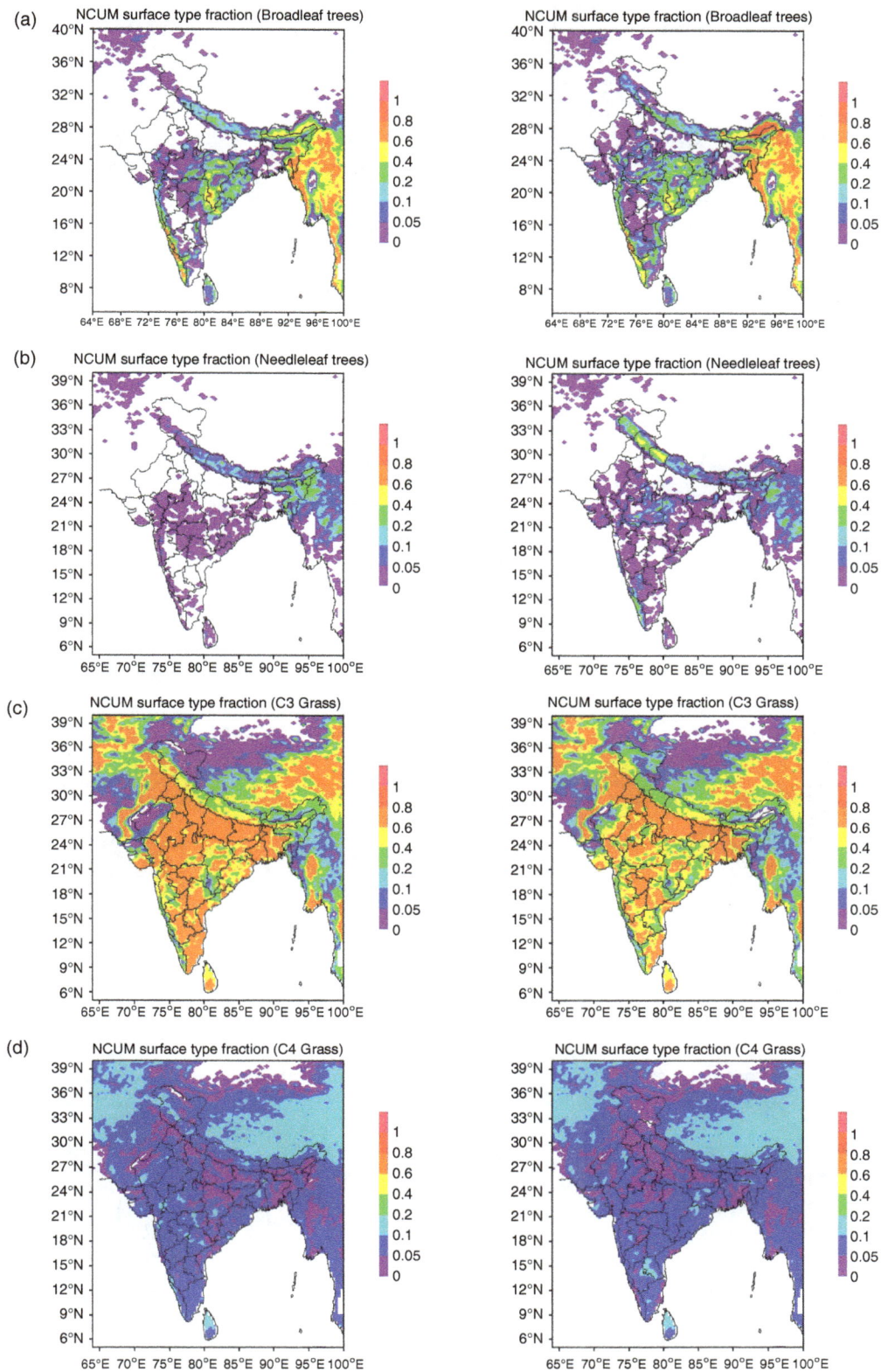

Figure 1. The surface fractions of (a) broad leaf trees, (b) needle leaf tree, (c) c3 grass, (d) c4 grass, (e) shrubs, (f) urban, (g) inland water and (h) bare soil fraction. Left panel shows the IGBP data and right panel shows NRSC ISRO data.

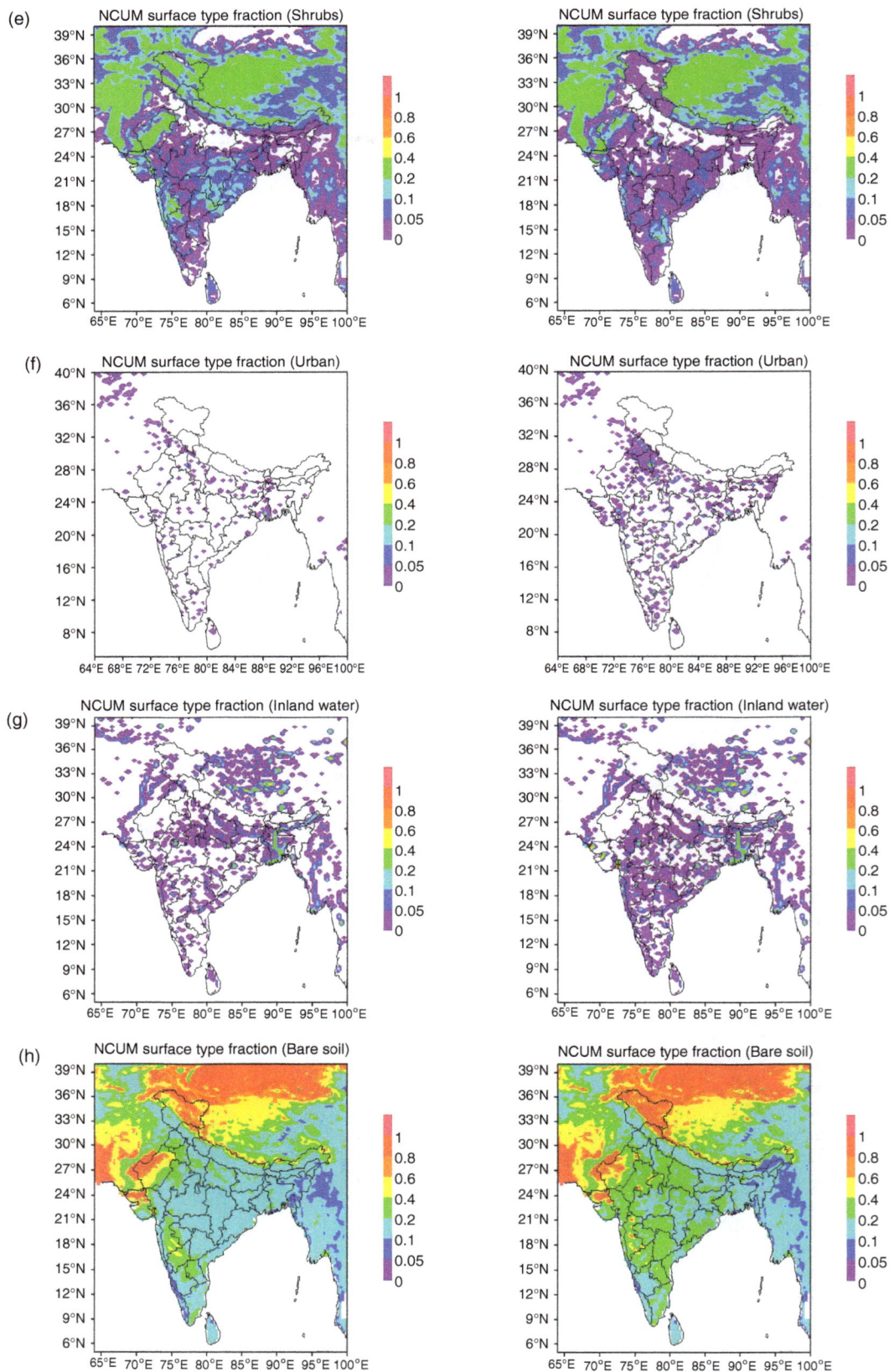

Figure 1. Continued.

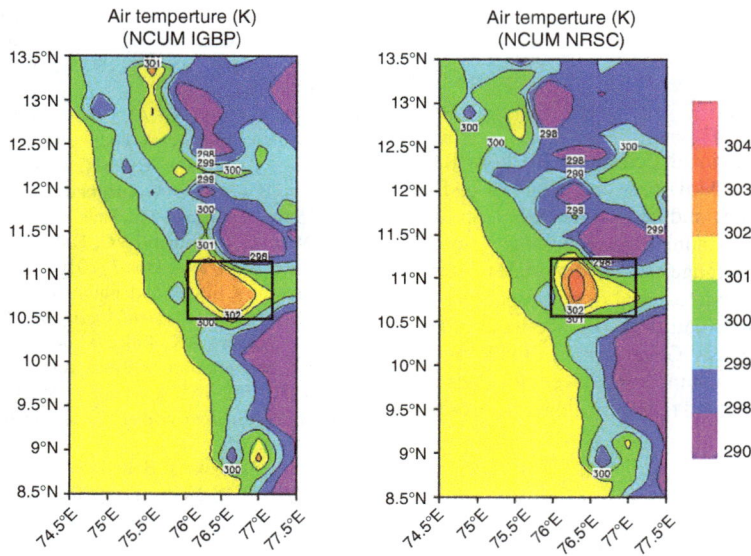

Figure 2. The comparison of 2 m temperature simulations using both Lu/Lc data sets. (a) The simulation with IGBP data and (b) simulation using ISRO data.

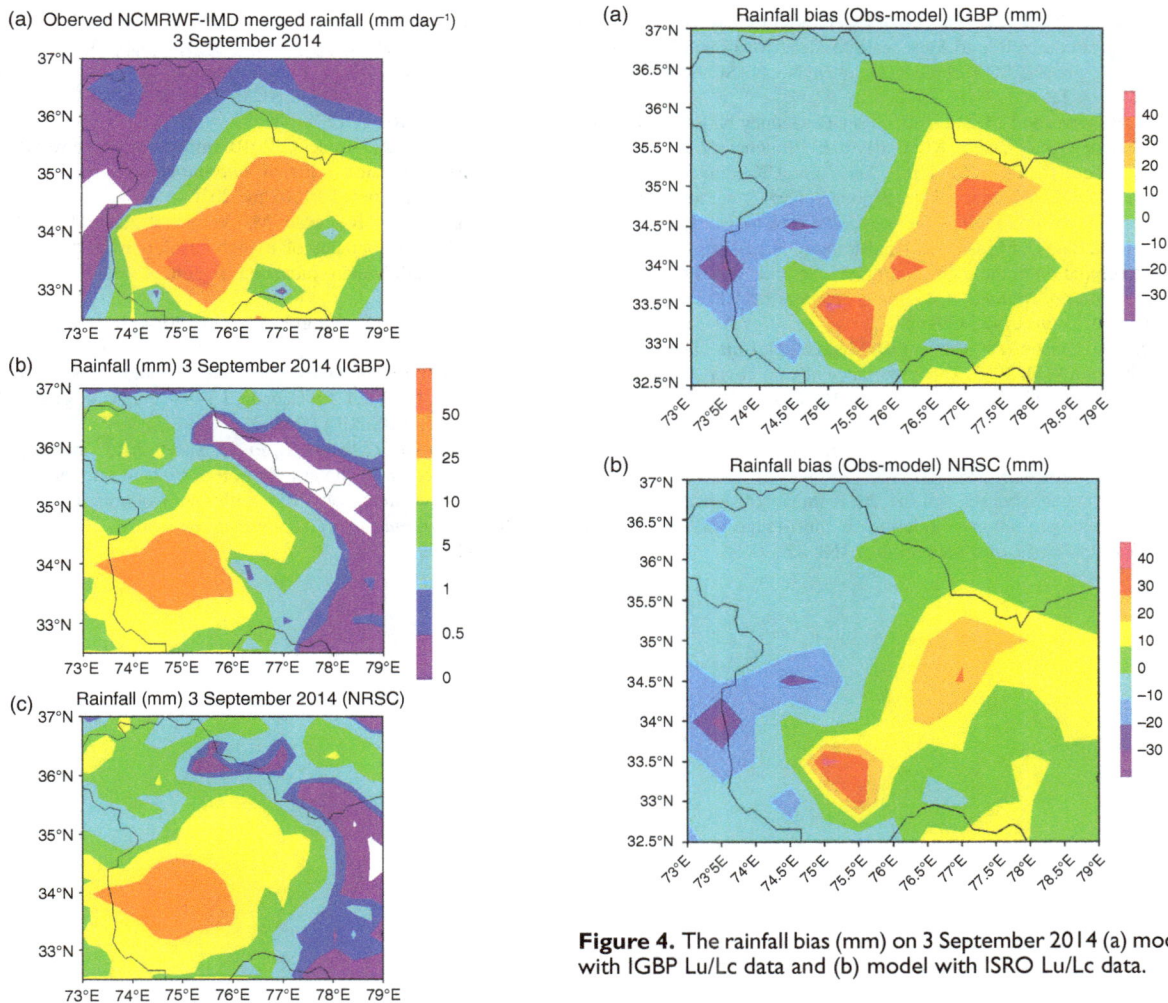

Figure 3. The comparison of model simulated rainfall (mm day^{-1}) with observation on 3 September 2014. (a) Observed rainfall, (b) model using IGBP data and (c) model using ISRO Lu/Lc data.

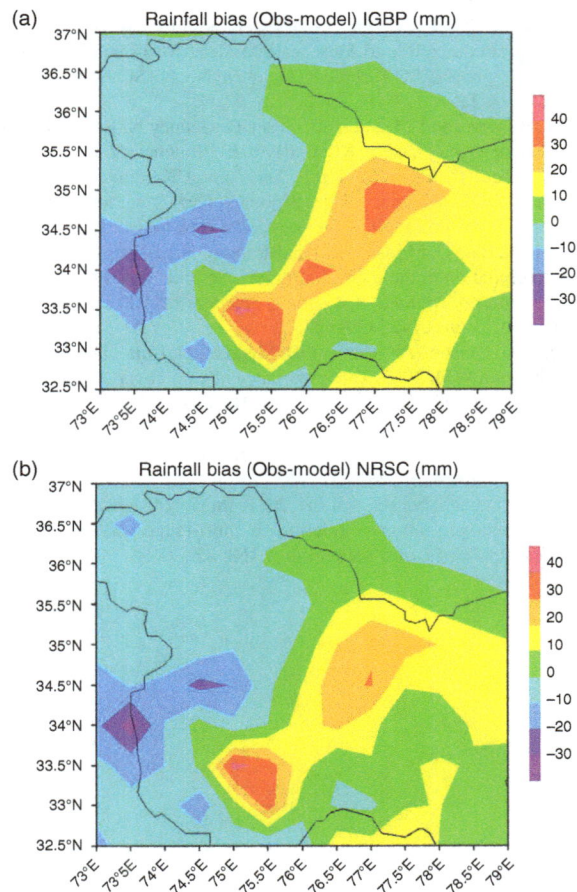

Figure 4. The rainfall bias (mm) on 3 September 2014 (a) model with IGBP Lu/Lc data and (b) model with ISRO Lu/Lc data.

This result is also consistent with previous experiment of Xie *et al.* (2014), who found that Lu/Lc can impact on precipitation simulation in WRF regional model over Beijing. Our preliminary results suggest that the use of more realistic Lu/Lc data can improve the weather predictions.

Acknowledgements

Unified Model is used at National Centre for Medium Range Weather Forecasting under a MoU between Ministry of Earth Sciences, Government of India and Met Office, UK to collaborate for developing a seamless numerical modelling system for prediction over different time ranges and spatial scales. Authors would like to thank all data sources used in this paper. We also thank IMD for providing the weather report. We would like to thank Ministry of Earth Sciences, Government of India for their support. Authors thank the anonymous reviewers for their constructive comments, which helped to improve the manuscript.

References

Best MJ, Pryor M, Clark DB, Rooney GG, Essery RLH, Ménard CB, Edwards JM, Hendry MA, Porson A, Gedney N, Mercado LM, Sitch S, Blyth E, Boucher O, Cox PM, Grimmond CSB, Harding RJ. 2011. The Joint UK Land Environment Simulator (JULES), model description – Part 1: Energy and water fluxes. *Geoscientific Model Development* **4**: 677–699.

Biswadip G. 2014. IRS-P6 AWiFS derived gridded land use/land cover data compatible to Mesoscale Models (MM5 and WRF) over Indian Region. NRSC Technical Document No. NRSC-ECSA-ACSG-OCT-2014-TR-651, 1–11.

Clark DB, Mercado LM, Sitch S, Jones CD, Gedney N, Best MJ, Pryor M, Rooney GG, Essery RLH, Blyth E, Boucher O, Harding RJ, Huntingford C, Cox PM. 2011. The Joint UK Land Environment Simulator (JULES), model description – Part 2: carbon fluxes and vegetation dynamics. *Geoscientific Model Development* **4**: 701–722.

Devaraju N, Bala G, Modak A. 2015. Effects of large-scale deforestation on precipitation in the monsoon regions: remote versus local effects. *Proceedings of National Academy of Sciences USA* **112**: 3257–3262, doi: 10.1073/pnas.1423439112.

Feddema JJ, Oleson KW, Bonan GB, Mearns LO, Buja LE, Meehl GA, Washington WM. 2005. The importance of land-cover change in simulating future climates. *Science* **310**: 1674–1678, doi: 10.1126/science.1118160.

Kalnay E, Cai M. 2003. Impact of urbanization and land-use change on climate. *Nature* **423**: 528–531.

Kumar P, Bhattacharya BK, Pal PK. 2013. Impact of vegetation fraction from Indian geostationary satellite on short-range weather forecast. *Agricultural and Forest Meteorology* **168**: 82–92.

Loveland TR, Belward AS. 1997. The IGBP-DIS global 1 km land cover data set. DISCover: first results. *International Journal of Remote Sensing* **18**(15): 3289–3295, doi: 10.1080/01431 1697217099.

Mahmood R, Quintanar AI, Conner G, Leeper R, Dobler S, Pielke RA Sr, Beltran-Przekurat A, Hubbard KG, Niyogi D, Bonan G, Lawrence P, Chase T, McNider R, Wu Y, McAlpine C, Deo R, Etter A, Gameda S, Qian B, Carleton A, Adegoke JO, Vezhapparambu S, Asefi S, Nair US, Sertel E, Legates DR, Hale R, Frauenfeld OW, Watts A, Shepherd M, Mitra C, Anantharaj VG, Fall S, Chang H-I, Lund R, Treviño A, Blanken P, Du J, Syktus J. 2010. Impacts of Land Use/Land Cover Change on Climate and Future Research Priorities. *Bulletin of the American Meteorological Society* **91**: 37–46.

Mitra AK, Bohra AK, Rajeevan MN, Krishnamurti TN. 2009. Daily Indian precipitation analyses formed from a merge of rain-gauge with TRMM TMPA satellite derived rainfall estimates. *Journal of Meteorological Society of Japan* **87A**: 265–279, doi: 10.2151/jmsj.87A.265.

Pielke RA, Adegoke J, Beltran-Przekurat A, Hiemstra CA, Lin J, Nair US, Niyogi D, Nobis TE. 2007. An overview of regional land-use and land-cover impacts on rainfall. *Tellus B* **59**: 587–601, doi: 10.1111/j.1600-0889.2007.00251.x.

Pielke RA, Pitman A, Niyogi D, Mahmood R, McAlpine C, Hossain F, Goldewijk KK, Nair U, Betts R, Fall S, Reichstein M, Kabat P, de Noblet N. 2011. Land use/land cover changes and climate: modeling analysis and observational evidence. *WIREs Climate Change* **2**: 828–850.

Rajagopal EN, Iyengar GR, George JP, Das Gupta M, Mohandas S, Siddharth R, Gupta A, Chourasia M, Prasad VS, Aditi, Sharma K, Ashish A. 2012. Implementation of unified model based analysis-forecast system at NCMRWF, NCMRWF Technical Report No. NMRF/TR/2/2012, 1–46.

Sertel E, Robock A, Ormeci C. 2010. Impacts of land cover data quality on regional climate prediction. *International Journal of Climatology* **30**: 1942–1953, doi: 10.1002/joc.2036.

Unnikrishnan CK, Rajeevan M, Vijaya Bhaskara Rao S, Kumar M. 2013. Development of a high resolution land surface dataset for the South Asian monsoon region. *Current Science* **105**(9): 1235–1246.

Xie Y, Shi J, Lei Y, Xing J, Yang A. 2014. Impacts of land cover change on simulating precipitation in Beijing area of China. IEEE International Geoscience and Remote Sensing Symposium (IGARSS), Quebec City, Canada, 13–18 July 2014, 4145–4148, doi: 10.1109/IGARSS.2014.694740.

Xu Z, Mahmood R, Yang ZL, Fu C, Su H. 2015. Investigating diurnal and seasonal climatic response to land use and land cover change over monsoon Asia with the Community Earth System Model. *Journal of Geophysical Research – Atmospheres* **120**: 1137–1152.

5

Analysis of Rocky Mountain mesoscale convective system initiation location clusters in the Arkansas-Red River Basin

Elisabeth Callen[1,*] and Donna F. Tucker[2]

[1]Department of Geological and Atmospheric Sciences, Iowa State University, Ames, IA, USA
[2]Department of Geography and Atmospheric Science, University of Kansas, Lawrence, KS, USA

*Correspondence to:
E. Callen, Department of
Geological and Atmospheric
Sciences, Iowa State University,
3014 Agronomy, Ames, IA
50011, USA.
E-mail: ecallen@iastate.edu

Abstract

Mesoscale convective systems (MCSs) are the largest precipitation producers in terms of total accumulated precipitation and are, therefore, the focus of this study. For this analysis, the MCSs forming within the Rocky Mountain portion of the Arkansas-Red River Basin were studied in April through September in 1996–2006. Once variables favourable for the initiation were determined, statistical analyses were performed on the variables. Presented are the general results of an analysis of the multiple linear regressions and principal component analyses showing that in each area where MCSs initiate, the most favourable conditions for initiation are unique.

Keywords: mesoscale convective system; statistical analysis; mountain meteorology

1. Introduction

Mesoscale convective systems (MCSs) and mesoscale convective complexes (MCCs; Maddox, 1980; Zipser, 1982) are prolific rain producers during the warm season (Fritsch *et al.*, 1986), potentially causing flash flooding (Doswell *et al.*, 1996 – hereafter D96; Schumacher and Johnson, 2005). The orogenic MCSs (Tripoli and Cotton, 1989) compound flooding issues typically associated with MCSs due to the limited terrain (e.g. valleys) the floodwaters can occupy. While, Tucker and Crook (1998) studied initiation characteristics for an orogenic MCS, which occurred just east of Denver, Colorado (CO), Maddox (1980) and Fritsch *et al.* (1986) determined a portion of MCCs developed on the Rocky Mountains' eastern slope. Also, Carbone *et al.* (2003) indicated that diurnal forcing in the Rocky Mountains plays an important role in the initiation and propagation of MCSs.

Many studies (Cotton *et al.*, 1983; Banta and Schaaf, 1987 – hereafter BS87; Banta, 1990 – hereafter B90; Tucker and Crook, 2001 – hereafter TC01; Tucker and Crook, 2005 – hereafter TC05) have examined thunderstorm mountain initiation and arrived at conclusions concerning the initiation locations and wind speeds and directions and are the starting point for MCS mountain initiation consideration. The discussions from Cotton *et al.* (1983), BS87, B90, TC01, and TC05 indicate that there is a potential for wind speed and direction at ridgetop height to predict the initiation location of a thunderstorm.

While not pertaining to orogenic MCSs in particular, several studies provide valuable information for variable selection. Jirak and Cotton (2007 – hereafter

JC07) and McAnelly and Cotton (1986 – hereafter MC86), among others, went into extensive detail on the MCS precursor environment. Rotunno *et al.* (1988 – hereafter R88) was pivotal in determining shear needed for squall lines (subset of MCSs). Johnson and Mapes (2001 – hereafter JM01) discussed that local barriers do affect time and place of initiation.

As a precursor to this analysis, Tucker and Li (2009 – hereafter TL09) studied the single cell systems, multicell systems, and MCSs within the entire Arkansas-Red River Basin (ARB), specifically looking at precipitation, size, and the number of each system type for the years 1996 through 2006. The study described here uses a small subset of the TL09 database when analysing the initiation (appearance of first precipitation) characteristics for the Rocky Mountain MCSs, which was not carried out in TL09. The study uses precipitation that initiated within the Rocky Mountains that will eventually become the MCSs. These Rocky Mountain MCSs were chosen since three fourths of the ARB precipitation fell from MCSs (TL09) and mountainous terrain can increase the chances of flooding from large precipitation events. Section 2 details the steps taken to determine the MCS initiation variables. Section 3 contains the resulting statistical runs of the multiple linear regressions (MLRs) and principal component analysis (PCAs). Trends for the most significant variables are included to show the overall change from 6 h prior to initiation through the initiation hour with no comparison against null cases. Conclusions follow in section 4. We hypothesize that a single set of variables cannot be applied to the Rocky Mountain portion, parts of CO and New Mexico (NM), of the ARB to determine MCS initiation location. The hypothesis is consistent with Moninger *et al.* (1991), who indicated that one set of criteria in a large domain for initiation does not provide meaningful results (no generalized conditions).

2. Data and methods

This study utilizes the multi-sensor precipitation data (stage 3; Young *et al.*, 2000) used in TL09's analysis. Delineation of the multi-sensor precipitation data into systems using MATLAB occurred as part of TL09. The MCS criteria, met within the ARB, were any system having continuous precipitation for at least 6 h and had a footprint (number of cells occupied) of at least 21 precipitation cells (TL09) at its maximum, where a precipitation cell is approximately 4 km by 4 km (ARM: Climate Research Facility, 2011).

For determining MCSs, only the ones initiating (first precipitation on the hourly digital precipitation array) west of 104°W, in the warm season (April to September), and in the years 1996–2006 in the ARB were used for mountain initiation. These MCSs were located in the westernmost area of the ARB. For these MCSs, only the appearance of the first precipitation (mountain initiation) had to occur by 104°W. The MCSs did not have to meet the MCS criteria by 104°W; the criteria had to be met within the TL09 domain. MATLAB was used to determine the MCSs, change the coordinate system into Polar Stereographic coordinates, and transform the local time in TL09's database into UTC. Once the transformations occurred, a cluster analysis, using the latitude and longitude of the first appearance of precipitation on the hourly digital precipitation array for each MCS (MCS initiation location), was run using between-groups linkage and Euclidean distance. Between-groups linkage is computed by calculating the smallest average distance between each set of points and then combining the closest points into a group until the desired clusters are obtained. The assumption made is that the MCSs close together in latitude and longitude have similar initiation characteristics leading to individual cluster domains centred on individual peaks/one specific area. Only clusters with 30 or more members were to be singled out for analysis, but the overall results (all the clusters combined) included the values from all clusters with 20 or more members. Figure 1 shows the clusters that were used in this analysis. The line at 104°W indicates the eastern boundary of the Rocky Mountains. No null cases were included due to the available information in the TL09 database.

Once the cluster analyses were completed, two North American Regional Reanalysis (NARR) model runs before and one model run after MCS initiation were downloaded for analysis (ex: initiation hour: 17Z; models downloaded/analysed: 12Z, 15Z, 18Z) from the NOAA National Operational Model Archive and Distribution System (NOMADS). NARR was used

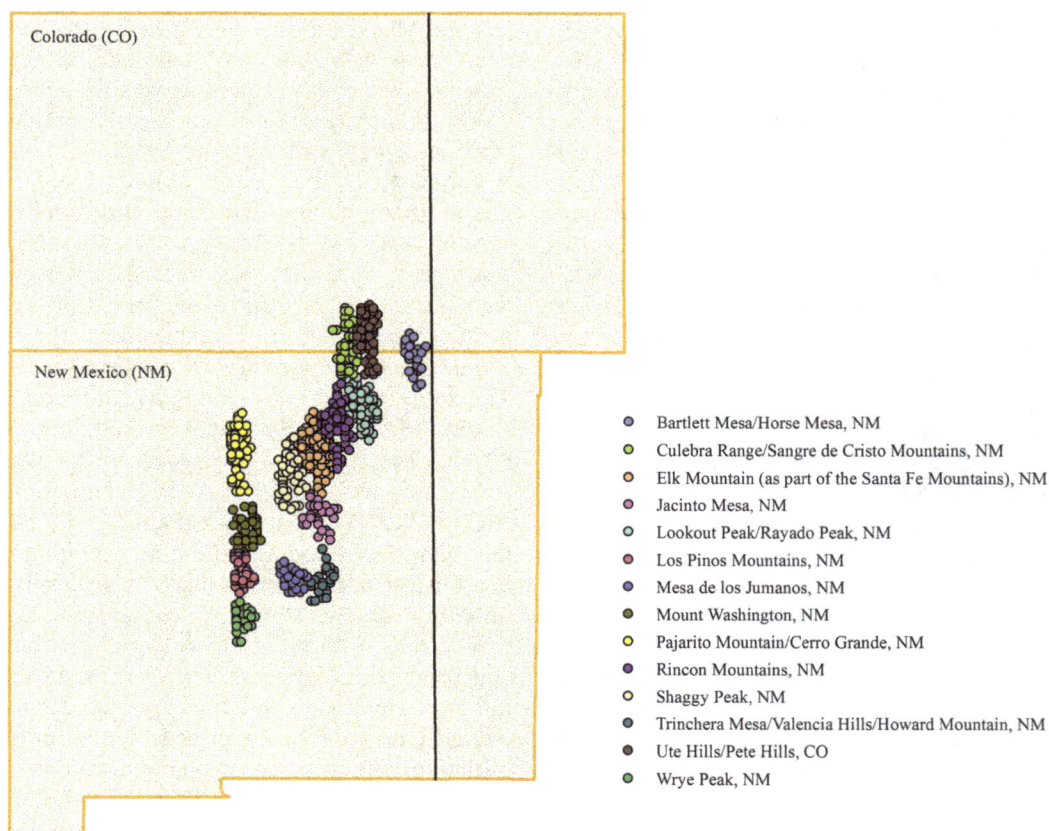

Colorado (CO)

New Mexico (NM)

- Bartlett Mesa/Horse Mesa, NM
- Culebra Range/Sangre de Cristo Mountains, NM
- Elk Mountain (as part of the Santa Fe Mountains), NM
- Jacinto Mesa, NM
- Lookout Peak/Rayado Peak, NM
- Los Pinos Mountains, NM
- Mesa de los Jumanos, NM
- Mount Washington, NM
- Pajarito Mountain/Cerro Grande, NM
- Rincon Mountains, NM
- Shaggy Peak, NM
- Trinchera Mesa/Valencia Hills/Howard Mountain, NM
- Ute Hills/Pete Hills, CO
- Wrye Peak, NM

Figure 1. Cluster locations. Black line indicates 104°W.

over other models as it contained no missing data and was at a constant resolution over the entire period. NARR was also run every 3 h since January 1979 and was run on 45 vertical layers with a horizontal grid spacing of 32 km (Mesinger et al., 2006). Even though the analysis is over a 7-h period and NARR is run every 3 h, there was no temporal interpolation performed to 'fill in' the hourly data between the NARR model runs and the variable values were not 'carried over' into the next hour to fill in the gaps. The NARR data were collected at the initiation location for each MCS.

Integrated Data Viewer (IDV, Murray, 2003) was used to analyse the NARR data to collect the NARR variables listed in Table 1. The polar stereographic coordinates of the systems were re-projected into the correct coordinate system. The derived variables, listed in Table 1, were calculated, using MATLAB, from the variable information (i.e. U and V component wind speed values) collected in IDV (using NARR). No filtering of the MCS initiation environments (e.g. modification by active convection, lifting by various types of fronts) occurred due to the nature of the analysis (observing clusters as a whole) and the size of the smaller

clusters. If the clusters were further broken down by MCS initiation environments, the sample size (especially with the smaller clusters) would be too small to glean any significant results.

MLRs and PCAs were performed on the non-standardized data sets using Statistical Packages for the Social Sciences (SPSS). The data were not standardized because of the variations present from cluster to cluster since the standardization of one cluster is not the same as the standardization of another cluster. MLRs and PCAs were applied to the 3 hourly data spanning the 6 h prior to initiation through the initiation hour at the initiation location, similar to the JC07 analysis. The systems would typically meet MCS criteria once the systems reached the Plains (east of the initiation locations). Fourteen statistical runs for each cluster were obtained – seven for MLRs and seven for PCAs on the 3 hourly data over the 7-h data period.

MLR was used as an abbreviated way to determine the variables that were needed to attain the MCS footprint within each cluster, even if it was not a linear fit. The assumption was the variables needed to produce the footprint were also the variables needed for initiation.

Table 1. Variable information including name, description, units, and the literature reference.

Name	Description	Units	Reference
NARR			
Thickness	1000–500 hPa thickness	gpm	JC07
PW	Precipitable water	mm	MC86
CAPE	Surface convective available potential energy	$J\,kg^{-1}$	JM01
CIN	Surface convective inhibition	$J\,kg^{-1}$	JM01
SRH	Storm relative helicity (0–3000 m)	$m^2\,s^{-2}$	JC07
GH600	Geopotential height, 600 hPa	gpm	JC07
GH500	Geopotential height, 500 hPa	gpm	JC07
GH300	Geopotential height, 300 hPa	gpm	JC07
GH200	Geopotential height, 200 hPa	gpm	JC07
SH850	Specific humidity, 850 hPa	$kg\,kg^{-1}$	D96
SH800	Specific humidity, 800 hPa	$kg\,kg^{-1}$	D96
SH600	Specific humidity, 600 hPa	$kg\,kg^{-1}$	D96
SH500	Specific humidity, 500 hPa	$kg\,kg^{-1}$	D96
SH300	Specific humidity, 300 hPa	$kg\,kg^{-1}$	D96
SH200	Specific humidity, 200 hPa	$kg\,kg^{-1}$	D96
UC600	U wind component, 600 hPa	$m\,s^{-1}$	B90, TC05
UC500	U wind component, 500 hPa	$m\,s^{-1}$	B90, TC05
UC300	U wind component, 300 hPa	$m\,s^{-1}$	B90, TC05
UC200	U wind component, 200 hPa	$m\,s^{-1}$	B90, TC05
VC600	V wind component, 600 hPa	$m\,s^{-1}$	B90, TC05
VC500	V wind component, 500 hPa	$m\,s^{-1}$	B90, TC05
VC300	V wind component, 300 hPa	$m\,s^{-1}$	B90, TC05
VC200	V wind component, 200 hPa	$m\,s^{-1}$	B90, TC05
T600	Temperature, 600 hPa	°C	JC07
T500	Temperature, 500 hPa	°C	JC07
T300	Temperature, 300 hPa	°C	JC07
T200	Temperature, 200 hPa	°C	JC07
Derived			
WD600	Wind direction, 600 hPa	°	BS87
WD500	Wind direction, 500 hPa	°	BS87
UWSS500	U component wind shear, 500 hPa to surface	$m\,s^{-1}$	R88
UWS600500	U component wind shear, 500 to 600 hPa	$m\,s^{-1}$	R88
UWSS600	U component wind shear, 600 hPa to surface	$m\,s^{-1}$	R88
VWSS500	V component wind shear, 500 hPa to surface	$m\,s^{-1}$	R88
VWS600500	V component wind shear, 500 to 600 hPa	$m\,s^{-1}$	R88
VWSS600	V component wind shear, 600 hPa to surface	$m\,s^{-1}$	R88

Footprint was used as the dependent variable because it is a part of the MCS criteria. The stepwise method was chosen with an entry value of 0.15 and an exit value of 0.20 with the exit of a variable occurring when the variable was too highly correlated to another variable. One issue was a lack of independence occurring in some of the MLRs, but this dependency was accepted because the variables could be coupled to one another. Cross validation was not performed due to the smaller clusters' sample sizes and due to how the variables were sampled from NARR.

PCA was used as a more in-depth tool than the MLR and also because of its dimension reduction. The variance considered in this analysis is from the overall variance accounted for by the components. The significance of a variable in the PCA was determined by the amount of variance accounted for throughout all components. Higher values of accounted for variance could indicate a stronger association between the variable and the likelihood for initiation.

3. Results

The MLR and PCA results, at the 95% confidence level, are presented for each cluster with 30 or more members, while Callen (2012) presented a cluster by cluster breakdown of all the variable results for clusters with 20 or more members. For the results of the MLRs and PCAs, Table 2 includes the best and worst fits for the analyses from the 3 hourly data over the 7-h data period. The results from Table 2 indicate that as the sample size decreases, the fits were better, which is expected since there were fewer data points to analyse. While, at times, the fits to the data were very poor, these analyses were still used to determine the best variables for potential MCS initiation in the ARB. A second cluster analysis was performed on the original Elk Mountain, NM cluster because the fits

were poor and it was determined the original cluster could be broken down by the 500 and 600 hPa wind directions.

An analysis was also performed on all the clusters together (listed as 'Overall') and was performed to show the individual cluster analyses were a better fit to the data than the overall results. The most significant aspect of the overall analysis was the very poor fits. The poor fits indicated that finding most likely MCS initiation conditions for the entire Rocky Mountain portion of the ARB would not likely succeed.

The individual cluster results could be used in an ingredients-based approach (Johns and Doswell, 1992, D96), with Table 3 showing the trends of the important variables in each cluster over the 7-h analysis window with the trends providing more information about the state of the atmosphere than variable values. Depending on the cluster, 85–90% of the cases have the indicated trends. The ingredients-based approach works since it generalizes the conditions needed for MCS initiation within the various clusters. The importance of a variable was determined by the number of times and how a variable was included in the MLRs and PCAs. Any positive or negative trend shown in Table 3 was considered significant enough to report. As can be seen in the Table 3 trends, no two clusters are the same overall. There are some similarities in the trends from 6 h prior to initiation through the initiation hour. These similarities include: no significant changes in the upper air temperatures, no significant changes in precipitable water (PW), and negative changes in convective inhibition (CIN) with most clusters. However, the other variables differ providing resistance to a common set of variable trends. It should be noted that WD500 and WD600 were not significant in any of the clusters.

Some useful results can be compiled from the information available in Table 3. For example, for the overall Elk Mountain, NM cluster, the first most important variables of a growing GH200, unchanging

Table 2. High and Low R^2 from the MLRs for each cluster and accounted for variance with eigenvalues of one or greater from the PCAs for each cluster. Also, the sample size is included.

Cluster name	N	Highest R^2	Lowest R^2	Highest variance	Lowest variance
Elk Mountain, NM	154	0.794	0.226	89.751	82.145
Elk Mountain wind direction group 1, NM	114	0.893	0.291	91.377	80.198
Elk Mountain wind direction group 2, NM	40	1.000	0.715	99.835	92.294
Ute Hills/Pete Hills, CO	90	0.906	0.226	94.217	86.300
Rincon Mountains, NM	76	0.995	0.457	92.545	86.017
Lookout Peak/Rayado Peak, NM	70	0.996	0.347	93.685	86.564
Pajarito Mountain/Cerro Grande, NM	61	1.000	0.869	94.912	87.970
Culebra Range/Sangre de Cristo Mountains, CO	56	0.996	0.298	93.926	88.737
Shaggy Peak, NM	51	0.999	0.610	95.357	88.736
Los Pinos Mountains, NM	43	1.000	0.687	98.639	89.600
Mount Washington, NM	38	1.000	0.246	96.947	93.801
Wrye Peak, NM	36	1.000	0.178	100.000	92.265
Mesa de los Jumanos, NM	34	1.000	0.613	97.647	93.769
Jacinto Mesa, NM	32	1.000	0.441	100.000	93.055
Bartlett Mesa/Horse Mesa, NM	31	1.000	0.864	100.000	92.515
Trinchera Mesa/Valencia Hills/Howard Mountain, NM	31	1.000	0.416	98.563	93.562
Overall	1165	0.272	0.093	82.979	76.747

Table 3. Trends for each cluster and variable from the 6 h prior to initiation to the initiation hour. P, significant positive trend; M, significant negative trend; N, no significant trend; I, variables are the most significant in that cluster; 2, variables are the second most significant in that cluster; 3, variables are the third most significant in that cluster. The variable's significance is independent of trend and a cluster can have multiple variables with the same significance.

Cluster	Thickness	PW	CAPE	CIN	SRH	GH600	GH500	GH300	GH200	SH850
Elk Mountain, NM	2 P						3 N	2 N	1 P	
Elk Mountain Wind Direction Group I, NM							3 N	2 N	3 P	
Elk Mountain Wind Direction Group 2, NM	3 P									
Ute Hills/Pete Hills, CO		3 P								
Rincon Mountains, NM	3 N	3 N					2 N			
Lookout Peak/Rayado Peak, NM	3 N	3 N					2 N	3 M	3 N	3 M
Pajarito Mountain/Cerro Grande, NM	2 N					1 N	3 N			
Culebra Range/Sangre de Cristo Mountains, CO	2 M					3 N			2 P	3 N
Shaggy Peak, NM	1 P	2 N								
Los Pinos Mountains, NM										
Mount Washington, NM	3 M	3 N			3 M	3 P	2 P	2 M	2 M	2 M
Wrye Peak, NM			1 P			1 N	2 N	2 M		
Mesa de los Jumanos, NM				3 M			3 P			
Jacinto Mesa, NM	1 M			2 N			2 N			
Bartlett Mesa/Horse Mesa, NM		3 N		3 M						3 M
Trinchera Mesa/Valencia Hills/Howard Mountain, NM	1 P	2 N		3 P		1 P	1 P	3 N	1 M	3 M

Cluster	SH800	SH600	SH500	SH300	SH200	UC600	UC500	UC300	UC200	VC600
Elk Mountain, NM	3 M						2 P			
Elk Mountain Wind Direction Group I, NM										
Elk Mountain Wind Direction Group 2, NM							2 M	3 P		3 P
Ute Hills/Pete Hills, CO										
Rincon Mountains, NM										
Lookout Peak/Rayado Peak, NM	3 M		3 N				1 N			
Pajarito Mountain/Cerro Grande, NM									3 P	
Culebra Range/Sangre de Cristo Mountains, CO							2 P			2 M
Shaggy Peak, NM										
Los Pinos Mountains, NM										
Mount Washington, NM						3 P	2 M	3 N		
Wrye Peak, NM							3 P	3 P	1 P	1 N
Mesa de los Jumanos, NM				2 P			3 M	2 P	3 N	
Jacinto Mesa, NM							1 N		2 P	
Bartlett Mesa/Horse Mesa, NM	3 N						3 P	1 N		3 P
Trinchera Mesa/Valencia Hills/Howard Mountain, NM	3 M						3 M	3 P		

Cluster	VC500	VC300	VC200	T600	T500	T300	T200	WD600	WD500	UWSS500
Elk Mountain, NM	1 M			3 N	3 N					1 N
Elk Mountain Wind Direction Group I, NM					3 N					1 N
Elk Mountain Wind Direction Group 2, NM						1 N				
Ute Hills/Pete Hills, CO										2 N
Rincon Mountains, NM										
Lookout Peak/Rayado Peak, NM										
Pajarito Mountain/Cerro Grande, NM										
Culebra Range/Sangre de Cristo Mountains, CO										
Shaggy Peak, NM										
Los Pinos Mountains, NM										
Mount Washington, NM				3 N				3 P		
Wrye Peak, NM										
Mesa de los Jumanos, NM										
Jacinto Mesa, NM										
Bartlett Mesa/Horse Mesa, NM	3 N	3 N		1 N		1 N			3 P	
Trinchera Mesa/Valencia Hills/Howard Mountain, NM	3 M	2 N	3 N	3 P	2 M	3 P	3 M	3 N	3 M	

Table 3. Continued.

	VC500	VC300	VC200	T600	T500	T300	T200	WD600	WD500	UWSS500
Rincon Mountains, NM	3 M			3 N						
Lookout Peak/Rayado Peak, NM				3 P	3 M	3 N				2 N
Pajarito Mountain/Cerro Grande, NM										
Culebra Range/Sangre de Cristo Mountains, CO										1 P
Shaggy Peak, NM										1 P
Los Pinos Mountains, NM		1 M								
Mount Washington, NM	2 M		1 P	2 M	3 N	2 N				3 N
Wrye Peak, NM	3 N		2 M							3 M
Mesa de los Jumanos, NM	2 M	3 M		3 N			3 N			
Jacinto Mesa, NM										
Bartlett Mesa/Horse Mesa, NM		2 M		3 N	3 N					3 P
Trinchera Mesa/Valencia Hills/Howard Mountain, NM	1 M	1 M	2 P	1 N	3 N	3 N				2 M

	UWS600500	UWSS600	VWSS500	VWS600500	VWSS600
Elk Mountain, NM			2 M		3 P
Elk Mountain Wind Direction Group 1, NM			2 M		
Elk Mountain Wind Direction Group 2, NM	2 N	2 P	1 N		
Ute Hills/Pete Hills, CO			3 M		1 N
Rincon Mountains, NM		1 N			
Lookout Peak/Rayado Peak, NM		3 N	2 P		
Pajarito Mountain/Cerro Grande, NM			3 N		
Culebra Range/Sangre de Cristo Mountains, CO	3 P		1 M		
Shaggy Peak, NM			1 M		
Los Pinos Mountains, NM			2 N		3 P
Mount Washington, NM	1 N	1 P	2 M	3 M	1 N
Wrye Peak, NM	2 M	3 M	2 P		2 N
Mesa de los Jumanos, NM	3 P	1 M	3 M		
Jacinto Mesa, NM					
Bartlett Mesa/Horse Mesa, NM	2 M				
Trinchera Mesa/Valencia Hills/Howard Mountain, NM	3 M	2 P	2 M		2 N

UWSS500, and decreasing VC500, indicate if these variables occur, then there is a greater likelihood for MCS initiation within the cluster domain. While more weight is given to the first most significant variables, the second and third most important variables need to be considered as well. This can be done for each cluster. The trends and significant variables show that no two sets of conditions needed for MCS initiation at different peaks are the same. These differences would cause issues for any forecaster trying to predict MCS initiation within the ARB portion of the Rocky Mountains.

For the individual clusters and generalized results, the most significant variables are the wind shear/wind speed, geopotential height, and specific humidity variables, which would be a starting point, although the specific variables differ from cluster to cluster. Observing the wind shear/wind speed, geopotential height, and specific humidity variables alone significantly reduces the number of variables needed for observation. Of the top six most used variables, four were wind shear variables with the wind shear between the surface and an upper level being the most significant. While the geopotential height variables are highly correlated in each cluster, the most significant one is not consistent through the clusters. The specific humidity could be a marker for the depth of the moist layer or for a pocket of moist air present at any level. PW should be considered since MCSs require relatively high amounts, but PW was not often considered one of the most important variables. Another ingredient to consider is T500, included relatively often, because changes could be a proxy for instability as it directly relates to Lifted Index (LI). The needed lift would be provided by the mountains. This variable combination into an ingredients-based approach further reinforces the fact that no two clusters are exactly the same and the ingredients needed for MCS initiation within each cluster domain are different.

The initiation variable differences observed in the clusters could be caused by a variety of factors including topography. The topography affects the flow of all variables causing some areas of have higher variable values than other areas. This effect can be seen especially in the wind variables were ridgetop height winds affect the potential for initiation. The topography-affected flow would cause certain places (i.e. specific peaks) to be more prominent areas of initiation over other areas.

4. Conclusion

While there were similarities present from cluster to cluster in that wind shear, specific humidity, and geopotential height variables occurred often, one variable, overall, is not considered most significant. The hour and cluster dependence of each significant variable is noticeable in Table 3 with the differing trends from 6 h prior to the initiation hour and differing significance of each variable in each cluster. Therefore, one set of criteria for MCS initiation within a large domain will not present meaningful results, consistent with Moninger

et al. (1991). The topography in the area contributes to the difficulty in determining generalized conditions over the Rocky Mountain portion of the ARB since different peaks require different initiation conditions.

In conclusion, there does appear to be certain conditions needed for MCS initiation with the individual clusters like wind shear, specific humidity, and geopotential height which can be used to help predict initiation. It is very difficult to predict MCS initiation within the Rocky Mountain portion of the ARB due to the conditions for MCS initiation in the Rocky Mountains being so diverse. Due to the diverseness of the MCS initiation conditions in the Rocky Mountain portion of the ARB, one forecast model could not be used to predict MCS initiation within this area. The non-generalized conditions would render any generalized forecast model unusable and would call for a much more sophisticated, complex forecast model to account for all the initiation characteristic differences. Many future steps need to be done to verify the analyses to predict the MCSs, including the inclusion of null cases.

References

ARM: Climate Research Facility. 2011. ARM XDC Datastreams. http://www.arm.gov/xdc/xds/abrfc (accessed 1 May 2011).

Banta RM. 1990. The role of mountains flows in making clouds. Atmospheric processes over complex terrain, meteorology monograph. *American Meteorological Society* **45**: 229–282.

Banta RM, Schaaf CB. 1987. Thunderstorm genesis zones in the Colorado Rocky Mountains as determined by traceback of geosynchronous satellite images. *Monthly Weather Review* **115**: 463–476.

Callen E. 2012. A statistical analysis of characteristics of mesoscale convective system mountain initiation location clusters in the Arkansas-Red River Basin. MS thesis, Department of Geography, University of Kansas, 489 pp.

Carbone RE, Tuttle JD, Ahijevych DA, Trier SB. 2003. Inferences of predictability associated with warm season precipitation episodes. *Journal of the Atmospheric Sciences* **59**: 2033–2056.

Cotton WR, George RL, Wetzel PJ, McAnelly RL. 1983. A long-lived mesoscale convective complex. Part 1: the mountain-generated component. *Monthly Weather Review* **111**: 1893–1918.

Doswell CA III, Brooks HE, Maddox RA. 1996. Flash flood forecasting: an ingredients-based methodology. *Weather and Forecasting* **11**: 560–581.

Fritsch JM, Kane RJ, Chelius CR. 1986. The contribution of mesoscale convective weather systems to warm-season precipitation in the United States. *Journal of Applied Meteorology* **25**: 1333–1345.

Jirak IL, Cotton WR. 2007. Observational analysis of the predictability of mesoscale convective systems. *Weather and Forecasting* **22**: 813–838.

Johns RH, Doswell CA III. 1992. Severe local storms forecasting. *Weather and Forecasting* **7**: 588–612.

Johnson RH, Mapes BE. 2001. Mesoscale processes and severe convective weather. Severe convective storms, meteorology monograph. *American Meteorological Society* **50**: 71–122.

Maddox RA. 1980. Mesoscale convective complexes. *Bulletin of the American Meteorological Society* **61**: 1374–1387.

McAnelly RL, Cotton WR. 1986. Meso-β-scale characteristics of an episode of meso-α-scale convective complexes. *Monthly Weather Review* **114**: 1740–1770.

Mesinger F, DiMego G, Kalnay E, Mitchell K, Shafran PC, Ebisuzaki W, Jovic D, Woollen J, Rogers E, Berbery EH, Ek MB, Fan Y, Grumbine R, Higgins W, Li H, Lin Y, Manikin G, Parrish D, Shi W. 2006. North American regional reanalysis. *Bulletin of the American Meteorological Society* **87**: 343–360.

Moninger WR, Bullas J, de Lorenzis B, Ellison E, Flueck J, McLeod JC, Lusk C, Lampru PD, Phillips RS, Roberts WF, Shaw R, Stewart TR, Weaver J, Young KC, Zubrick SM. 1991. Shootout-89, a comparative evaluation of knowledge-based systems that forecast severe weather. *Bulletin of the American Meteorological Society* **72**: 1339–1354.

Murray D. 2003. The Integrated Data Viewer – a web-enabled application for scientific analysis and visualization. In 19th International Conference on Interactive Information Processing Systems, American Meteorological Society, Long Beach, CA, USA.

Rotunno R, Klemp JB, Weisman ML. 1988. A theory for strong, long-lived squall lines. *Journal of the Atmospheric Sciences* **45**: 463–485.

Schumacher RS, Johnson RH. 2005. Organization and environmental properties of extreme-rain-producing mesoscale convective systems. *Monthly Weather Review* **133**: 961–976.

Tripoli GJ, Cotton WR. 1989. Numerical study of an observed orogenic mesoscale convective system. Part I: simulated genesis and comparison with observations. *Monthly Weather Review* **117**: 273–304.

Tucker DF, Crook NA. 1998. The generation of a mesoscale convective system from mountain convection. *Monthly Weather Review* **127**: 1259–1273.

Tucker DF, Crook NA. 2001. Favored regions of convective initiation in the Rocky Mountains. In 9th Conference on Mesoscale Processes, American Meteorological Society, Fort Lauderdale, FL, USA.

Tucker DF, Crook NA. 2005. Flow over heated terrain. Part II: generation of convective precipitation. *Monthly Weather Review* **133**: 2565–2582.

Tucker DF, Li X. 2009. Characteristics of warm season precipitating storms in the Arkansas-Red River Basin. *Journal of Geophysical Research* **114**: D13108, doi: 10.1029/2008JD011093.

Young CB, Bradley AA, Krajewski WF, Kruger A. 2000. Evaluating NEXRAD multisensory precipitation estimates for operational hydrologic forecasting. *Journal of Hydrometeorology* **1**: 241–254.

Zipser EJ. 1982. Use of a conceptual model of the life of mesoscale convective systems to improve very-short-range forecasts. In *Nowcasting*, Browning K (ed). Academic Press; 191–204.

6

Estimation of sea level pressure fields during Cyclone Nilam from Oceansat-2 scatterometer winds

Ch. Purna Chand,[1] M. V. Rao,[1] I. V. Ramana,[1] M. M. Ali,[1,*] J. Patoux[2] and M. A. Bourassa[3]

[1] National Remote Sensing Centre, ISRO, Hyderabad, India
[2] Department of Atmospheric Sciences, University of Washington, Seattle, WA, USA
[3] Center for Ocean-Atmospheric Prediction Studies, The Florida State University, Tallahassee, FL, USA

*Correspondence to:
M. M. Ali, National Remote
Sensing Centre, ISRO,
Hyderabad, India.
E-mail: mmali110@gmail.com

Abstract

Sea level pressure observations are rarely available over most of the oceans and no remote sensing instrument can measure it directly. Here, we explore the estimation of sea level pressure fields using surface wind observations from OSCAT, the scatterometer onboard Oceansat-2, during the 27 October–2 November 2012 passage of Cyclone Nilam over the Bay of Bengal. The estimated pressure is validated with moored-buoy observations. Scatterometer-derived pressure differences are closer to *in situ* observations than those in examined numerical model estimations. The estimated pressure is more accurate than European Centre for Medium-range Weather Forecasting re-analysis values in location, track, and intensity.

Keywords: scatterometer; sea level pressure; OSCAT

1. Introduction

The sea level pressure pattern is one of the most important meteorological parameters related to several aspects of severe weather. Horizontal gradients in pressure are extremely useful for analyzing weather patterns and weather features. Winds driven by pressure gradients in turn drive waves; cyclone storm surge is dependent on these winds and on the inverse barometer effect. *In situ* pressure measurements over the ocean are relatively scarce compared to measurements available over land. One approach to obtaining sea level pressure fields over the ocean is to contour these limited observations from buoys or ships to obtain sea level pressure maps. A second approach is to use numerical weather forecast model analyses in which *in situ* and scatterometer winds are assimilated (e.g. http://www.knmi.nl/publications/fulltexts/oceansat_cal_val_report_final_copy1.pdf). A third alternative is to infer the sea level pressure fields indirectly from remote sensing observations of winds. Brown and Levy (1986), Harlan and O'Brien (1986), Hsu *et al.* (1997), Hsu and Liu (1996), and Zierden *et al.* (2000) demonstrated how scatterometer winds can be used to estimate sea level pressure. The resulting surface pressure fields estimated from scatterometer observations are more accurate in terms of gradients compared to isolated measurements obtained from platforms on buoys, islands, and ships.

Here, we apply the procedure described by Patoux *et al.* (2003) to estimate sea level pressure fields over the Indian Ocean using Oceansat-2 scatterometer (OSCAT) observations. OSCAT is onboard the Indian Oceansat-2 satellite along with two other sensors, an ocean-color monitoring sensor and a radio occultation sounder for the atmosphere. OSCAT is a Ku-band, pencil-beam scatterometer similar in design to the NASA SeaWinds instrument onboard QuikSCAT.

The Ku-band, pencil-beam scatterometer is an active microwave radar operating at 13.515 GHz. It consists of a parabolic, 1-m diameter dish antenna, which rotates at 20 revolutions per minute using a direct current motor and generates two beams. The inner beam (1400 km swath) operates with HH polarization and the outer beam (1800 km swath) operates with VV polarization. The energy of the transmitted radio frequency (RF) pulse backscattered by the ocean surface is received at the antenna and, after onboard 'range compression', is digitized and transmitted to the ground station. The normalized radar cross section, referred to as sigma-naught (σ_0), is calculated from this echo data and a geophysical model function inverts σ_0 to wind vectors, resulting in a 1800 km wide swath of 50×50 km vector winds. The accuracy of wind speed is found to be $1.4\,\mathrm{m\,s^{-1}}$ [root mean square error (RMSE)] when compared to European Centre for Medium-Range Weather Forecasts (ECMWF) analysis winds and $1.7\,\mathrm{m\,s^{-1}}$ (RMSE) when compared to National Centers for Environmental Prediction (NCEP) wind analysis. In the case of wind direction, the RMSE is $17.2°$ with ECMWF analysis and $18.8°$ with NCEP analysis. These results are well within the OSCAT mission goal of $2\,\mathrm{m\,s^{-1}}$ wind speed accuracy and $20°$ in direction, in the range of 4–$24\,\mathrm{m\,s^{-1}}$ wind speed (Chakraborty *et al.*, 2013).

In this paper, we retrieve sea level pressure fields from OSCAT winds using the University of Washington planetary boundary layer (UWPBL)

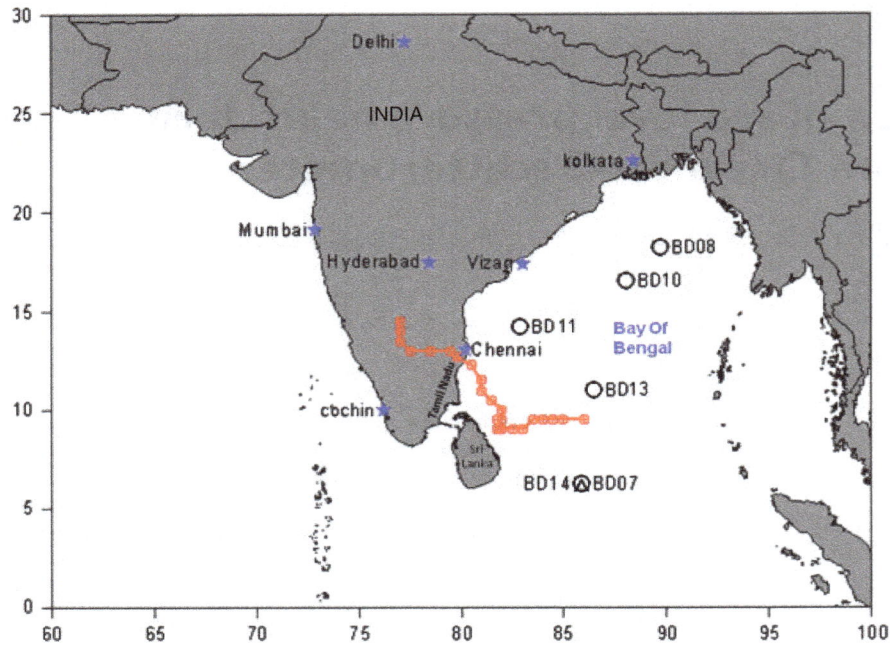

Figure 1. Location of the moored buoys (black circles and a triangle in the case of BD14) used in the study, along with the cyclone track. Red circles indicate positions of the cyclone as estimated by the India Meteorological Department.

model of Patoux *et al.* (2003) during the passage of Cyclone Nilam [India Meteorological Department (Cyclone Warning Division, 2012) Preliminary report on cyclonic storm – www.imd.gov.in]. Cyclone Nilam formed in the Southwest Bay of Bengal at 1130 IST on 28 October 2012 near 09.50°N and 86.00°E. It moved westward and intensified into a deep depression on the morning of 29 October. By the morning of 30 October, the cyclone was over the Southwest Bay of Bengal off the coast of Sri Lanka. It moved north-northwestward and crossed the North Tamilnadu coast near Mahabalipuram, South of Chennai, between 1600 and 1700 IST on 31 October 2012. After landfall the storm moved west-northwestward and weakened into a depression on the morning of 1 November 2012. Before making landfall, the cyclone had a wind maximum speed of 23.15 m s^{-1} (reported by IMD). It caused approximately \$56.7 million in damage, with a death toll of up to 75 (en.wikipedia.org/wiki/Cyclone_Nilam).

We will show that the OSCAT-derived pressure compares well with individual buoy observations and the central locations of the cyclone, as determined from pressure fields, are better placed in the scatterometer-derived pressure fields than in the ECMWF re-analysis (ERA) pressure values.

2. Data and methods

In this study, we used 27 October–2 November 2012 OSCAT Level 2B winds from the National Remote Sensing Centre (www.nrsc.gov.in); 27, 29, and 31 October and 2 November 2012 were descending passes

(around 0630 GMT), and 28 and 30 October and 1 November 2012 were ascending passes (around 1830 GMT) in the Bay of Bengal region. We used only passes that covered the cyclone area.

Wind measurements severely contaminated by rain were discarded. Rain flagging made use of pre-retrieval and post-retrieval rain probabilities, indicated by values given in the retrieval cost function for the sixth and fifth wind vector solutions, respectively. The threshold values of 0.9 and 0.5 were used for pre-retrieval and post-retrieval rain flagging. The information on cyclone track and intensity was obtained from IMD (www.imd.gov.in). *In situ* pressure observations (at 3-h interval) were obtained from the moored-buoy record available at Indian National Centre for Ocean Information Services (INCOIS) (http://www.incois.gov.in). The locations and names of the moored buoys are shown in Figure 1 along with the cyclone track. We selected the buoy observations of 0600 and 1800 GMT to coincide with the satellite descending and ascending passes. We used different integration schemes to compute pressure from OSCAT using the UWPBL model. Then the Pearson's correlation coefficient, r, was computed between the model estimated values and the *in situ* observations. We also computed uncertainties in the slope (Ω_{slope}) and intercept ($\Omega_{\text{y-int}}$) following Taylor (1997),

$$\Omega_{\text{slope}} = \sqrt{n\Omega_y^2/\Delta} \qquad (1)$$

and

$$\Omega_{\text{y}-int} = \sqrt{\Omega_y^2 \left(\sum_{i=1}^{n} x_i^2\right)/\Delta} \qquad (2)$$

where

$$\Delta = n \left(\sum_{i=1}^{n} x_i^2 \right) - \left(\sum_{i=1}^{n} x_i \right)^2 \qquad (3)$$

and

$$\Omega_y = \sqrt{ \frac{1}{n-2} \sum_{i=1}^{n} (y_i - mx_i - b)^2 } \qquad (4)$$

where $x_i = $ in situ observation, $y_i = $ model pressure, $m = $ slope, $b = $ y-intercept, and $n = $ number of points.

3. The model

The UWPBL model computes two-dimensional fields of geostrophic wind gradients from the corresponding fields of surface wind vectors using a combination of a two-layer similarity model in the mid latitudes and a mixed layer model in the tropics. It then fits a pressure pattern by integrating the geostrophic equation via least-squares minimization. This model requires air temperature, sea surface temperature (SST), humidity, and the horizontal temperature gradient. The integration to determine the pressure field is off by a constant of integration; therefore, in situ sea level pressure observations can be useful to eliminate this offset (Patoux, 2000). If air temperature and SST are not used, then the code defaults to constant values. In particular, it defaults to the same potential temperature for air and sea, such that the potential temperature difference is zero (neutral stratification). The model also requires a background guess for the winds. We used climatological data for 925 hPa winds in the tropics to solve the force balance in the mixed layer model. The mean 925 hPa winds were calculated from the NCEP/National Center for Atmospheric Research (NCAR) re-analysis (Kalnay et al., 1996). In situ sea level pressures were taken from the buoy observations shown in Figure 1.

4. Anchor pressure values

During the process of integrating from pressure gradients to pressure, the constant of integration can be set to either a constant value or a value that minimizes the difference between the pressure field and in situ pressure measurements at the time of scatterometer observations. We used different constants of integration in our analysis, based on the available moored-buoy observations in the Bay of Bengal (Figure 1), to compute the pressure fields: (1) a constant pressure of 1013 hPa (global average) throughout the study period, (2) pressure observations from only one buoy (**BD13**), (3) a daily average (one value) of observations from three buoys (**BD08**, **BD13**, and **BD14**), (4) a daily average (one value) of all six buoy observations, and (5) all available daily individual pressure values (six values). The pressure observations (average and individual values) used in the integration scheme on different dates during the cyclone period (27 October–2 November 2012) are shown in Table I.

5. Validation

By considering each individual buoy measurements (see Section on Data and Methods), we have 42 comparison points from six buoy locations in 7 days. When we use a pressure observation from a single buoy to set the constant of integration, we do not use values from this buoy in our comparison statistics; hence, we have 35 points in this case. But considering the average of three buoys or six buoys gives us 42 comparison points because these average values are different from any buoy observation. The statistical analysis comparing the pressures obtained from OSCAT with buoy measurements and with a constant pressure value of 1013 hPa is summarized in Table II. Among all comparisons, estimation obtained by using all six buoy pressure values yields the best results with a bias of −0.06, a root mean square difference (RMSD) of 0.67, an R^2 (coefficient of determination) of 0.90, a scatter index (SI: defined as the ratio between RMSD and the in situ data mean), and a standard deviation ratio (SDR: defined as the ratio between the standard deviation error and the in situ data standard deviation) of 0.33. Very poor results were obtained when a constant pressure of 1013 hPa was used.

The results obtained by using (1) one buoy observation, (2) the average of three buoy observations, (3)

Table I. Different pressure values used as the initial guess during 27 October–02 November 2012.

Date	One buoy value (BD13)	Average of three buoys (BD08, BD13, BD14)	Average of all six buoys	Buoy pressure values (set of six) (BD14/BD13/BD08/BD10/BD11/BD07)
27 October 2012	1005	1007	1008	1006.8/1005.6/1010.4/1010.0/1008.9/1007.4
28 October 2012	1007	1007	1008	1005.2/1007.9/1010.6/1010.3/1008.8/1006.0
29 October 2012	1008	1008	1008	1005.7/1008.2/1010.4/1010.1/1008.8/1005.4
30 October 2012	1006	1007	1007	1005.9/1006.5/1010.6/1010.1/1005.7/1007.1
31 October 2012	1008	1009	1008	1007.9/1008.7/1010.9/1010.1/1006.0/1008.4
1 November 2012	1011	1011	1011	1011.2/1011.0/1012.3/1011.7/1009.8/1010.9
2 November 2012	1010	1010	1010	1010.6/1010.6/1011.5/1011.3/1010.2/1010.4

Table II. Statistical analysis of the comparison between *in situ* and model-derived pressure during 27 October–2 November 2012 using different constants of integration.

Pressure criteria	Bias	RMSD	Coefficient of determination (R^2)	Scatter index (SI)	Standard deviation ratio (SDR)	Uncertainty in slope	Uncertainty in intercept
1013	4.77	4.97	0.54	0.00493	0.686	0.08	87.88
One buoy value	−0.29	1.43	0.72	0.00142	0.667	0.11	114.81
Average of three buoys	0.20	1.13	0.79	0.00112	0.536	0.08	84.29
Average of six buoys	0.34	0.84	0.88	0.00083	0.369	0.06	58.10
All buoys individual pressure values	−0.06	0.67	0.90	0.00067	0.325	0.05	50.86

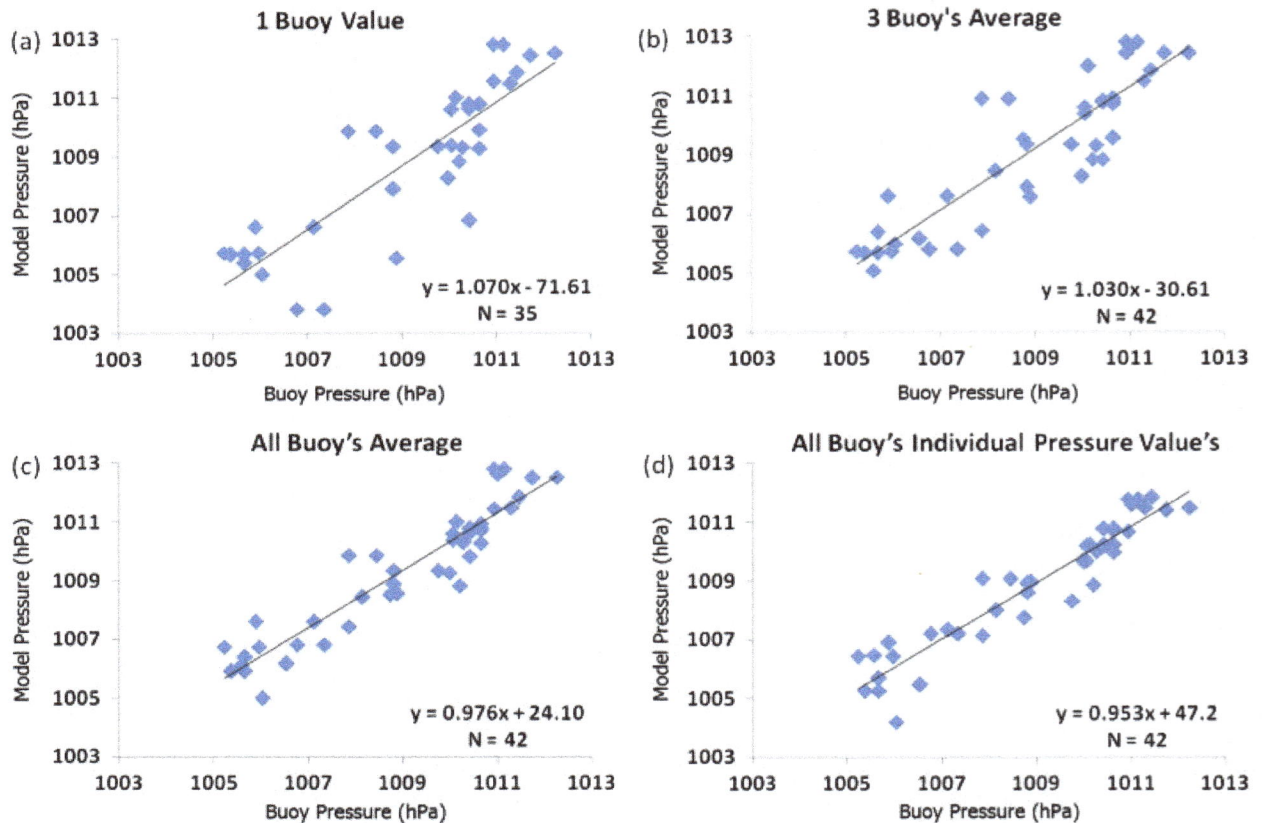

Figure 2. The scatter between the *in situ* and model-derived pressure during 27 October–2 November 2012. Four different initial pressure values are given for the 7 days: (a) one pressure value from Buoy-BD13, (b) one average pressure value from three buoys (BD08, BD13, BD14), (c) one average pressure from all six buoys, and (d) pressure values from the six individual buoys.

the average of six buoy observations, and (4) the individual values of all six buoys are shown in Figure 2. (The results obtained by using a constant pressure value of 1013 hPa are not shown.) Of the integration methods we adopted, considering all individual buoy measurements yielded the best result (Figure 2(d) and Table II). The regression has a good fit, with a slope of 0.95 and an intercept of 47 hPa when all the individual pressure values are considered. From this comparison, we can conclude that using more *in situ* pressure measurements to determine the constant of integration will yield better absolute values of pressure from scatterometer winds. From this point onward in the study, we used the pressure estimated by using observations from all six buoys. The low uncertainties in the slope and *y*-intercept are another indication of

the excellent quality of our pressure values. We then conducted one more analysis in which we estimated the UWPBL pressure on the first day over the entire satellite pass using the six individual buoy observations to determine the constant of integration. For the buoy locations, we then obtained the mean difference between these scatterometer-derived values for the first day and the buoy observations from the following day. This mean difference was then treated as a bias, and it was applied to all the UWPBL pressure values of the first day; these bias-corrected pressure values were then given as the first guess on the second day. In a similar way, the mean difference between the UWPBL-estimated pressure of the second day and the buoy values of the third day is applied to the estimated pressure of the second day. This approach is used to

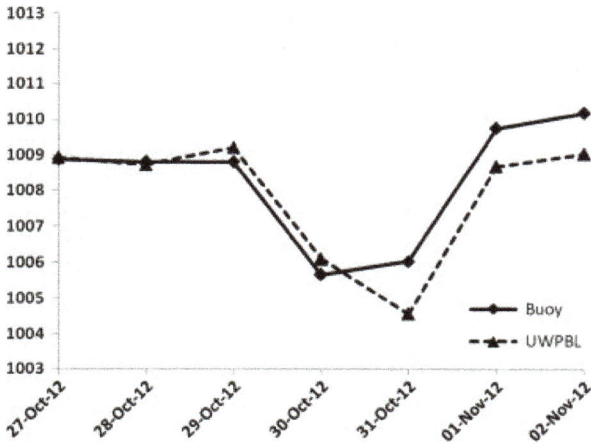

Figure 3. Comparison of scatterometer-derived pressure at Buoy-BD11 with the observed pressure.

Besides validating the scatterometer-derived pressure fields with individual buoy observations, we compared the pressure difference between buoy pairs. Thus, by using different buoy combinations, we obtained 15 pressure differences for 1 day and 105 values for 7 days from the six buoys. Similarly, we calculated the corresponding pressure differences from the scatterometer-derived pressure fields at the same buoy pair locations. The pressure fields estimated by using all six individual buoy observations (Table II) correlate relatively well with *in situ* measurements. There is an R^2 of 0.81, a slope of 0.9, an uncertainty in slope of 0.04, and an uncertainty in intercept of 0.11 hPa (Figure 4); thus, the scatterometer-derived pressure fields also provide relatively accurate pressure gradients.

determine the pressure field's initial guess to run the model for each remaining day. In this approach, the actual buoy observations are given only on the first day run. Based on a comparison of buoy pressures to these scatterometer-derived pressures, these estimations are more accurate than those using a constant value of 1013, one buoy value, three buoy or six buoy average observations as the first guess (first three approaches listed in Table II). However, the estimations are not as accurate as those obtained by individually using all six buoy values as the initial guess.

To compare the model estimations with an individual buoy whose values were not used in the model, we estimated pressure by using only five buoy observations and the resulting estimated pressure values were compared with *in situ* values of the sixth buoy (BD11: Figure 3). This comparison had a correlation of 0.90 and an RMSD of 0.84 hPa.

6. Spatial structure of pressure fields

On 27 October at 1205 IST, the pressure derived from scatterometer winds was about 1008 hPa at the center of the cyclone (figure not shown). This low pressure developed into a well-formed depression with 1004 hPa central pressure (at 'X' in Figure 5) on 28 October, and was centered at about 9°N and 83°E. (Because of the limitation of the swath we do not have observations beyond 90°E.) The cyclone then continued northwestward and the central pressure decreased to 1002 hPa.

The scatterometer-derived sea level pressure fields well describe the movement and intensity of the cyclone. Except on 29 October, the lowest pressure values were observed on the right side of the track. Generally, as the translational vector of the cyclone adds to the system's rotational winds, a Northern Hemisphere cyclone's winds will be maximum on the

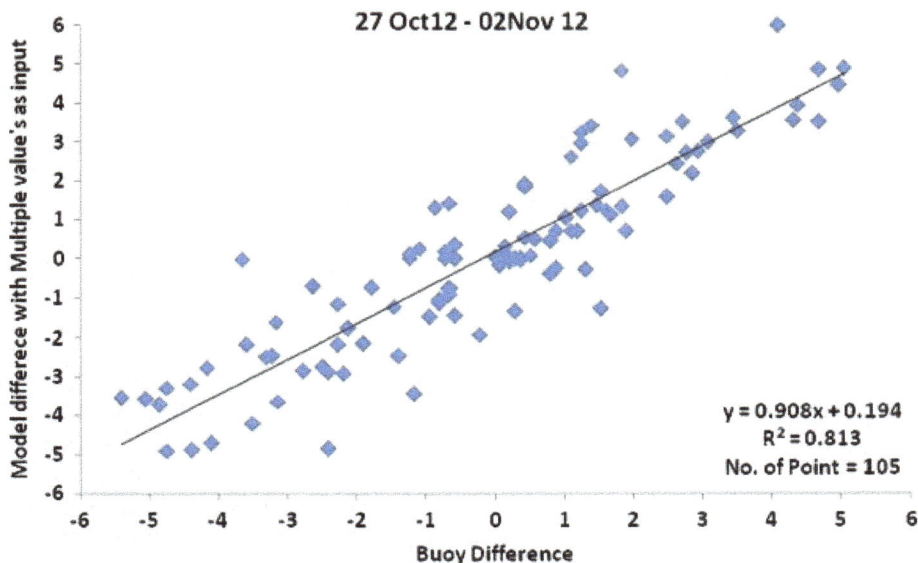

Figure 4. Scatter between the pressure differences between the buoy observations and the pressure differences estimated from scatterometer winds.

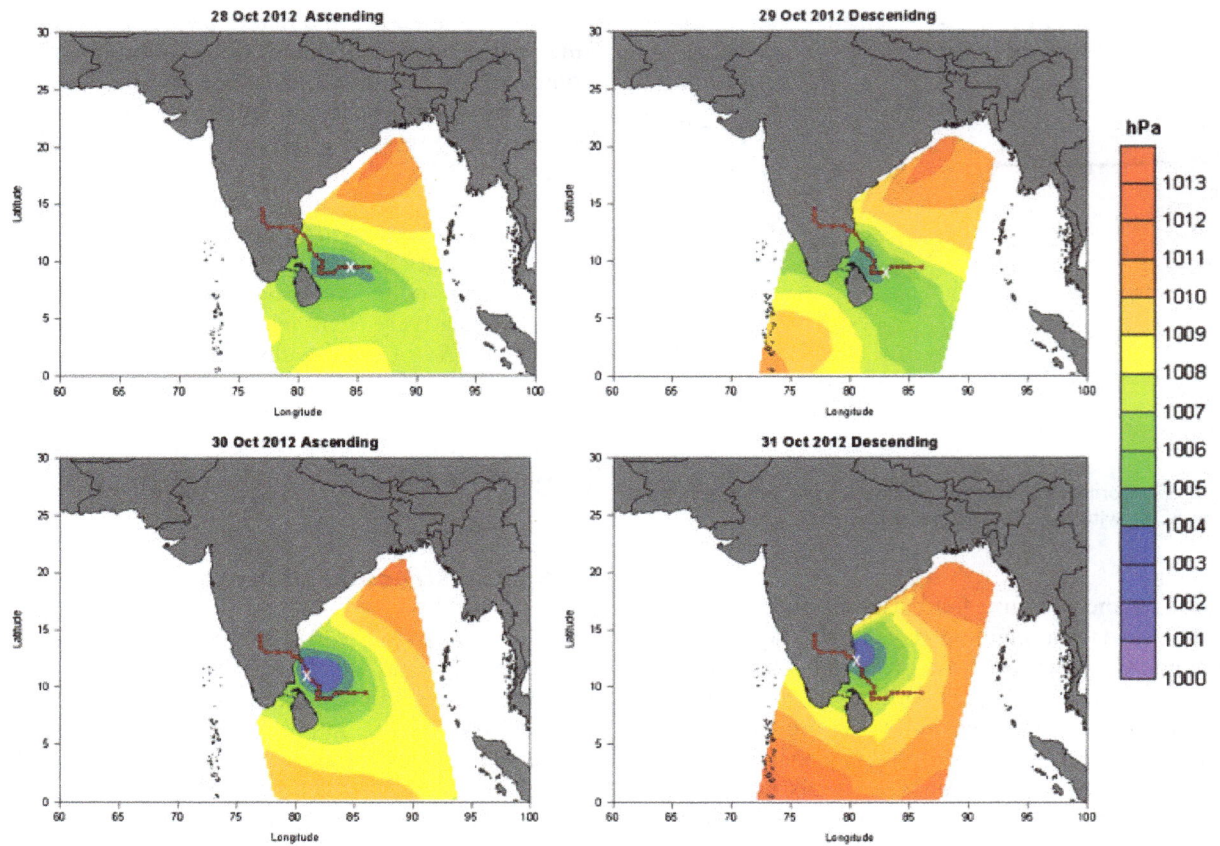

Figure 5. Distribution of sea level pressure fields along with cyclone tracks. The center of the cyclone location is marked as 'X' on the figures.

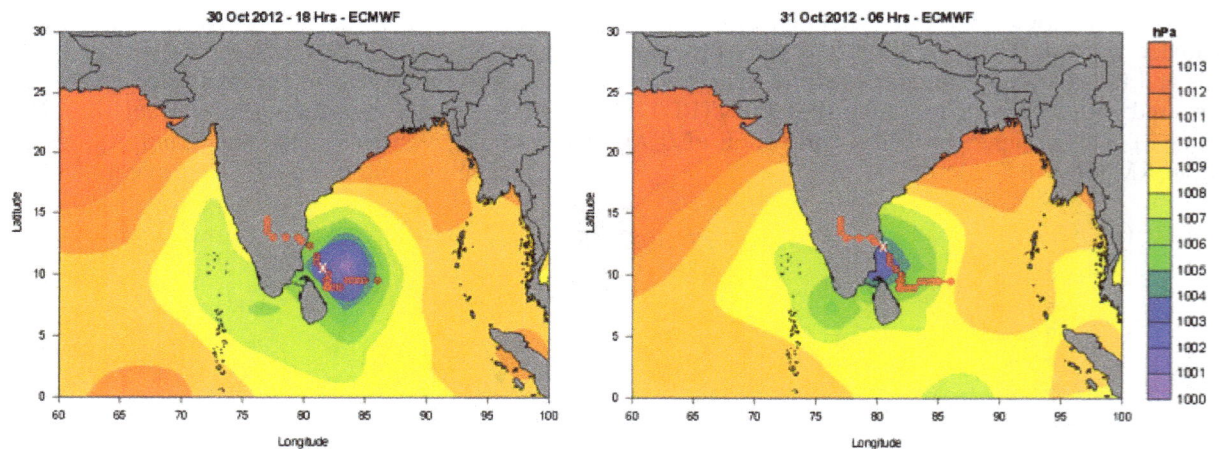

Figure 6. Distribution of sea level pressure fields from ECMWF re-analysis along with cyclone tracks. The center of the cyclone location is marked as 'X' on the figures.

right side of the track. The contours became more circular on 30 and 31 October as the cyclone intensified.

7. Comparison with ECMWF re-analyzed fields

To confirm the superiority of the scatterometer-derived pressures, we compared the ECMWF re-analyzed

pressure values with the buoy observations. The best fit line (figure not shown) between the ECMWF and *in situ* (scatterometer and in situ) has a slope of 0.72, an R^2 of 0.88, and an RMSD of 0.84 hPa, whereas these statistics between scatterometer-derived pressure and *in situ* observations are 0.95, 0.90, and 0.67. Thus, scatterometer-derived pressure fields correlate better with the *in situ* measurements. The spatial distribution of ECMWF re-analyzed pressure fields on 30 and 31 October 2012 are shown in Figure 6. The lowest pressure obtained from the scatterometer is closer to the observed center of the cyclone than found for the cyclone center in the ECMWF re-analysis pressure fields. In addition, the values are closer to the values reported by IMD. Hence, we can conclude that the central locations of the cyclone on these 2 days are better represented in the scatterometer-derived pressure fields than in the ECMWF re-analyze pressure values.

8. Summary and conclusions

We estimated the sea level pressure fields during Cyclone Nilam (27 October–2 November 2012) in the Bay of Bengal, using OSCAT wind measurements and the University of Washington planetary boundary layer model. We compared the resulting sea level pressure values with buoy measurements. Best results, an RMSD of 0.67 hPa and R^2 of 0.90, were obtained when all individual buoy measurements were integrated into the model as the first guess. Very small values of the scatter index (0.00067), standard deviation ratio (0.33), and very low uncertainty in fitting parameters show the accuracy of the estimations. Besides the individual pressure values, gradients are also well estimated by the model. The OSCAT-derived pressure fields describe the evolution of the cyclone well, both in track and intensity, compared to the ECMWF re-analysis fields.

Acknowledgements

The authors thank the respective organizations for the encouragement and support given during the progress of this work. This work is carried out as part of the Oceansat-2 Announcement of Opportunity Project of Space Applications Centre, ISRO, and with National Remote Sensing Centre (NRSC), ISRO and NASA support. Pressure data from the buoy observations were obtained from the INCOIS website and the cyclone tracks from IMD. The OSCAT wind vectors were provided by NSRC/ISRO. The authors thank C-DAC, Pune for their help in understanding the code. The authors have no conflict of interest in any way. [Correction added 29 January 2014 after original online publication: the Acknowledgements section has been amended.]

References

Brown RA, Levy G. 1986. Ocean surface pressure fields from satellite sensed winds. *Monthly Weather Review* **114**: 2197–2206.

Chakraborty A, Deb SK, Sikhakolli R, Gohil BS, Kumar R. 2013. Intercomparison of OSCAT winds with numerical-model-generated winds. *IEEE Geoscience and Remote Sensing Letters* **10**(2): 260–262.

Harlan J Jr, O'Brien JJ. 1986. Assimilation of scatterometer winds into surface pressure fields using a variational method. *Journal of Geophysical Research* **91**: 7816–7836.

Hsu CS, Liu WT. 1996. Wind and pressure fields near tropical cyclone Oliver derived from scatterometer observations. *Journal of Geophysical Research* **101**: 17021–17027.

Hsu CS, Wurtele MG, Cunningham GF, Woiceshyn PM. 1997. Construction of marine surface pressure fields from scatterometer winds alone. *Journal of Applied Meteorology* **36**: 1249–1261.

Cyclone Warning Division (2012). A Preliminary Report on cyclonic storm, NILAM over Bay of Bengal, India Meteorological Department, New Delhi. (www.imd.gov.in).

Kalnay E, Kanamitsu M, Kistler R, Collins W, Deaven D, Gandin L, Iredell M, Saha S, White G, Woollen J, Zhu Y, Chelliah M, Ebisuzaki W, Higgins W, Janowiak J, Mo KC, Ropelewski C, Wang J, Leetmaa A, Reynolds R, Jenne R, Joseph D. 1996. The NCEP/NCAR 40-year reanalysis project. *Bulletin of the American Meteorological Society* **77**(3): 437–471.

Patoux J. 2000. UWPBL 3.0, The University of Washington Planetary Boundary Layer (UWPBL) Model, Technical Note, University of Washington, 54 pp.

Patoux J, Foster RC, Brown RA. 2003. Global pressure fields from scatterometer winds. *Journal of Applied Meteorology* **42**: 813–826.

Taylor J. 1997. An Introduction to Error Analysis: The Study of Uncertainties in Physical Measurement, University Science Books, ISBN: 093570275X (ISBN13: 9780935702750).

Zierden DF, Bourassa MA, O'Brien JJ. 2000. Cyclone surface pressure fields and frontogenesis from NASA scatterometer (NSCAT) winds. *Journal of Geophysical Research* **105**: 23967–23981.

7

Atmospheric water parameters measured by a ground-based microwave radiometer and compared with the WRF model

Federico Cossu,[1,2]* Klemens Hocke,[1,2] Andrey Martynov,[2,3] Olivia Martius[2,3] and Christian Mätzler[1,2]

[1]Institute of Applied Physics, University of Bern, Switzerland
[2]Oeschger Centre for Climate Change Research, University of Bern, Switzerland
[3]Institute of Geography, University of Bern, Switzerland

*Correspondence to:
F. Cossu, Institute of Applied Physics, University of Bern, Sidlerstrasse 5, Bern 3012, Switzerland.
E-mail:
federico.cossu@iap.unibe.ch

Abstract

The microwave radiometer TROWARA measures integrated water vapour (IWV) and integrated cloud liquid water (ILW) at Bern since 1994 with a time resolution of 7 s. In this study, we compare TROWARA measurements with a simulation of summer 2012 in Switzerland performed with the Weather Research and Forecasting (WRF) model. It is found that the WRF model agrees very well with TROWARA's IWV variations with a mean bias of only 0.7 mm. The ILW distribution of the WRF model, although similar in shape to TROWARA's distribution, overestimates the fraction of clear sky periods (83% compared to 60%).

Keywords: microwave radiometer; integrated water vapour; integrated cloud liquid water; WRF model

1. Introduction

The TROpospheric WAter RAdiometer (TROWARA) continuously monitors vertically integrated water vapour (IWV) and vertically integrated cloud liquid water (ILW) at Bern since 1994.

Water vapour and clouds are essential components of the climate system for their role in the water cycle and in the Earth's radiation energy budget. Water vapour, being the most important natural greenhouse gas, contributes to the warming of the atmosphere. Clouds, instead, play a dual role: the cloud droplets, whose mean radius ranges from 4 to 24 μm depending on cloud type (Mason, 1971), reflect the shortwave radiation from the sun (cooling effect) and at the same time they absorb and re-emit the infrared long-wave radiation from the surface (warming effect).

Uncertainties in the representation of cloud processes in climate models explain much of the spread in modelled climate sensitivity, leading to regional errors on cloud radiative effect (CRE) of several tens of watts per square metre (Flato et al., 2013). Komurcu et al. (2014) showed that global annual mean fields of cloud liquid water path (LWP, identical to ILW) of six different global climate models (GCMs) can differ by more than a factor of two and that the GCMs' mean LWP is more than two times larger than the highest value observed by satellites.

The fractional coverage of clouds and their liquid water content are two important properties which need to be further investigated in order to understand the biases in CRE in models. A small change of 0.1 mm (1 mm is equal to 1 kg m^{-2}) in ILW produces a variation of several hundred watts per square metre in the downward shortwave flux at the surface (Turner et al., 2007). Given the scarcity of high-quality ground-based cloud liquid water measurements, TROWARA's observations are valuable for a more in-depth analysis of those properties and their comparison with regional climate models (RCMs).

Until now, only TROWARA's water vapour measurements have been considered and a trend analysis study of IWV has been published by Hocke et al. (2011). Here we present for the first time TROWARA measurements of integrated cloud liquid water (ILW). The aim of this study is to compare data of IWV and ILW from TROWARA with coincident data from a RCM simulation of summer 2012 in Switzerland performed with the Weather Research and Forecasting (WRF) model.

The outline of this article is as follows. In Section 2, we provide detailed information about TROWARA and the WRF model simulation. In Sections 3 and 4 the time series of IWV and ILW and the characteristics of the statistical distributions of ILW are analysed for TROWARA and for the WRF model. Section 5 contains an outlook on future work. The conclusions about the comparison are presented in Section 6, together with a discussion about the microphysical schemes influence on the WRF model simulation.

2. Data

2.1. TROWARA data

The TROWARA microwave radiometer has been operated at the University of Bern (46.95°N, 7.44°E, 575 m a.s.l.) since 1994. To improve the measurement accuracy during rainy periods, in 2002 TROWARA was moved inside a temperature-controlled room, looking towards southeast through a styrofoam window which is transparent to microwaves. Due to the southeast exposure, most rain events do not deposit any rain on the window. Therefore emission of most rain events can be accurately measured.

The instrument consists of two microwave channels at 21.4 GHz (bandwidth 100 MHz) and 31.5 GHz (bandwidth 200 MHz). The lower frequency is more sensitive to microwave emission from water vapour, while the higher frequency is more sensitive to liquid water. In addition, TROWARA has a channel in the thermal infrared ($\lambda = 9.5 - 11.5\,\mu m$) which is required for the estimation of the cloud temperature. The standard elevation angle is 40° and the main lobe of TROWARA's antenna pattern has a full width of 4° at half power.

The measurement principle of TROWARA is based on passive remote sensing of the microwave emission from water vapour and liquid cloud droplets. The zenith opacities at 21 and 31 GHz are derived from the observed brightness temperatures in the two microwave channels. Then, IWV and ILW are derived from a linear combination of the opacities. The retrieval process requires auxiliary information such as surface temperature, pressure and relative humidity which are provided by the local meteorological station.

Compared to the satellite-based microwave radiometers, their surface-based counterparts like TROWARA have the advantage of a well-known cosmic background whose brightness temperature is very isotropic and much less than any atmospheric temperature. Peter and Kämpfer (1992) stated that for TROWARA the error of IWV is about 3% of the measured value. This error is due to the variable water vapour sensitivity of the instrument with altitude. For ILW Peter and Kämpfer (1992) stated an error of 10–20% of the measured value. This error is due to uncertainties in the dielectric constant of water at the unknown cloud temperature. In the currently adopted refined retrieval of Mätzler and Morland (2009), the cloud temperature is estimated using an infrared radiometer, thus reducing the ILW error. Furthermore, progress has been achieved in dielectric models of water (Mätzler et al., 2010). In addition to the percent error, there was a zero bias on the order of 0.01–0.02 mm. Improvements have also been achieved by refined physical retrievals. The present instrument uses a correction to reduce this error to 0.001 mm (Mätzler and Morland, 2009). The remaining uncertainty is limited to rainy conditions when the emission is enhanced by Mie effects.

During rain, the microwave emission of the rain droplets is much larger than that of the cloud droplets. As we are focusing on the estimation of the cloud droplets only (suspended hydrometeors), the emission from the large precipitating hydrometeors (i.e. rain droplets) is a contamination to our measurements.

Therefore, in these conditions, a proper determination of cloud liquid water is not feasible. To overcome this problem, the retrieval algorithm has been designed to allow the user to choose a threshold for ILW, beyond which all further emission is attributed to rain. Mätzler and Morland (2009) found that the presence of rain droplets is likely for ILW values exceeding 0.4 mm (or $0.4\,kg\,m^{-2}$). Thus our study concentrates on the analysis of ILW measurements below the threshold of 0.4 mm. We applied the retrieval algorithm to the year 2012 and obtained two time series of IWV and ILW with a time resolution of about 7 s.

2.2. WRF simulation

We performed a simulation of the summer 2012 (June, July and August) in Switzerland with the WRF model version 3.4.1 (Skamarock et al., 2008). The simulation domain extends from 42.72° to 49.91°N and from 4.14° to 12.09°E (Figure S1, Supporting Information) with a horizontal resolution of 2.14 km and 35 vertical levels (top level at 50 hPa). The time step is 4 s and the output is saved every 60 min. The initial state and the update of the boundary conditions (every 6 h) are specified with ECMWF analysis data.

The simulation was repeated using four microphysical schemes to find the scheme which produces the most realistic distributions of water vapour and clouds. The schemes considered in this study are the WRF single-moment 6-class scheme (WSM6) (Hong and Lim, 2006), the new Thompson et al. scheme (NT) (Thompson et al., 2008), the Milbrandt-Yau double-moment 7-class scheme (MY) (Milbrandt and Yau, 2005) and the Morrison double-moment scheme (MO) (Morrison et al., 2009). It is known that simulations with the same configuration and initial conditions but with different microphysical schemes can produce different results (Jankov et al., 2005, 2009; Otkin and Greenwald, 2008; Mercader et al., 2010; Awan et al., 2011). A comparison of 13 microphysical schemes for an idealized WRF model simulation can be found in Cossu and Hocke (2014).

For the model comparison with TROWARA measurements, we used the closest grid point to the instrument location (grid point at 46.949°N, 7.432°E). Every microphysical scheme of the WRF model partitions atmospheric water in water vapour and in several hydrometeor classes, such as cloud liquid water, cloud ice, rain droplets and so on. We computed IWV and ILW from the mass mixing ratio vertical profiles of water vapour and cloud liquid water.

3. IWV comparison

Summer months in Switzerland are characterized by the highest amounts of IWV throughout the year which can reach values between 40 and 45 mm. Water vapour is generally transported from the Atlantic Ocean to

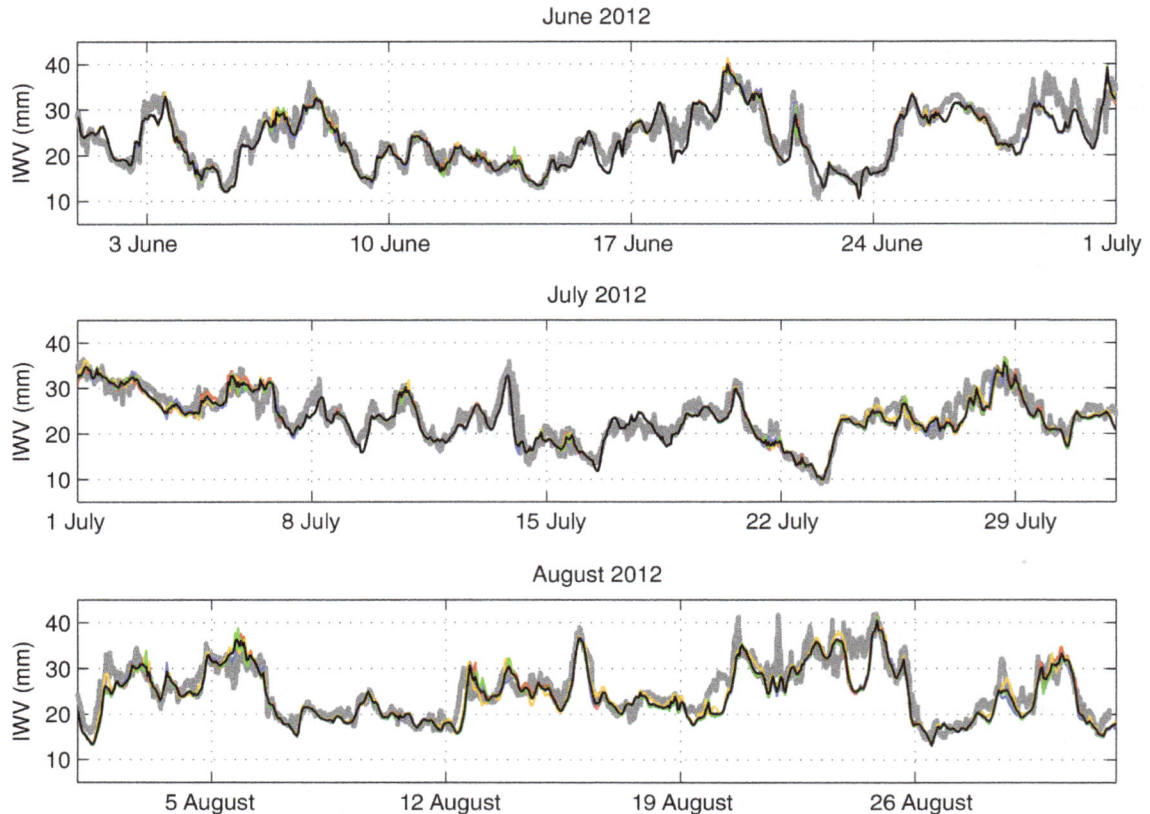

Figure 1. Integrated water vapour (IWV) measured by TROWARA in summer 2012 (grey points) and simulated by the WRF model with four different microphysical schemes (blue, red, yellow and green lines with their mean in black).

Switzerland, but local production of significant quantities of water vapour may occur during hot summer days by evaporation from lakes, rivers, vegetation and moist soil (Sodemann and Zubler, 2010).

Figure 1 shows the time series of IWV measured by TROWARA (grey points) and simulated by the WRF model with the four different microphysical schemes (blue, red, yellow and green lines with their mean in black). TROWARA measurements range from about 9 mm on 23 July to about 42 mm on 24 August with an average of about 24 mm.

The microphysical schemes (WSM6, NT, MY and MO) do not deviate too much from each other. The maximum mean deviation is between MY and MO, with a mean difference IWV_{MY} minus IWV_{MO} of 0.19 mm and a standard deviation of 0.84 mm. The minimum variability between the schemes is found during periods without clouds, for example between 14 and 19 June (a zoom on this period is provided in Figure S2).

The average value of the microphysical schemes, $IWV_{<WRF>}$, follows the TROWARA time series well. The mean difference $IWV_{TROWARA}$ minus $IWV_{<WRF>}$ is 0.7 mm and its standard deviation is 2.3 mm. IWV is simulated with great accuracy being $IWV_{<WRF>}$ on average only 2.4% lower than $IWV_{TROWARA}$.

The good agreement of the model with the measurements can also be seen in the scatter plot of Figure 2, where we show $IWV_{<WRF>}$ versus $IWV_{TROWARA}$. The points above the identity line (overestimation)

are shown in magenta colour, while the points below the identity line (underestimation) are shown in cyan colour. The linear regression line is shown in black and has a correlation coefficient of 0.915, indicating the good agreement between the WRF model and TROWARA. The regression line lies for the most part under the identity line, meaning that the underestimation is stronger than the overestimation. By analysing the occurrence times and the coincident ILW values, we found that the overestimation occurs most often in the late evening and in more cloudy conditions (average ILW = 0.062 mm), while the underestimation occurs most often in the early afternoon and in less cloudy conditions (average ILW = 0.025 mm) (occurrence times distributions in Figure S3).

4. ILW comparison

ILW is the vertical integral of the cloud liquid-water content (kg m^{-3}), which depends on the drop size distribution (DSD) of the cloud droplets. Because of the complex microphysical and dynamical processes within clouds, DSDs are highly variable in time and space (Westwater et al., 2005), causing ILW to vary as well. Other processes that determine the variability of DSD and ILW are geographical location and cloud height, local convection and fluxes from the surface, regional and large-scale dynamics.

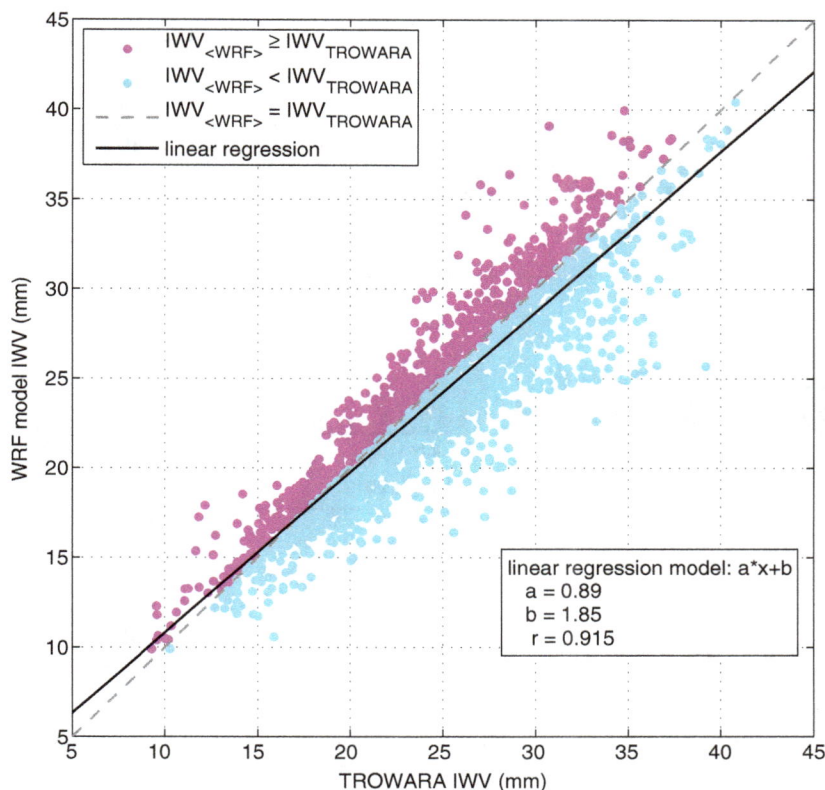

Figure 2. Scatter plot of IWV simulated by the WRF model (average of four different microphysical schemes) versus IWV measured by TROWARA. The dashed grey line is the identity line, the black line is the linear regression line, the magenta points are the cases in which the WRF model overestimates TROWARA measurements and the cyan points the cases of underestimation.

The variability of ILW is larger compared to the variability of IWV. The relative variation of IWV and ILW (computed as the ratio of standard deviation to mean value) for an interval of about 10 min provides a good measure of the variability of the two parameters. For TROWARA measurements in summer 2012, the mean relative variation is about 1% for IWV and about 51% for ILW. The frequency distribution of the relative variation of IWV and ILW is provided in Figure S4.

4.1. Data overview

Figure 3 gives an overview of the ILW data sets of TROWARA and the WRF model and their agreement or disagreement. As for Figure 1, the grey points indicate TROWARA measurements and the coloured lines indicate the four WRF model microphysical schemes with their mean in black. The simulated ILW does not match the radiometer measurements as well as in the case of IWV. The mismatch is caused by the model poor sampling (1 record every hour) and by the difficulty to represent cloud-scale processes in models.

For periods characterized by a more frequent and prolonged occurrence of clouds, as for example between 3 and 14 June, the WRF model performs better, showing a similar probability in the occurrence of clouds to TROWARA. Further, in extended periods dominated by fair weather, as between 14 and 17 June, TROWARA and the WRF model agree in showing no clouds.

Another difference between TROWARA and the WRF model is the range of ILW values. TROWARA values are limited to 0.4 mm, as explained in Section 2.1, while the WRF model values can be higher. The maximum values, which are not displayed in Figure 3, are 1.41 mm (WSM6), 1.49 mm (NT), 2.16 mm (MY) and 3.21 mm (MO).

4.2. ILW distribution

In the following sections we change our focus from the absolute ILW values to their occurrence probability which allows to characterize their distributions. To compute the ILW distributions, we divide the ILW values into three groups: the first group contains the values smaller than 0.01 mm, which represent, ignoring ice clouds (TROWARA is not sensitive to ice particles and we ignored ice clouds in the WRF model as well), clear sky conditions; the second group is subdivided into 10 intervals between 0.01 and 0.4 mm; the third group includes the values greater than or equal to 0.4 mm. The probability of occurrence of each bin is computed by counting the number of values that fall in that bin divided by the total number of points. The computed ILW distributions for TROWARA and for the WRF model are shown in Figure 4.

In summer 2012 at Bern, TROWARA measures a clear sky fraction of 60%, while it is 83% on average for the WRF model simulations. The surplus of clouds

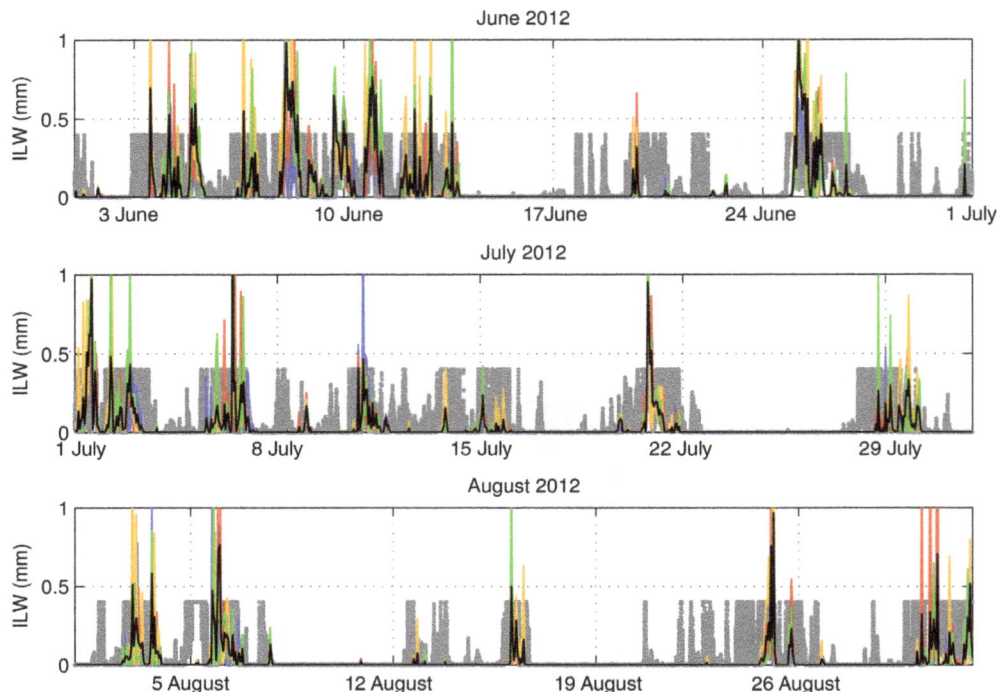

Figure 3. Integrated cloud liquid water (ILW) measured by TROWARA in summer 2012 (grey points) and simulated by the WRF model with four different microphysical schemes (blue, red, yellow and green lines with their mean in black).

in TROWARA data compared to that of the WRF model can also be clearly seen in the time series of Figure 3. There are often periods in which TROWARA measures a positive ILW and the WRF model a null ILW. For these periods, there is an easy way to verify whether TROWARA or WRF is correct by looking at the archive images of the local webcam. A cloudy day is easily recognizable and it would rule out the null ILW value simulated by the WRF model. By looking at the webcam images of the Climate and Environmental Physics Institute of the University of Bern, we could confirm the validity of TROWARA measurements. One of these images is shown in Figure S5.

Between 0.01 and 0.4 mm the ILW distributions of both TROWARA and the WRF model are clearly decreasing, meaning that thick (thin) clouds and/or clouds with a high (low) liquid water content occur less (more) frequently. Owing to the high sensitivity of TROWARA and to the large number of available points, it is possible to increase the binning in this range and to better define the shape of the distribution, as shown in Figure 5. The probability distribution between 0.01 and 0.4 mm seems to follow a power law function and by fitting the data with a two-term power model we obtain the following equation:

$$P(\text{ILW}) = a\text{ILW}^b + c \qquad (0.01 \leq \text{ILW} < 0.4) \quad (1)$$

where $P(\text{ILW})$ is the probability density function of the continuous variable ILW between 0.01 and 0.4 mm, $a = 0.0022$, $b = -0.71$ and $c = -0.0028$. The coefficient of determination R^2 between the measured ILW distribution and the computed probability density function is 0.996, indicating that the ILW distribution in this range resembles well a power law function.

The last bin from 0.4 mm up, albeit having a width of 0.01 mm, includes all the values ≥ 0.4 mm, the threshold value. The occurrence probability of the last bin is lower in the WRF model (0.03 on average) than in TROWARA (0.11) partially because most of the WRF model values are in the first bin, causing the part of the distribution with ILW ≥ 0.01 mm to be lower.

The ILW distribution of the WRF model varies slightly from scheme to scheme and it is similar to TROWARA. In fact, the first bin is the highest one, the bins between 0.01 and 0.4 mm decrease in size and the last bin is higher than the previous ones. A zoom of the occurrence probability between 0 and 0.1 for the WRF model ILW distributions is provided in Figure S6.

There is a substantial difference in the number of points used to compute the distributions of TROWARA and of the WRF model: more than one million points for TROWARA compared to 2209 points for the WRF model. A larger number of points for the WRF model would allow not only to draw more robust statistical conclusions on the clear sky occurrence probability but also to study in more detail the shape of the distribution between 0.01 and 0.4 mm, as done with TROWARA, and to enhance the differences between the microphysical schemes. To increase the number of available points the output frequency must be increased or the simulation period must be extended. It was not possible for us to save the simulation data with such a high frequency because of hard disk space limitations.

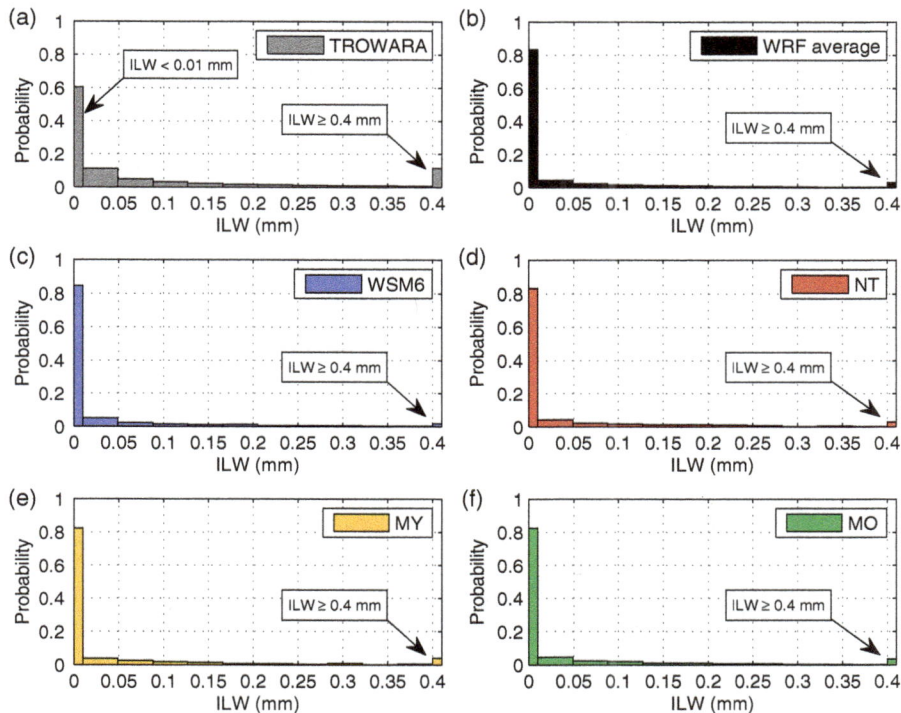

Figure 4. (a) ILW distribution of the values measured by TROWARA in summer 2012. The first bin on the left represents all the values smaller than 0.01 mm (even small negative values), while the last bin on the right represents all the values equal to or greater than 0.4 mm. (b) Average WRF ILW distribution computed by averaging the ILW distributions of the four microphysical schemes. (c–f) ILW distribution of the values simulated by the WRF model using four different microphysical schemes. The last bin on the right in each distribution represents all the values equal to or greater than 0.4 mm.

5. Future work

In future, we will investigate whether the statistics of the simulated ILW can be improved by analysing the output from several model grid-points in the surroundings of Bern. The average ILW distribution of the nearby grid-points could reduce the point-to-point internal variability of the numerical model and improve the model performance at this location.

Simulations with a higher output frequency will allow to better define the shape of the ILW distributions and to enhance the differences between the schemes.

By considering a longer time period, the seasonal behaviour of TROWARA ILW will be analysed to discover possible changes in the clear sky occurrence probability and in the shape of the ILW distribution over the past few years.

Because microwave radiometry is a well established and possibly the most accurate method for retrieving IWV and ILW (Westwater *et al.*, 2005), TROWARA measurements are also valuable for validation of satellite measurements, as in the study of Roebeling *et al.* (2008).

6. Conclusions

In this study, we presented a comparison of IWV and ILW measured in summer 2012 at Bern by the microwave radiometer TROWARA and simulated with the WRF model using different microphysical parameterizations.

We have found that over a 3-month simulation the WRF model agrees very well with TROWARA's IWV variations. The mean bias $IWV_{TROWARA}$ minus $IWV_{<WRF>}$ is only 0.7 mm and the standard deviation is 2.3 mm. The variation in IWV between the microphysical schemes was minimal and the schemes that differed the most were MY and MO with a maximum mean deviation IWV_{MY} minus IWV_{MO} of 0.19 mm and a standard deviation of 0.84 mm.

For ILW, the main focus of our comparison was on the probability density function of ILW which is especially relevant for climate research, since even small differences in cloud liquid water greatly modify the atmospheric radiative fluxes (Turner *et al.*, 2007). By computing the ILW distributions, we found only a partial agreement between TROWARA and the WRF model. In both ILW distributions the highest occurrence probability is for clear sky periods and the shape of both distributions is gradually decreasing between 0.01 and 0.4 mm. The WRF model, however, overestimates the clear sky fraction (83%) compared to TROWARA (60%).

As in the case of IWV, we did not find substantial differences in the ILW distributions simulated with the selected microphysical schemes. All the four WRF model ILW distributions have in fact almost the same clear sky occurrence probability and a decreasing probability for ILW ≥0.01 mm. The relatively small

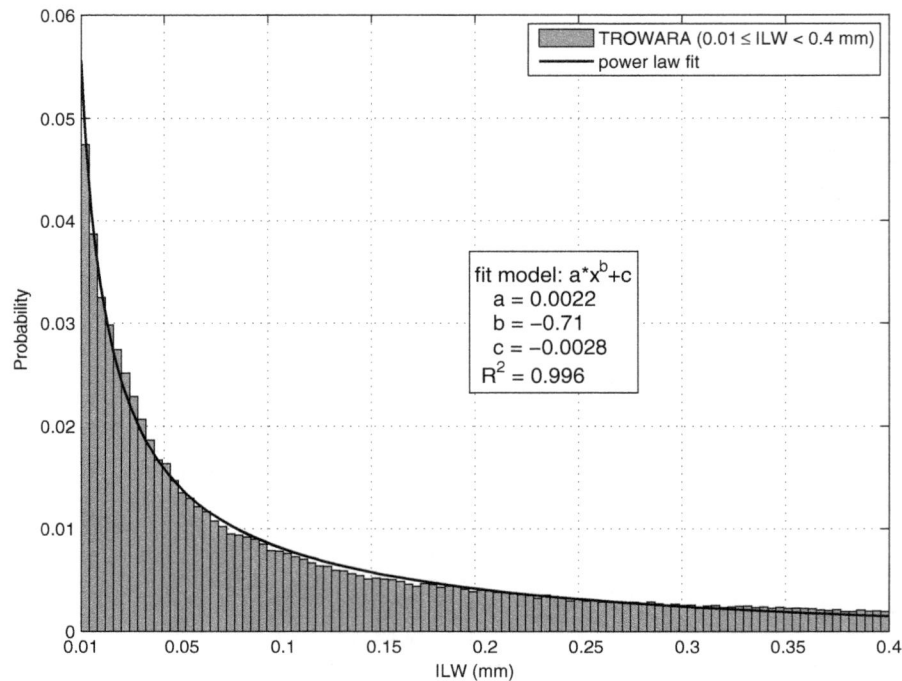

Figure 5. Finer binning and power law fit of the central part of TROWARA ILW distribution between 0.01 and 0.4 mm.

number of available points has however hindered a more detailed comparison of the distributions.

Additionally, we characterized TROWARA's ILW distribution between 0.01 and 0.4 mm, finding that it is well represented by a power law function with an exponent of − 0.71 (Equation 1).

Acknowledgements

The authors would like to thank the Oeschger Centre for Climate Change Research for providing the funding, the technicians and engineers of the Institute of Applied Physics for keeping TROWARA operative throughout the years and the WRF model developers for providing the freely available numerical model for the simulations.

Supporting information

The following supporting information is available:

Figure S1. The WRF simulation domain extends from 42.72° to 49.91°N and from 4.14° to 12.09°E. The white dot indicates the location of Bern (46.95°N, 7.44°E) where TROWARA is measuring.

Figure S2. IWV variability, expressed as difference between the microphysical schemes (upper panel), is lower during periods without clouds (null ILW, lower panel).

Figure S3. Dependence on local time for the cases in which the WRF model overestimates TROWARA's IWV (upper panel) and for the cases of underestimation (lower panel). The y-axis represents the number of events per hour. The overestimation occurs most often in the late evening, while the underestimation occurs most often in the early afternoon.

Figure S4. Absolute frequency distributions of the relative variations of IWV (top panel) and ILW (bottom panel) measured by TROWARA in summer 2012. The relative variations are computed as the ratio of standard deviation to mean value for intervals of 10 min.

Figure S5. Webcam image from the archive of the Climate and Environmental Physics institute of the University of Bern showing stratus clouds on 24 August 2012 at 0800 UTC. At that time, TROWARA data correctly show a positive ILW value, while the WRF model simulations have a null ILW value. TROWARA is installed near the webcam, pointing in the same direction and with an elevation angle of 40°.

Figure S6. Zoom of the occurrence probability between 0 and 0.1 for the ILW distributions simulated by the WRF model using four different microphysical schemes.

References

Awan NK, Truhetz H, Gobiet A. 2011. Parameterization-induced error characteristics of MM5 and WRF operated in climate mode over the Alpine region: an ensemble-based analysis. *Journal of Climate* **24**: 3107–3123, doi: 10.1175/2011JCLI3674.1.

Cossu F, Hocke K. 2014. Influence of microphysical schemes on atmospheric water in the Weather Research and Forecasting model. *Geoscientific Model Development* **7**: 147–160, doi: 10.5194/gmd-7-147-2014.

Flato G, Marotzke J, Abiodun B, Braconnot P, Chou S, Collins W, Cox P, Driouech F, Emori S, Eyring V, Forest C, Gleckler P, Guilyardi E, Jakob C, Kattsov V, Reason C, Rummukainen M. 2013. Evaluation of climate models. In *Climate Change 2013: The Physical Science Basis. Contribution of Working Group I to the Fifth Assessment Report of the Intergovernmental Panel on Climate Change*, Stocker T, Qin D, Plattner G-K, Tignor M, Allen S, Boschung J, Nauels A, Xia Y, Bex V, Midgley P (eds). Cambridge University Press: Cambridge, UK and New York, NY.

Hocke K, Kämpfer N, Gerber C, Mätzler C. 2011. A complete long-term series of integrated water vapour from ground-based microwave radiometers. *International Journal of Remote Sensing* **32**: 751–765, doi: 10.1080/01431161.2010.517792.

Hong S-Y, Lim J-OJ. 2006. The WRF single-moment 6-class microphysics scheme (WSM6). *Journal of Korean Meteorological Society* **42**: 129–151.

Jankov I, Gallus WA, Segal M, Shaw B, Koch SE. 2005. The Impact of different WRF model physical parameterizations and their interactions on warm season MCS rainfall. *Weather Forecasting* **20**: 1048–1060, doi: 10.1175/WAF888.1.

Jankov I, Bao J-W, Neiman PJ, Schultz PJ, Yuan H, White AB. 2009. Evaluation and comparison of microphysical algorithms in ARW-WRF model simulations of atmospheric river events affecting the California coast. *Journal of Hydrometeorology* **10**: 847–870, doi: 10.1175/2009JHM1059.1.

Komurcu M, Storelvmo T, Tan I, Lohmann U, Yun Y, Penner JE, Wang Y, Liu X, Takemura T. 2014. Intercomparison of the cloud water phase among global climate models. *Journal of Geophysical Research: Atmospheres* **119**: 3372–3400, doi: 10.1002/2013JD021119.

Mason B. 1971. *The Physics of Clouds*. Oxford Monographs on Meteorology. Clarendon Press: Oxford, UK.

Mätzler C, Morland J. 2009. Refined physical retrieval of integrated water vapor and cloud liquid for microwave radiometer data. *IEEE Transactions on Geoscience and Remote Sensing* **47**: 1585–1594, doi: 10.1109/TGRS.2008.2006984.

Mätzler C, Rosenkranz PW, Cermak J. 2010. Microwave absorption of supercooled clouds and implications for the dielectric properties of water. *Journal of Geophysical Research: Atmospheres (1984–2012)* **115**, doi: 10.1029/2010JD014283.

Mercader J, Codina B, Sairouni A, Cunillera J. 2010. Results of the meteorological model WRF-ARW over Catalonia, using different parameterizations of convection and cloud microphysics. *Journal of Weather and Climate of the Western Mediterranean* **7**: 75–86, doi: 10.3369/tethys.2010.7.07.

Milbrandt J, Yau M. 2005. A multimoment bulk microphysics parameterization. Part I: analysis of the role of the spectral shape parameter. *Journal of the Atmospheric Sciences* **62**: 3051–3064, doi: 10.1175/JAS3534.1.

Morrison H, Thompson G, Tatarskii V. 2009. Impact of cloud microphysics on the development of trailing stratiform precipitation in a simulated squall line: comparison of one- and two-moment schemes.

Monthly Weather Review **137**: 991–1007, doi: 10.1175/2008MWR2556.1.

Otkin JA, Greenwald TJ. 2008. Comparison of WRF model-simulated and MODIS-derived cloud data. *Monthly Weather Review* **136**: 1957–1970, doi: 10.1175/2007MWR2293.1.

Peter R, Kämpfer N. 1992. Radiometric determination of water vapor and liquid water and its validation with other techniques. *Journal of Geophysical Research: Atmospheres* **97**: 18173–18183, doi: 10.1029/92JD01717.

Roebeling R, Deneke H, Feijt A. 2008. Validation of cloud liquid water path retrievals from SEVIRI using one year of CloudNET observations. *Journal of Applied Meteorology and Climatology* **47**: 206–222, doi: 10.1175/2007JAMC1661.1.

Skamarock W, Klemp J, Dudhia J, Gill D, Barker D, Duda M, Huang XY, Wang W. 2008. A description of the advanced research WRF version 3. Technical report, Mesoscale and Microscale Meteorology Division, National Center for Atmospheric Research, Boulder, CO. http://nldr.library.ucar.edu/repository/collections/TECH-NOTE-000-000-000-855 (accessed 13 December 2012).

Sodemann H, Zubler E. 2010. Seasonal and inter-annual variability of the moisture sources for Alpine precipitation during 1995–2002. *International Journal of Climatology* **30**: 947–961, doi: 10.1002/joc.1932.

Thompson G, Field PR, Rasmussen RM, Hall WD. 2008. Explicit forecasts of winter precipitation using an improved bulk microphysics scheme. Part II: implementation of a new snow parameterization. *Monthly Weather Review* **136**: 5095–5115, doi: 10.1175/2008MWR2387.1.

Turner D, Vogelmann A, Austin R, Barnard J, Cady-Pereira K, Chiu JC, Clough S, Flynn C, Khaiyer M, Liljegren J, Johnson K, Lin B, Long C, Marshak A, Matrosov S, McFarlane S, Miller M, Min Q, Minnis P, O'Hirok W, Wang Z and Wiscombe W. 2007. Thin liquid water clouds: their importance and our challenge. *Bulletin of the American Meteorological Society* **88**: 177–190, doi: 10.1175/BAMS-88-2-177.

Westwater ER, Crewell S, Matzler C. 2005. Surface-based microwave and millimeter wave radiometric remote sensing of the troposphere: a tutorial. *IEEE Geoscience and Remote Sensing Society Newsletter* **134**: 16–33.

8

The extreme forecast index at the seasonal scale

E. Dutra,* M. Diamantakis, I. Tsonevsky, E. Zsoter, F. Wetterhall, T. Stockdale, D. Richardson and F. Pappenberger

European Centre for Medium Range Weather Forecasts, Reading, UK

*Correspondence to:
E. Dutra, European Centre for
Medium Range Weather
Forecasts, Reading, UK.
E-mail:
emanuel.dutra@ecmwf.int*

Abstract

The extreme forecast index (EFI) concept has been applied to the European Centre for Medium-Range Weather Forecasts (ECMWF) seasonal forecasts (S4) of 2-m temperature (T2M) and total precipitation (TP) using a novel semi-analytical technique. Results derived from synthetic data highlight the importance of large ensemble sizes to reduce the EFI calculation uncertainty due to sampling. This new diagnostic complements current diagnostics as exemplified for the 2012 warm summer in south central and eastern Europe. The EFI provides an integrated measure of the difference between a particular seasonal forecast ensemble and the underlying model climate which can be used as an early warning indicator.

Keywords: seasonal forecasts; extreme forecast index; predictability

1. Introduction

The quality of ensemble forecast systems has steadily improved over the last decades and several measures of extreme or anomalous situations that can potentially provide the forecast users with early warning systems have been developed. The extreme forecast index (EFI, Lalaurette, 2003; Zsoter, 2006), developed at ECMWF, is an example of an index that was designed to identify situations where the medium-range ensemble prediction system (EPS) forecasts are detecting extreme situations. Detection of extremes can be accomplished by comparing model forecasts to the underlying model climatology (Lalaurette, 2003; Thielen-del Pozo et al., 2009; Bartholmes et al., 2009; Cloke et al., 2010; Alfieri et al., 2011). The major advantage of such an approach is that it can be applied everywhere including in areas where observations are sparse or unavailable and that it inherently accounts for the need of forecast calibration as it is based on relative difference between forecasts and model climatology. This does not overcome the problems associated with sparse or unavailable observations, in particular, surface observations of temperature and precipitation which are a limitation in any evaluation of the model skill.

At the seasonal time scales (up to 6-month lead time) comparable extreme indexes have not been applied. Operational detection of 'extremes' has been mainly focused at the probability of exceeding percentiles. Seasonal forecasting is not a weather forecast: weather can be considered as a snapshot of continually changing atmospheric conditions. Seasonal forecasts provide a range of possible climate changes that are likely to occur in the season ahead. Owing to the chaotic nature of the atmospheric circulation, it is not possible to predict the daily weather variations at a specific location months in advance. However, in some parts of the world, and in some circumstances, it may be possible to give a relatively narrow range within which weather values are expected to occur. Such forecast can easily be understood and acted upon, some of the forecasts associated with strong El Nino events fall into this category (e.g. Stockdale et al., 2011). More typically, the probable ranges of the atmospheric conditions differ only slightly from year to year. Forecasts of these modest shifts might be useful for some users, for example as a first warning, and could support further decision making involved in drought risks and water resources management.

In this article, we describe the development and application of the EFI methodology toward seasonal forecasts (at the monthly to seasonal time scales). While in the medium range, the EFI provides information on extreme events at a daily/local scale (e.g. storms), at the seasonal scale it will measure how the mean model climate and the actual forecast differ from a monthly/large scale. The extraction and analysis of large volumes of ensemble data is a complex and difficult task. The EFI is a possible and efficient way of summarizing the available information by scaling the ensemble forecast with respect to the model climate. The real advantage of using the EFI lies in the fact that it is an integral measure referenced to the model climate that contains all the information regarding variability of a parameter in location and time. Therefore, the user can recognize anomalous situations without defining different space- and time-dependent thresholds. This will summarize the probabilistic forecasts and can highlight potential anomalies in the long range that should be analyzed in detail by the user/forecast provider using the full range of the ensembles forecasts. In the following section the data and methods are presented followed by the results; the main conclusions are summarized in the last section.

2. Data and methods

2.1. ECMWF seasonal forecasts

ECMWF seasonal forecasts, based on an atmosphere–ocean coupled model, were used for the EFI calculations. We evaluate the recently implemented System 4, which became available in November 2011 (S4; Molteni et al., 2011). The horizontal resolution of the atmospheric model is about 0.7° in the grid-point space (spectral truncation TL255) with 91 vertical levels in the atmosphere. The ocean model has 42 vertical levels with a horizontal resolution of approximately 1°. The seasonal forecasts consist of a 51-member ensemble with 7-month lead time, including the month of issue, referred as 0-month lead time. S4 also has a set of re-forecasts starting on the first of every month for the years 1981–2010. These hindcasts are identical to the real-time forecasts in every way, except that the ensemble size if only 15 rather than 51. The data from these hindcasts create the 'model climate' that can be used for the calibration of forecasts products, and are used here to define the model climate from which the EFI forecasts are calculated. Molteni et al. (2011) present an overview of the model biases and forecast scores of S4.

The EFI calculations were performed for 2-m temperature (T2M) and total precipitation (TP) considering the model climate as the full hindcast period with 450 values (30 years × 15 ensemble members). Prior to the calculations, the T2M and TP fields were spatially interpolated from TL255 resolution to a regular grid of 2.5° × 2.5° (mass conservative for TP and bilinear for T2M), reducing the amount of data to process and smoothing the spatial fields.

2.2. EFI calculation

The EFI formulation scale departures from the reference climate cumulative distribution function (CDF) and is defined as:

$$\text{EFI} = \frac{2}{\pi} \int_0^1 \frac{p - F(p)}{\sqrt{p(1-p)}} \, dp \qquad (1)$$

$F(p)$ is a function denoting the proportion of ensemble members lying below the p quantile of the climate record. The term $1/\sqrt{p(1-p)}$, which takes its minimum for $p = 0.5$ and its maximum at both ends of the probability range, is used to give more weight to the tails of the distribution. This can also be interpreted as using the statistical Anderson–Darling (Anderson and Darling, 1952) test as a modification of the known Kolmogorov–Smirnov test. Given that $0 \leq F(p) \leq 1$, EFI values will lie in the same interval with unit values obtained when all the ensemble members are above (positive) or below (negative) the climate distribution.

The numerical integration of Equation (1), where $F(p) \in [0,1]$, is problematic near 0 or 1 where the integrated function $\to \infty$. To circumvent this problem,

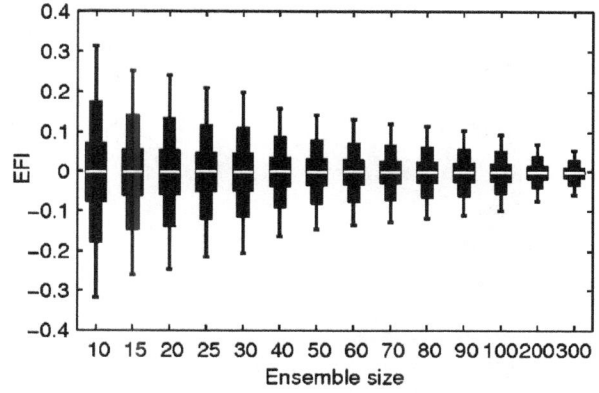

Figure 1. EFI values distribution comparing 100 000 forecasts with different ensemble sizes (horizontal axis) sampled from the same distribution as the model climate (with 450 values). The boxplots represent the percentiles 10, 30, 50 (white line), 70 and 90 and the lines extend from percentiles 1 to 99.

the endpoints can be excluded from the integration interval and the function is integrated in a slightly smaller domain $[\varepsilon, 1-\varepsilon]$. However, this has several disadvantages: (1) loss of accuracy, (2) EFI calculation is sensitive to the chosen numerical integration method and (3) the EFI values can overshoot/undershoot the interval $[-1, 1]$.

To deal effectively with these problems, a new technique, semi-analytical in nature, has been developed which is stable and produces results in the desired interval $[-1, 1]$. The main idea is to do an analytical integration thus avoiding the numerical problems associated with the singularity of the integrand. However, for an analytical calculation, a continuous representation of the model data-dependent function $F(p)$ is needed. This can be performed using a monotone interpolation formula such as linear interpolation:

$$F(p_i) \cong \widehat{F}(p_i) = f_i + \frac{p - p_i}{p_{i+1} - p_i}(f_i - f_{i+1}),$$
$$p \in [p_i, p_{i+1}], \quad f_i \equiv F(p_i) \qquad (2)$$

where the index i refers to the sorted climate record within a set of N samples.

The resulting EFI formula is a second order accurate finite series expansion:

$$\begin{aligned}
\text{EFI} = \frac{2}{\pi} \sum_{i=0}^{N-1} &\left\{ \left[(2f_i - 1) - (2p_i - 1)\frac{\Delta f_i}{\Delta p_i} \right] \right. \\
&\times \left[\arccos\left(\sqrt{p_i + 1}\right) - \arccos\left(\sqrt{p_i}\right) \right] \\
&\left. + \left(\sqrt{p_{i+1}(1 - p_{i+1})} - \sqrt{p_i(1 - p_i)} \right) \frac{\Delta f_i}{\Delta p_i} \right\}
\end{aligned}$$
$$(3)$$

where $\Delta f_{i+1} = f_{i+1} - f_i$; $\Delta p_i = p_{i+1} - p_i$; $p_0 = 0$; $p_N = 1$

In the results that follow, the EFI calculations were performed using Equation (3). The numerical accuracy of Equation (3) is mainly dependent on the number

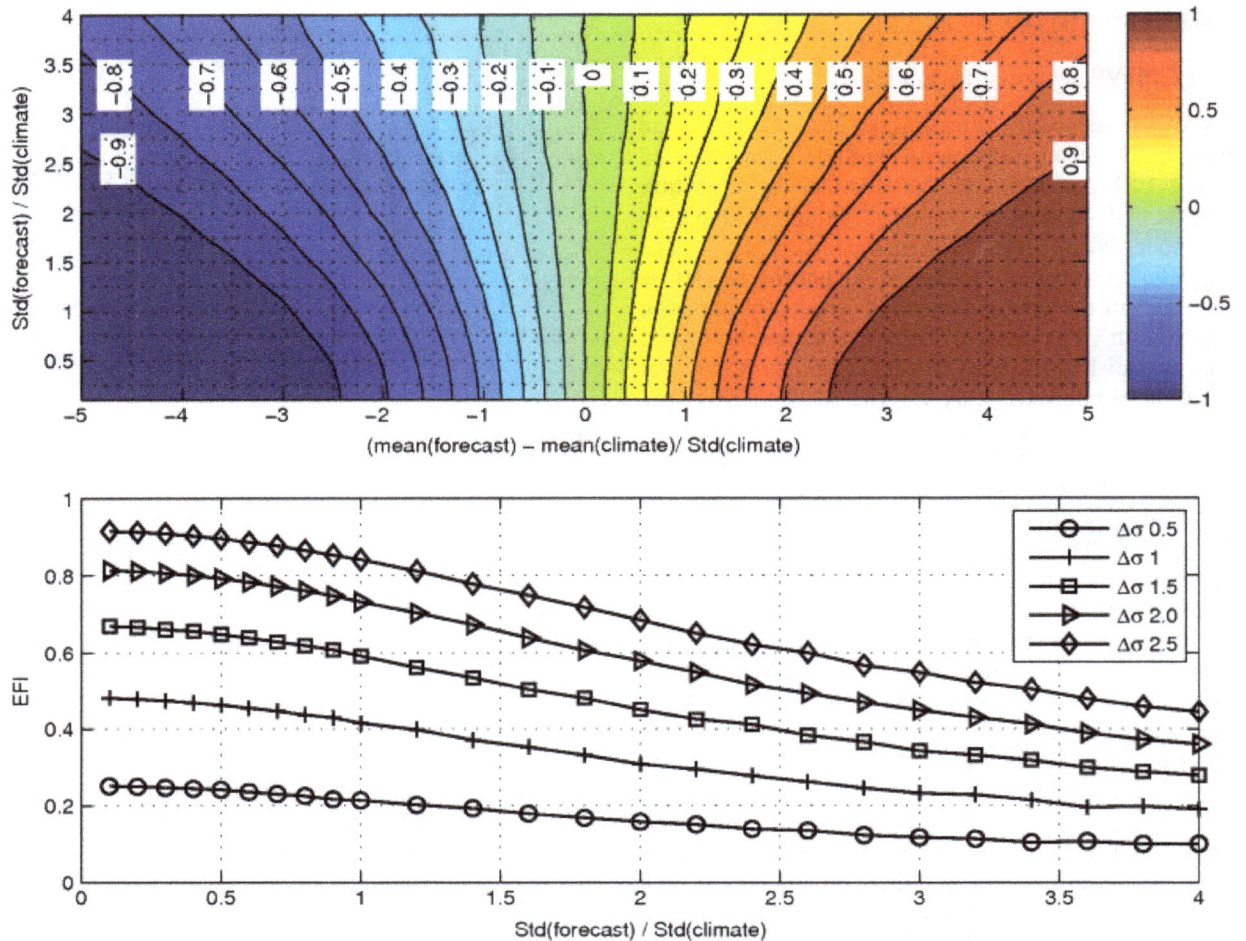

Figure 2. Top: EFI values as a function of changes in the ensemble mean and standard deviation of the forecast in respect to the climate. The changes in the mean (horizontal axis) are rescaled as the climate standard deviation anomalies added/subtracted to the climate mean, and the changes in the standard deviation (vertical axis) are given as the ration between the forecast and the climate standard deviation. Bottom: sub-sample of the contours in the top panel of the EFI for changes in the forecast ensemble mean (plus or minus) of 0.5 (circles), 1 (plus), 1.5 (square), 2 (right triangle) and 2.5 (diamond) standard deviations of the climate as a function of the changes in the standard deviation (horizontal axis).

of samples N in the climate distribution. Numerical experiments show that for a setup similar to the one used by ECMWF seasonal forecasts ($N = 450$ samples: 15 ensemble members × 30 years), the absolute value of the numerical error will be less than 10^{-2}. The good numerical accuracy of Equation (3) can be attributed to two facts: (1) the formula integrates the singular part of the integral exactly and (2) the piecewise linear approximation used for $F(p)$ turns to be accurate given that $F(p)$ is smooth and monotonically increasing at each interval $[p_i, p_{i+1}]$.

3. Results

3.1. Idealized EFI

In this section, we present the EFI sensitivity to the forecast ensemble size and the EFI relation with the changes in the forecast ensemble mean and standard deviation using synthetic data, which allows a broad testing and understanding of the EFI behavior.

3.1.1. EFI sensitivity for forecast ensemble size

Each seasonal forecast of S4 is composed of 51 ensemble members in real time and 15 ensemble members in the hindcast period. The decision on the number of ensemble members is mainly constrained by the available computational resources. As the real-time forecasts of S4 are only available since May 2011, EFI calculations previous to that period will be comparing CDF from the model climate with 450 samples against forecasts of only 15 samples. The reduced size in the hindcast period will impact the uncertainty of the EFI values. To evaluate the impact of the ensemble size on the EFI calculation, we performed a sensitivity analysis by producing synthetic data with the following characteristics:

- Model climate: 100 000 samples each with 450 values (to represent 30 years with 15 ensemble members) randomly sampled from a normal distribution;
- Forecasts: 100 000 forecasts with ensemble sizes ranging from 10 to 300 ensemble members. The

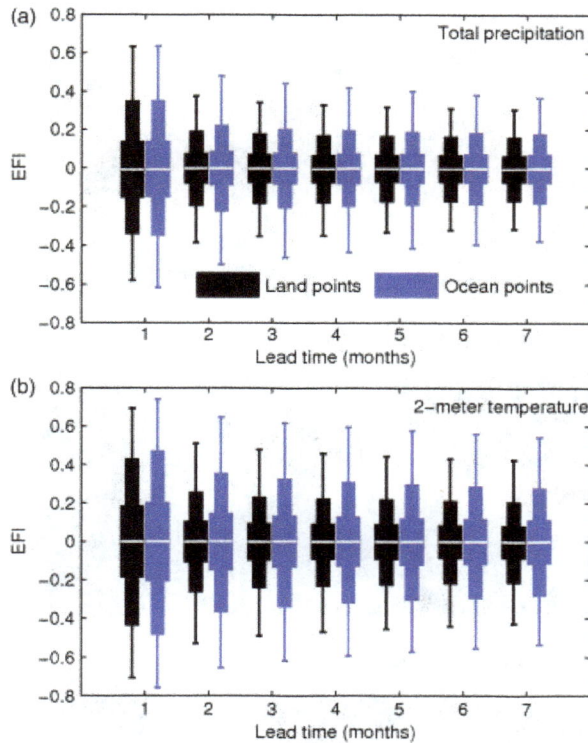

Figure 3. TP (a) and T2M (b) EFI distribution over all land points (black) and ocean points (blue) for the full S4 hindcast period (1981–2010, all months) as a function of lead time.

forecasts were generated by randomly sampling data from the model climate.

As the forecasts are the samples of the model climate, that is drawn for the same distribution, if the ensemble size would not have impact, the EFI values should be very close to zero. Figure 1 displays EFI values for each ensemble size, where the boxplots represent the EFI values distribution of the 100 000 forecasts with the same ensemble size. As expected, the median of the EFI values is zero because the forecasts are subsamples of the model climate. The vertical extension of the boxplots (between the 10‰ and 90–80% of the EFI values) gives an indication of the associated uncertainty of the EFI calculation due to the ensemble size, or sampling. For forecasts with 100 members the uncertainty is ±0.05, dropping to ±0.03 with 300 members. However, for ensemble sizes of 50 members the uncertainty is ±0.08 increasing to 0.14 for 15 ensemble members. Similar results were found when increasing the sample sizes, using a uniform random distribution, or changing the mean of the forecast (mean EFI is different from zero, but the uncertainty bounds remained the same). Although these results were derived from synthetic data they highlight the importance of large ensemble sizes to properly capture the forecast distribution and to allow the EFI calculation with a low uncertainty. The EFI sensitivity to the ensemble size is not caused by the numerical calculation (Equation (3)), but by the sampling errors of the forecast distribution ($F(p)$ in

Equation (1)) (due to small ensemble sizes). Similar results were also found when analyzing the fraction of ensemble members below or above the lower or upper tercile, which is a common metric used in seasonal forecasts. This should be considered when analyzing the EFI fields of S4 for dates previous to May 2011 (based on 15 ensemble members) that will have almost twice the uncertainty of the real-time EFI fields (based on 51 ensemble members).

3.1.2. EFI relation with the forecast mean and standard deviation

The EFI measures departures of the forecast CDF with respect to the climate CDF. These values are not intuitive, apart from 1 (or −1) when all the forecast ensemble members are above (or below) the climate distribution, or close to zero when the forecast distribution is very similar to the climate. To evaluate the relation between the EFI values and the changes in the ensemble mean and standard deviation of the forecast in the climate, we performed the following EFI synthetic calculations:

- Model climate: 1000 samples each with 450 values randomly sampled from a normal distribution of mean X and standard deviation Y;
- Forecasts: 1000 random forecasts with 51 ensemble members with normal distribution of mean $X + \Delta \times Y$ (with Δ varying from −5 to 5) and standard deviation $\Delta \times Y$ (with Δ varying from 0.2 to 4).

The changes in the EFI due to changes in the ensemble mean and standard deviation are represented in Figure 2. The results show that EFI is sensitive to changes in both the ensemble mean and spread: the same change in the ensemble mean in a sharper forecasts will results in a higher EFI. Note that in practice it is unlikely that a forecast will have a larger uncertainty than the climate, i.e. the scaled forecast uncertainty can be expected to be less than or approximately equal to one. A forecast with the same standard deviation as the climate, with changes in the ensemble mean of 0.5, 1, 1.5, 2 and 2.5 standard deviation of the climate will have 0.2, 0.42, 0.60, 0.74 and 0.85 of EFI values, respectively (see bottom panel of Figure 2). Different computations were performed (changing the baseline model climate, from normal to uniform distribution and sample sizes) and the results were similar. The information in Figure 2 can be used as a simple lookup table to connect EFI values to the changes in the ensemble mean/spread that can be further analyzed by examining in detail the different ensemble members.

3.2. EFI in the seasonal forecasts

Following the previous results using idealized distributions of the model climate and forecasts, this section presents the behavior of the EFI applied to the ECMWF S4 seasonal forecasts. This article is only

(a)T2M, 1 month lead time

(b)TP, 1 month lead time

(c)T2M, 2 months lead time

(d)TP 2 months lead time

(e)T2M, 3 months lead time

(f)TP, 3 months lead time

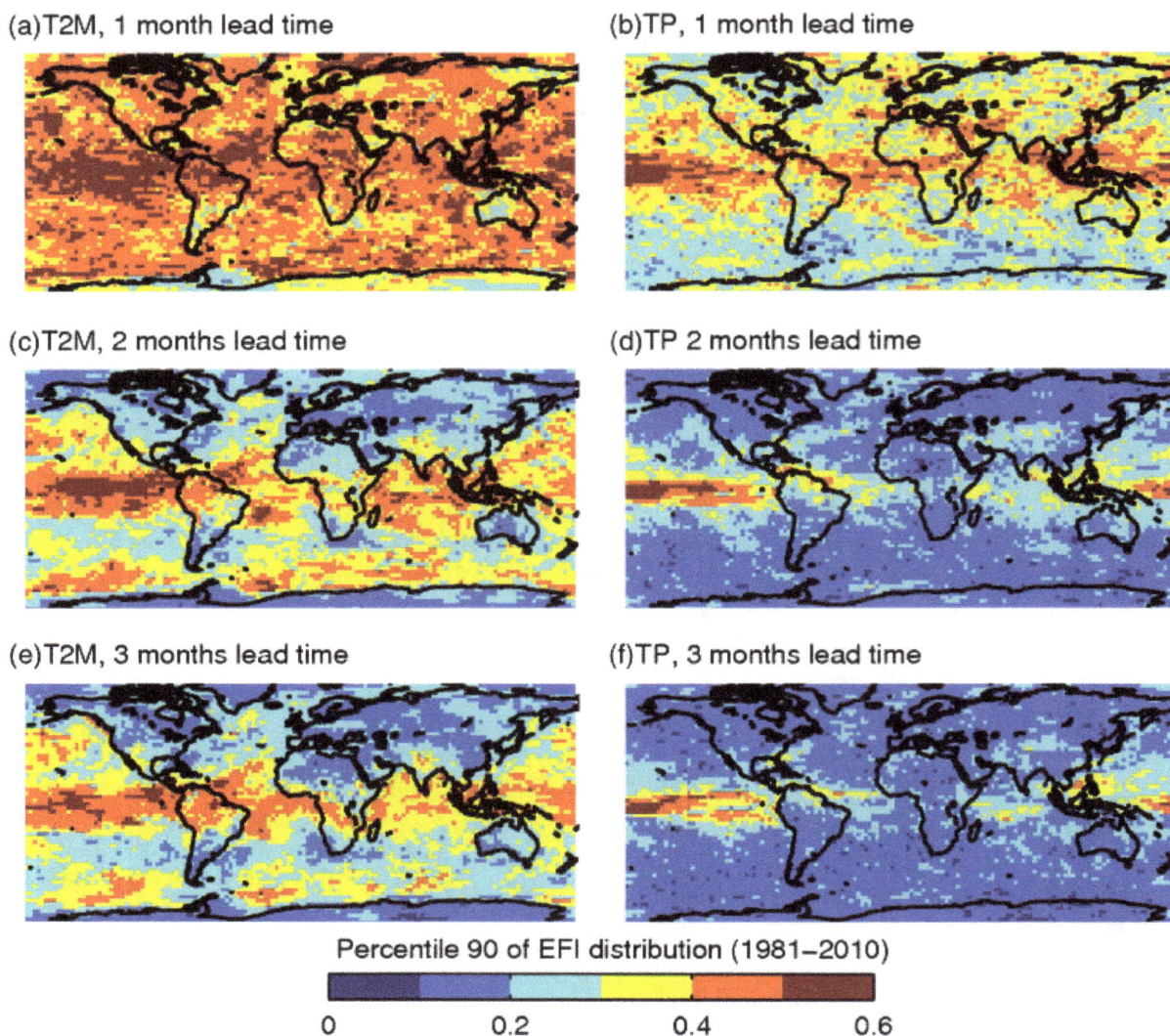

Percentile 90 of EFI distribution (1981–2010)

0   0.2   0.4   0.6

Figure 4. Spatial distribution of the 90 percentile of TP (b, d, f) and T2M (a, c, e) EFI calculated between 1981 and 2010 for the forecasts initialized in January for lead times of 1 (a, b), 2 (c, d) and 3 (e, f) months.

focused on the EFI development of seasonal forecasts and its behavior, whereas the forecasts skill is not addressed. For a particular application, the skill of the forecasts, i.e. comparing with actual observations should also be performed. The EFI will only have potential benefit to users in case the forecasts are skillful.

The distribution of EFI over all land points and ocean points for T2M and TP is represented in Figure 3 as a function of forecast month lead time. For both TP and T2M (and also land/ocean points), there is a decrease in the number of high (or low) values of the EFI with lead time. This decrease is mainly from the first/second month of forecast to the remaining forecast months. This behavior can be primarily attributed to the loss of predictability. While the first month of forecast still has some predictability in the medium range associated with the initial conditions, with increasing lead time the predictability is reduced along with the forecasts' sharpness. This is similar to what was shown in the previous Section 3.12, where

an increase of the forecasts' standard deviation for the same change in the forecast mean leads to a reduction of the EFI. Note that in the later months, the EFI is not much bigger than that the sampling errors would give for an identical forecast/climate distribution. There is also a remarkable difference between TP and T2M, where the EFI of the latter has a larger range in particular over the ocean. To further investigate this feature, Figure 4 represents the 90th percentile of TP and T2M EFI forecasts issued in January for different lead times from the hindcast period (1981–2010). The spatial distribution of the 90th percentile highlights the differences between the first forecast month and the remaining as well as the differences between T2M and TP and over land and ocean, resembling a map of predictability – higher 90th percentile values associated with higher predictability. The EFI in the medium-range forecasts is typically used as a warning for severe weather when its values are close to 1 (or −1). Applied to these long-range forecasts, the thresholds to define warnings cannot be so extreme, because

Forecasts initialized May 2012 valid for JJA 2012

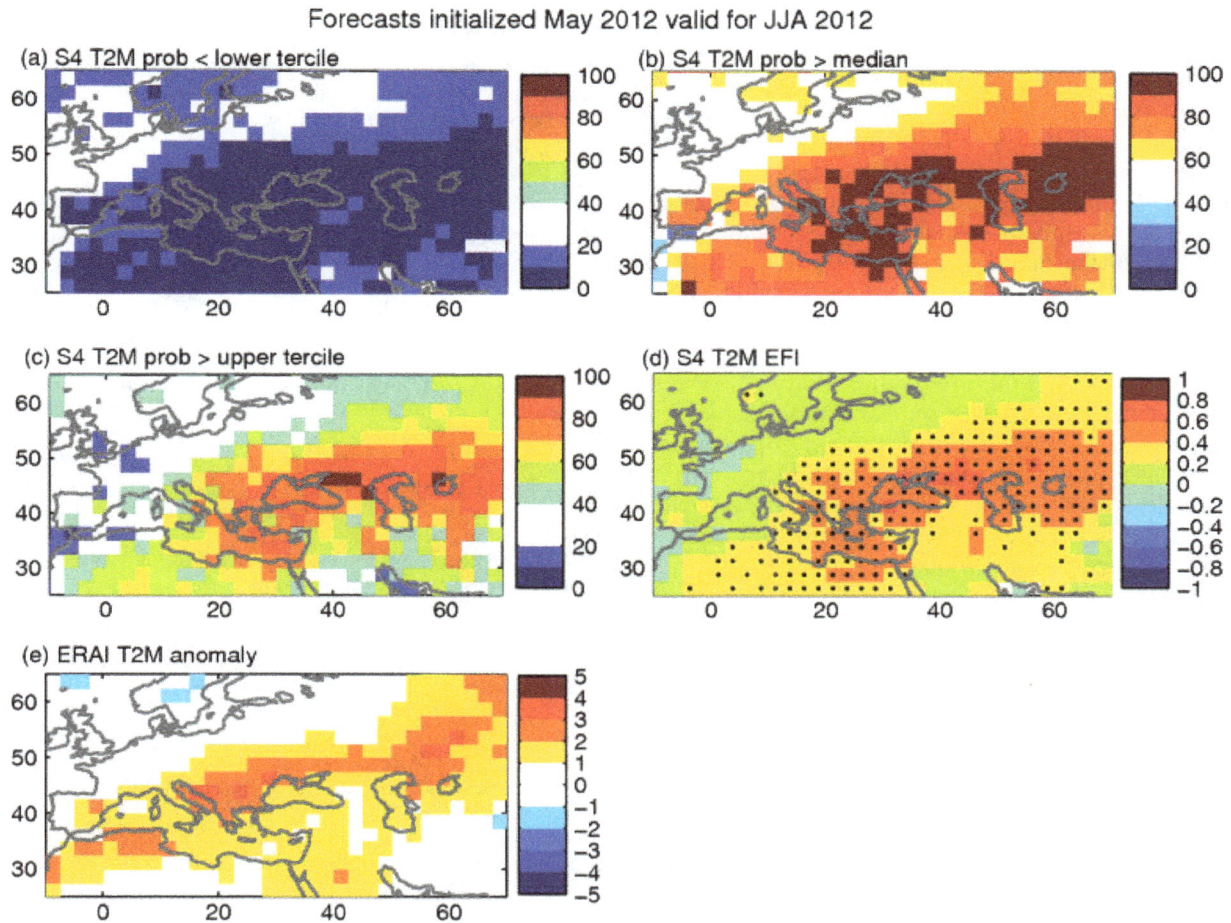

Figure 5. S4 T2M forecasts valid for JJA 2012 initialized in May 2012. (a) probability of T2M below the lower tercile; (b) probability of T2M exciding the median; (c) probability of T2M above the upper tercile; (d) T2M EFI and; (e) ERA-Interim (ERAI) JJA 2012 T2M anomaly. In panel (d) the black dots indicate EFI values above the 90th percentile of the EFI values between 1981 and 2010.

it will be very unlikely for a forecast distribution to differ greatly for the baseline climatology. An option is to use the hindcasts EFI to calculate different thresholds based on the percentiles, varying spatially, for each initial forecast data and lead time. The example shown in Figure 4 could be used to define warning levels as well as to identify areas and lead times where the EFI will have limitations, for example if the 90th percentile is below the sampling uncertainty, estimated as 0.08 for 50 ensemble members in Section 3.11.

Figure 5 compares three standard products: probabilities below the lower tercile, and above the median and upper tercile with the EFI for the S4 T2M forecasts initialized in May 2012 and valid for June,July,August (JJA) 2012. Summer 2012 had above normal temperature anomalies in south central and eastern Europe (Figure 5(e)). This warm anomaly was partially detected by S4 forecasts issued in May 2012, with high probabilities of T2M above the median and upper tercile, low probabilities of T2M below the lower tercile, and positive values of the EFI. An example of the use of thresholds to define EFI warning levels is presented in Figure 5(d) were the grid points with EFI values above the 90th percentile

are highlighted (Figure 5(d)). The EFI map resumes the information contained in the other three products. Additionally, user defined warning levels (e.g. values above below a certain percentile based on the hindcasts) could be used as early warning/detection of forecasts anomalies that should be further analyzed using remaining diagnostics.

4. Conclusions

The EFI concept, mainly used in medium-range ensemble forecasts, has been extended and implemented on seasonal forecasts. A new semi-analytical formulation is presented that allows an accurate calculation of the EFI. An assessment of the EFI behavior using synthetic data showed the variation of EFI with changes in the ensemble mean and spread: similar changes in the ensemble mean will result in higher or lower EFI values for low and high ensemble spreads, respectively. This information can be used as a guide for the interpretation of the EFI. Furthermore, we also show the importance of large ensembles to reduce the uncertainty of the EFI due to the sampling errors of the forecast distribution.

The EFI was applied to the ECMWF seasonal forecasts of monthly means of T2M and TP up to 6-month lead time. It was found that the EFI distribution changes with lead time, with a reduction in the occurrence of high/low values. This is associated with smaller changes of the ensemble mean with respect to the model climate, and to an increase of the ensemble spread. In this situation, the distribution of the forecast is similar to the underlying model climate. This was mainly visible for TP over land. On the other hand, the EFI of monthly T2M in the tropical regions shows a higher range, even on long lead times. These results are associated with the low predictability of TP over land when compared with sea surface temperature in the tropical regions. These results are coherent with the synthetic data tests showing that an increase in the ensemble spread (can be associated with a reduction of predictability) leads to a decrease of the EFI for similar changes in the ensemble mean.

We have successfully implemented the EFI applied to seasonal prediction, and examined its behavior. It is clearly sensitive to ensemble size, which makes a detailed study difficult with only 15 member hindcasts. Our results do not include an evaluation of the skill of the EFI, because it is not possible to derive an observed EFI for verification. Such skill assessment should be performed on the original fields, and the EFI can be evaluated on a case study basis, as it was shown for the 2012 summer in southern Europe. With the currently available data, the EFI does not bring additional information to the standard products, such as tercile or median probabilities, but resumes such information in a single indicator, complementing currently used diagnostics. Further investigations can be carried out when larger sample sizes become available (which is planned for selected start dates).

Acknowledgements

We thank one anonymous reviewer for the valuable comments and suggestions. This work was funded by the FP7 EU projects GLOWASIS (http://www.glowasis.eu) and DEW-FORA (http://www.dewfora.net).

References

Alfieri, L, Velasco, D, Thielen, J. 2011. Flash flood detection through a multi-stage probabilistic warning system for heavy precipitation events. *Advances in Geosciences* **29**: 69–75, DOI: 10.5194/adgeo-29-69-2011, www.adv-geosci.net/29/69/2011/.

Anderson T, Darling D. 1952. Asymptotic theory of certain goodness of fit criteria based on stochastic processes. *The Annals of Mathematical Statistics* **23**: 193–212.

Bartholmes JC, Thielen J, Ramos MH, Gentilini S. 2009. The European Flood Alert System EFAS - Part 2: Statistical skill assessment of probabilistic and deterministic operational forecasts. *Hydrology and Earth System Sciences* **13**: 141–153.

Cloke HL, Jeffers C, Wetterhall F, Byrne T, Lowe J, Pappenberger F. 2010. Climate impacts on river flow: projections for the Medway catchment, UK, with UKCP09 and CATCHMOD. *Hydrological Processes* , DOI: 10.1002/hyp.776.

Lalaurette F. 2003. Early detection of abnormal weather conditions using a probabilistic extreme forecast index. *Quarterly Journal of The Royal Meteorological Society* **129**(594): 3037–3057.

Molteni F, Stockdale T, Balmaseda M, Balsamo G, Buizza R, Ferranti L, Magnunson L, Mogensen K, Palmer T, Vitart F. 2011. The new ECMWF seasonal forecast system (System 4). *ECMWF Technical Memorandum* **656**: 49 pp.

Stockdale T, Anderson D, Balmaseda M, Doblas-Reyes F, Ferranti L, Mogensen K, Palmer T, Molteni F, Vitart F. 2011. ECMWF seasonal forecast system 3 and its prediction of sea surface temperature. *Climate Dynamics* **37**(3): 455–471, DOI: 10.1007/s00382-010-0947-3.

Thielen-del Pozo J, Bartholmes J, Ramos M-H, de Roo A. 2009. The European Flood Alert System. Part 1: concept and development. *Hydrology and Earth System Sciences* **13**: 125–140.

Zsoter E. 2006. Recent developments in extreme weather forecasting. *ECMWF Newsletter* **107**(107): 8–17.

9

An examination of potential seasonal predictability in recent reanalyses

Xia Feng* and Paul Houser

Geography and Geoinformation Science, George Mason University, Fairfax, VA, 22030, USA

*Correspondence to:
X. Feng, George Mason
University, Fairfax, VA 22030,
USA.
E-mail: xfeng@gmu.edu

Abstract

This study examined potential seasonal predictability of precipitation and 2-m temperature in the recent global reanalyses from 1979 to 2012. The reanalyses being investigated are R1, R2, ERA-40, JRA25, ERA-I, MERRA, CFSR and 20CR. When compared against Global Precipitation Climatology Project (GPCP) precipitation, ERA-I and CFSR provide the best estimates of potential predictability, MERRA, R1 and R2 overestimate predictability and ERA-40 and JRA25 are unrealistic in the tropics. Predictability estimates of 2-m temperature from 20CR, JRA25, R1, MERRA, CFSR and R2 exhibit better agreement with the reanalysis ensemble mean than ERA-40 and ERA-I, which identify less predictability.

Keywords: seasonal potential predictability; reanalysis; precipitation; 2-m temperature

1. Introduction

Estimates of the potential predictability of seasonal means are usually based on partitioning total interannual variability into 'weather noise' variability that is unpredictable on seasonal time scales (Lorenz, 1963), and 'potentially predictable signal' arising from slowly varying boundary conditions (Palmer and Anderson, 1994), external forcings and possibly internal atmospheric dynamics (Frederiksen and Zheng 2007). The potential predictability is measured by the degree to which interannual variability exceeds unpredictable weather noise. Several studies have proposed the statistical methods for estimating the potential predictability of seasonal means from a single realization of daily time series, such as a spectral method in Madden (1976), an analysis of variance method in Shukla and Gutzler (1983), a first-order Markov chain model in Katz (1983), a bootstrap procedure in Feng *et al.* (2011) and an analysis of covariance (ANOCOVA) method in Feng *et al.* (2012). The performance of these methods in estimating predictability of surface air temperature and precipitation was assessed by Feng *et al.* (2013a, 2013b). Their studies indicate that ANOCOVA produces good predictability estimate for temperature and the best estimate for precipitation in comparison with several other methods. We will therefore use the ANOCOVA method to examine the quality of the representation of potential seasonal predictability within the reanalyses.

During the past several decades, reanalyses have increasingly become an important tool for studying weather and climate systems by providing complete spatial and temporal data coverage over long time periods. They use a wide range of available observations to constrain a numerical weather prediction model through a state-of-art assimilation system to produce continuous and consistent estimate of the state of the atmosphere and land surface. One of the appealing features of reanalyses is to fill the observational gaps in space and time using data assimilation routines but in a manner consistent with model physics and dynamics. However, uncertainties in underlying models, input observations and assimilation technology could lead to uncertainty in the analyzed fields. There are a number of studies that identified biases and uncertainties in reanalysis products for representing precipitation (Bosilovich *et al.*, 2008; Kim and Alexander, 2013), tropical cyclones (Schenkel and Hart, 2012), atmospheric circulation (Nguyen *et al.*, 2013), land surface and ocean fluxes (Wang and Zeng, 2012; Chaudhuri *et al.*, 2013) and so on.

Given the impact of seasonal variability on climate prediction and subsequent socioeconomy, the question is raised as to whether the reanalyses can be used for investigating climate processes and predictability, particularly on seasonal time scales. This study aims to evaluate and intercompare the recent global reanalyses in representing potential seasonal predictability of precipitation and 2-m surface air temperature and identify strengths and deficiencies in each reanalysis product. The similarities and discrepancies between reanalysis estimates will be explored to quantify the uncertainties in the representation of seasonal predictability in reanalyses.

2. Data

Eight global reanalysis datasets examined in this study are the National Centers for Environmental Prediction–National Center for Atmospheric Research (NCEP/NCAR) 40-year Reanalysis (R1) (Kalnay *et al.*, 1996), the NCEP–Department of Energy (DOE) Atmospheric Model Intercomparison

Project 2 reanalysis (R2) (Kanamistu *et al.*, 2002), the European Centre for Medium-Range Weather Forecasts (ECMWF) 40-year Reanalysis (ERA-40) (Uppala *et al.*, 2005), Japan Meteorological Agency (JMA) 25-year Reanalysis Project (JRA-25) (Onogi *et al.*, 2007), the ECMWF Interim Reanalysis (ERA-I) (Dee *et al.*, 2011), the National Aeronautics and Space Administration (NASA) Modern-Era Retrospective Analysis for Research and Applications (MERRA) (Rienecker *et al.*, 2011), the NCEP Climate Forecast System Reanalysis (CFSR) (Saha *et al.*, 2010), and the National Oceanic and Atmospheric Administration–Cooperative Institute for Research in Environmental Sciences (NOAA/CIRES) 20th Century Reanalysis Version 2 (20CR) (Compo *et al.*, 2011).

To evaluate the reanalysis precipitation, we use the merged satellite–gage product from the Global Precipitation Climatology Project (GPCP) (Adler *et al.*, 2003). This dataset provides precipitation estimates at 2.5° lat/lon grid spacing over the entire 5-day intervals. Long-term daily record of global observation or ground truth of surface 2-m temperature is virtually not available. We thus generate an ensemble mean of 2-m temperature from the eight reanalysis products and use it as the reference.

All of the reanalysis datasets were regridded to a common horizontal resolution of 2.5° × 2.5°. We use daily average for 2-m temperature and 5-day average for precipitation, respectively over the time period 1979–2012 except for ERA-40 ending at 2002 (1979–2002). The annual cycle is removed from the averaged data by regressing out the first three annual harmonics as discussed in Feng *et al.* (2011). The derived anomaly is then divided into four seasonal time series: December–January–February (DJF), March–April–May (MAM), June–July–August (JJA) and September–October–November (SON). Each season defined in this study is 18 pentads for precipitation and 90 days for 2-m temperature.

3. Method

Potential seasonal predictability is estimated using ANOCOVA developed by Feng *et al.* (2012), who combined autoregressive model (AR) to account for temporal dependence, and the classical analysis of variance to take into account the variation of seasonal means

$$X_{d,y} = \phi_1 X_{d-1,y} + \phi_2 X_{d-2,y} + \cdots + \phi_p X_{d-p,y} + \mu_y + \epsilon_{d,y},$$ (1)

where $X_{d,y}$ denotes a daily time series of climate variable on the dth day in the yth year for $d = 1, \ldots, D$, $y = 1, \ldots, Y$, where D denotes number of days in a season, Y indicates number of years, the parameters $\phi_1, \phi_2, \ldots, \phi_p$ are autoregressive coefficients, p is an unknown order, μ_y is a constant value in year y and represents potential predictable signal, ϵ_{dy} is independent identically distributed random with mean 0 and

variance σ_ϵ^2. In terms of the model (1), testing no seasonal potential predictability is equivalent to testing the null hypothesis that the term μ_y is constant for every year

$$H_0 : \mu_1 = \mu_2 = \cdots = \mu_Y,$$ (2)

If this hypothesis is true, that is, $\mu_y = \mu$ for all y, the ANOCOVA model (1) reduces to

$$X_{dy} = \phi_1 X_{d-1y} + \phi_2 X_{d-2y} + \cdots + \phi_p X_{d-py} + \mu + \epsilon_{dy}.$$ (3)

Model (3) is a standard AR model and will be called the 'reduced' model, whereas model (1), consisting of both weather fluctuations and seasonal mean variations, will be called the 'full' model. H_0 can be tested by comparing

$$F_{ANOCOVA} = \frac{(SSR_{reduced} - SSR_{full})/(Y-1)}{SSR_{full}/(DY - Y - p)},$$ (4)

with critical value from the F distribution with the degrees of freedom of $(Y-1)$ and $(DY - Y - p)$, where $SSR_{reduced}$ and SSR_{full} stand for the sum of the squared residual errors for the reduced and full models, respectively.

The fraction of predicable variance (FPV) is defined as

$$FPV_{ANOCOVA} = \frac{\hat{\sigma}_S^2}{\hat{\sigma}_S^2 + \hat{\sigma}_N^2},$$ (5)

where caret symbol ^ indicates a sample quantity, $\hat{\sigma}_S^2$ and $\hat{\sigma}_N^2$ denote signal and noise variances, respectively, according to the model (1). Their formulations and further details on ANOCOVA can be found in Feng *et al.* (2012).

4. Results

4.1. Precipitation

In this section, we evaluate the realism of potential predictability in precipitation as depicted by the reanalyses against GPCP data using the ANOCOVA method. Since observed precipitation is not assimilated into the present global reanalysis, GPCP data is an independent reference. Note 20CR has high number of missing data in its 5-day precipitation average and hence is omitted from the evaluation here. We apply ANOCOVA to the remaining seven reanalyses and obtain FPV ratios during four seasons in comparison with GPCP as shown in Figure 1. The colored regions denote the null hypothesis of no seasonal potential predictability is rejected at 5% significance level.

In Figure 1, ERA-I and CFSR show the best performance in reproducing the observed pattern: potential predictability is high in the tropical Pacific ocean, low in extratropics and undergoes a distinct seasonal variation. Although R1, R2 and MERRA agree well with GPCP, they substantially overestimate predictability with statistically significant FPV ratios occurring over tropical

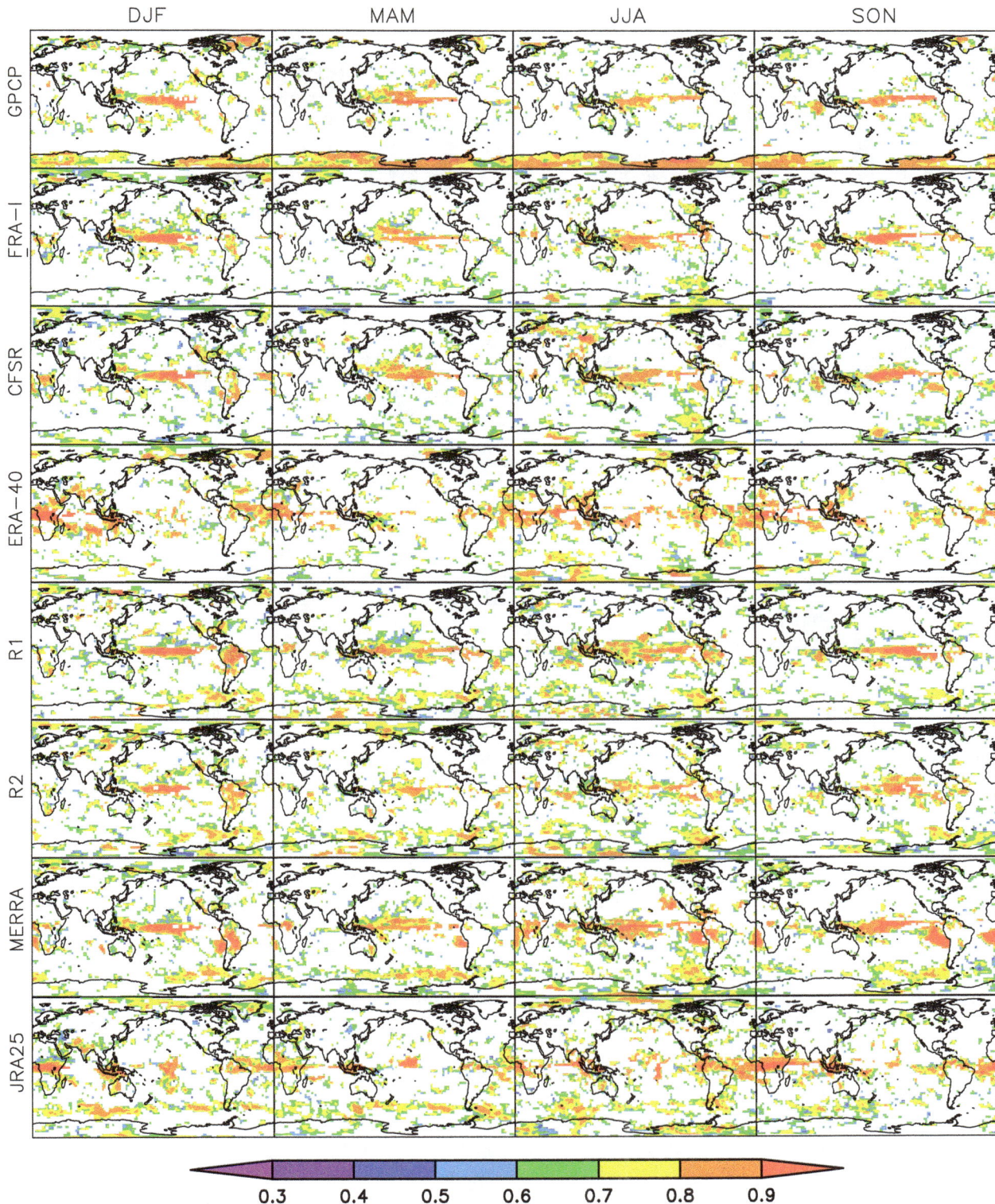

Figure 1. The fraction of predictable variance of precipitation estimated from GPCP, ERA-I, CFSR, ERA-40, R1, R2, MERRA and JRA25 in DJF, MAM, JJA and SON (left to right) over 1979 to 2012. The regions in which null hypothesis of no potential predictability is insignificant at the 5% level are masked out.

Indian and Atlantic oceans, Africa and southern ocean as well as a small portion of land areas over northern high latitude. Relatively good agreement is also found between ERA-40, JRA25 and GPCP over extratropical regions. However, the estimates are unrealistic in

the tropics with spotty predictability in the Pacific ocean and high predictability over Africa, Indian and Atlantic oceans. Figure 2 shows the fraction of area that is deemed potentially predictable by each individual reanalysis dataset. Both the reanalyses and GPCP

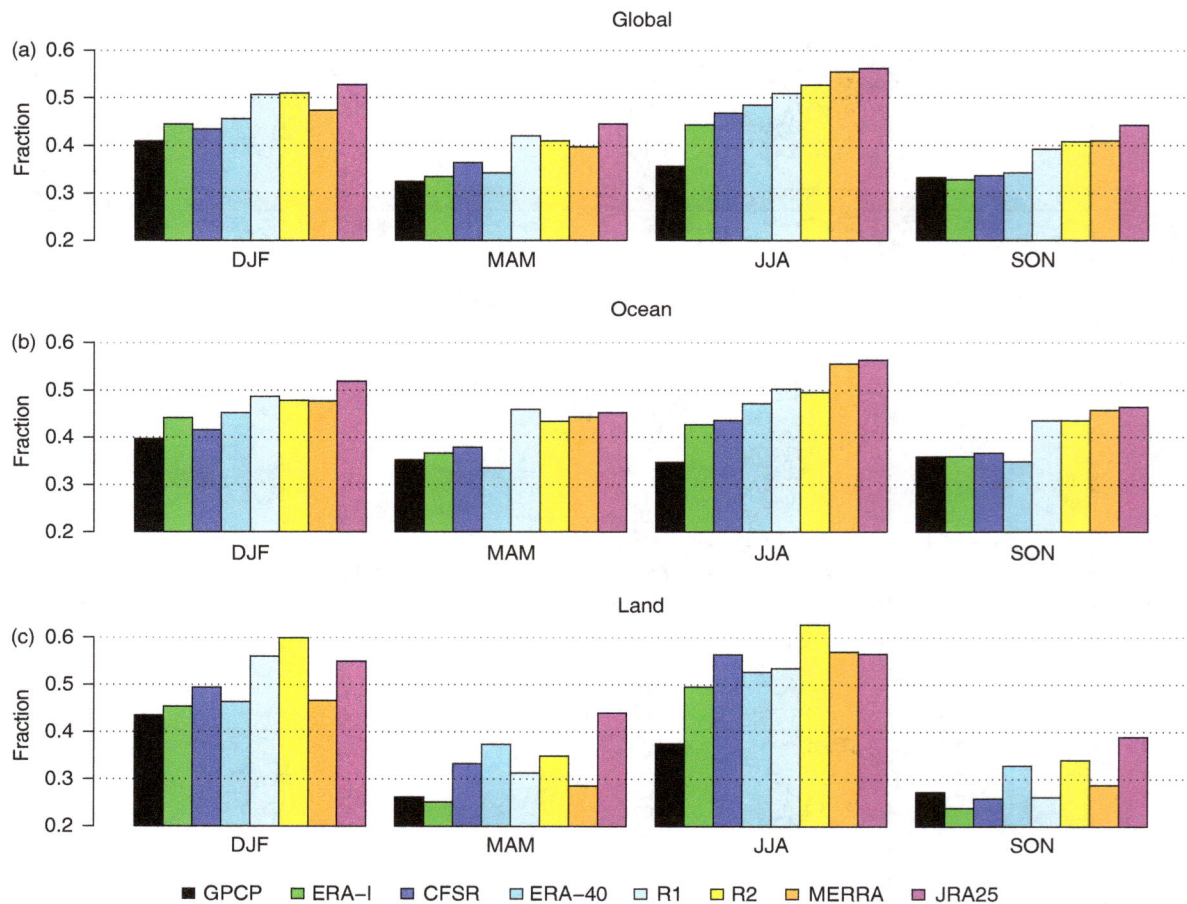

Figure 2. Histograms for fraction of regions where the null hypothesis of no potential predictability is rejected over (a) globe, (b) ocean and (c) land identified by GPCP, ERA-I, CFSR,ERA-40, R1, R2, MERRA and JRA25 in DJF, MAM, JJA and SON for precipitation over 1979 to 2012.

agree that predictability is higher in DJF and JJA than in MAM and SON, and more persistent over ocean than over land. The most discrepancies among reanalyses are found over land than over ocean.

Generally speaking, the new reanalyses perform better relative to the old reanalyses, probably due to the advances in the data assimilation system and observing system, improved model parameterizations and higher horizontal and vertical resolutions. For instance, the improvements in model physics and moisture analysis in ERA-I have eliminated some issues in the representation of the hydrological cycle in ERA-40 (Andersson *et al.*, 2005), which may contribute to the better performance of ERA-I here. Compared to R1 and R2, CFSR incorporates numerous improvements, including the use of the coupled model with finer resolution, assimilation of satellite radiances and direct forcing of land surface model with observed precipitation. The use of observed precipitation in foricng land surface model eliminates the biases associated with model estimated precipitation, subsequently improving precipitation output in CFSR.

The reanalyses are known to be prone to inhomogeneities and artificial trends due to the changes in the satellite-observing system. The introduction of

the Advanced Microwave Sounding Unit (AMSU) produces dramatic shifts in MERRA water and energy fluxes between 1999 and 2000 (Robertson *et al.*, 2011). These shifts would contribute to the detected seasonal signal and hence inflate the estimated predictability. In JRA-25, precipitation shift in 1987 coincides with the availability of the Special Sensor Microwave Imager (SSM/I) (Bosilovich *et al.*, 2008), which may explain the predictability overestimation over some of the tropical regions. The changes in the observing system also afflict CFSR (Saha *et al.*, 2010) and ERA-I (Bromwich *et al.*, 2011). However, in this study, they have no clear impact on the predictability estimates in CFSR and ERA-I.

To further look into the consistency and disagreement between the reanalyses and observation, we show an 'indicator function' plot in Figure 3, where 1 or −1 indicate that all reanalyses are consistent with observation in determining whether seasonal precipitation is insignificantly or significant potentially predictable, where 0 represents reanalyses disagree with the observation whether potential predicability is significant or not. Note ERA-40 and JRA25 are excluded from this analysis due to their misrepresentations in the tropics (Figure 1). The remaining reanalyses and observation

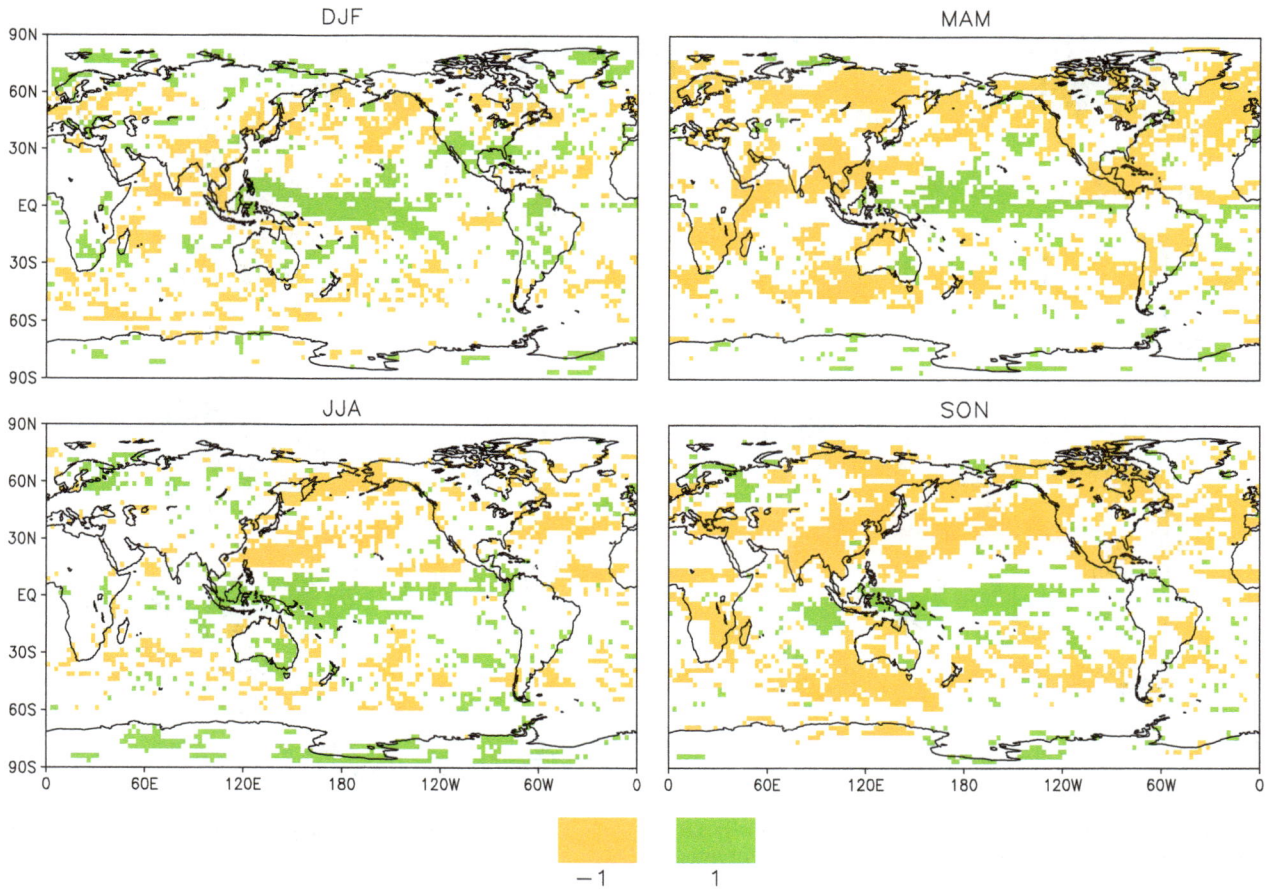

Figure 3. Indicator function of the consistency of ERA-I, CFSR, R1, R2, MERRA and GPCP for identifying the significance of potential predictability of precipitation in DJF, MAM, JJA and SON over 1979 to 2012. The indicator function is defined to be −1 or 1, representing whether reanalyses agree with GPCP that precipitation is insignificantly or significantly predictable, respectively, and is masked out otherwise.

reveal the most agreement in significance of predictability occur in MAM and SON over roughly 38% of the globe where more than 8% of the globe is significantly predictable predominated in the eastern and mid-tropical Pacific oceans, about 30% of the globe is insignificantly predictable scattered over extratropics, leaving 62% of the globe with inconsistent predictability. In contrast, DJF and JJA are the most discordant seasons where significant and insignificant predictability occupies 12% and 16% of the land, respectively, while inconsistent predictability dominate the remaining 72% of the land.

4.2. Temperature

The reanalysis ensemble mean 2-m temperature is used as reference to explore agreement and discrepancy between the reanalyses. The ensemble mean is the average among the eight reanalyses and hence is not a true independent reference. Using ANOCOVA, we derive potential predictability of 2-m temperature for eight reanalyses (Figure S1) and the reanalysis ensemble mean during all the four seasons as shown in Figure 4. The values of FPV are displayed only over regions where the null hypothesis of no predictability is rejected

at 5% significance level. The ensemble mean and reanalyses exhibit a pronounced tropical–extratropical contrast with higher FPV in the tropical oceans than the extratropical land areas.

To further gain insight on the similarities and differences among all datasets, we show in Figure 5, the fraction of predictable areas from each dataset over global, ocean and land during four seasons. The statistically significant FPV values from the ensemble mean and reanalyses occupy more than 90% of the globe with a seasonally varying predictability pattern. Specifically, there are more regions identified as potentially predictable in MAM and SON over ocean, but over land there are more predictable areas in MAM and JJA. Moreover, ocean displays more persistent predictable pattern than land during all of the seasons. Overall, estimates from JRA25, 20CR, R1, MERRA, CFSR and R2 are more comparable to those from the reanalysis ensemble mean, large discrepancies are found in ERA-40 and ERA-I, which have less predictable areas. The reanalyses exhibit better agreement over ocean than over land. The most discrepancy occurs in SON over land, where predictable areas range from 71% in ERA-40 to 84% in ERA-I. The small discrepancy over ocean is due to the fact that 2-m temperature is primarily

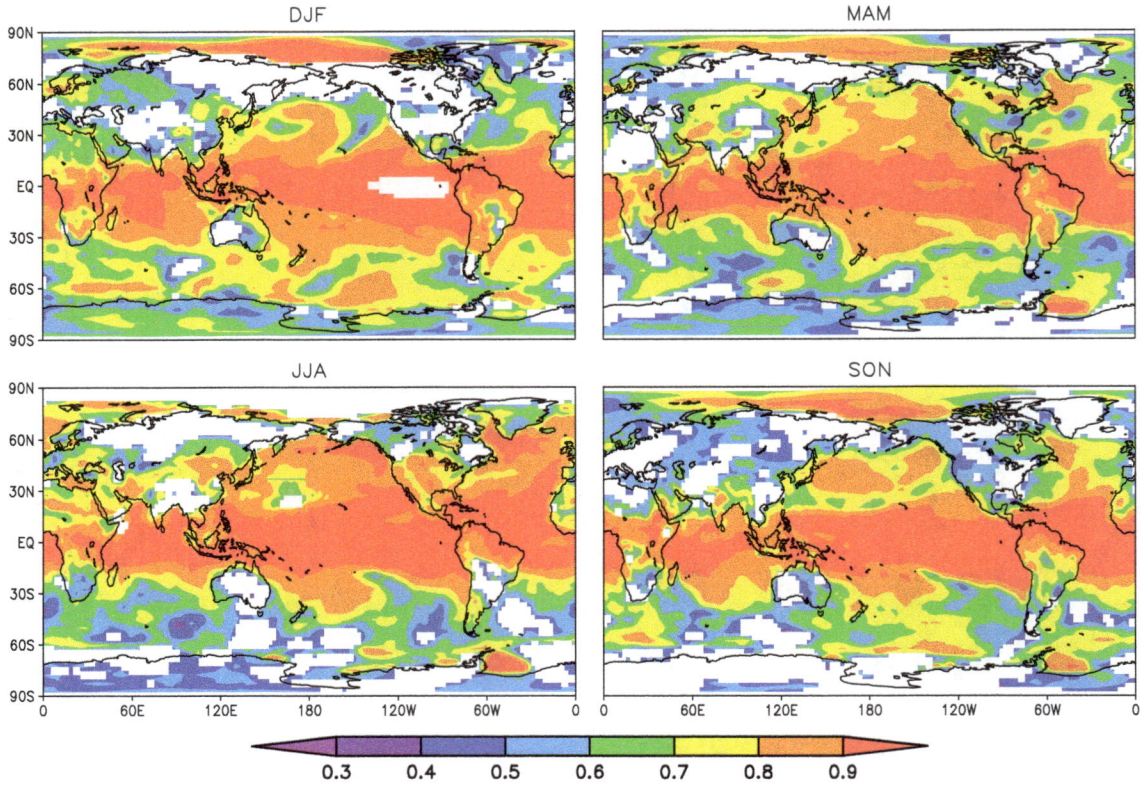

Figure 4. The fraction of predictable variance of reanalysis ensemble mean 2-m temperature in DJF, MAM, JJA and SON over 1979 to 2012. The regions in which null hypothesis of no potential predictability is insignificant at the 5% level are masked out.

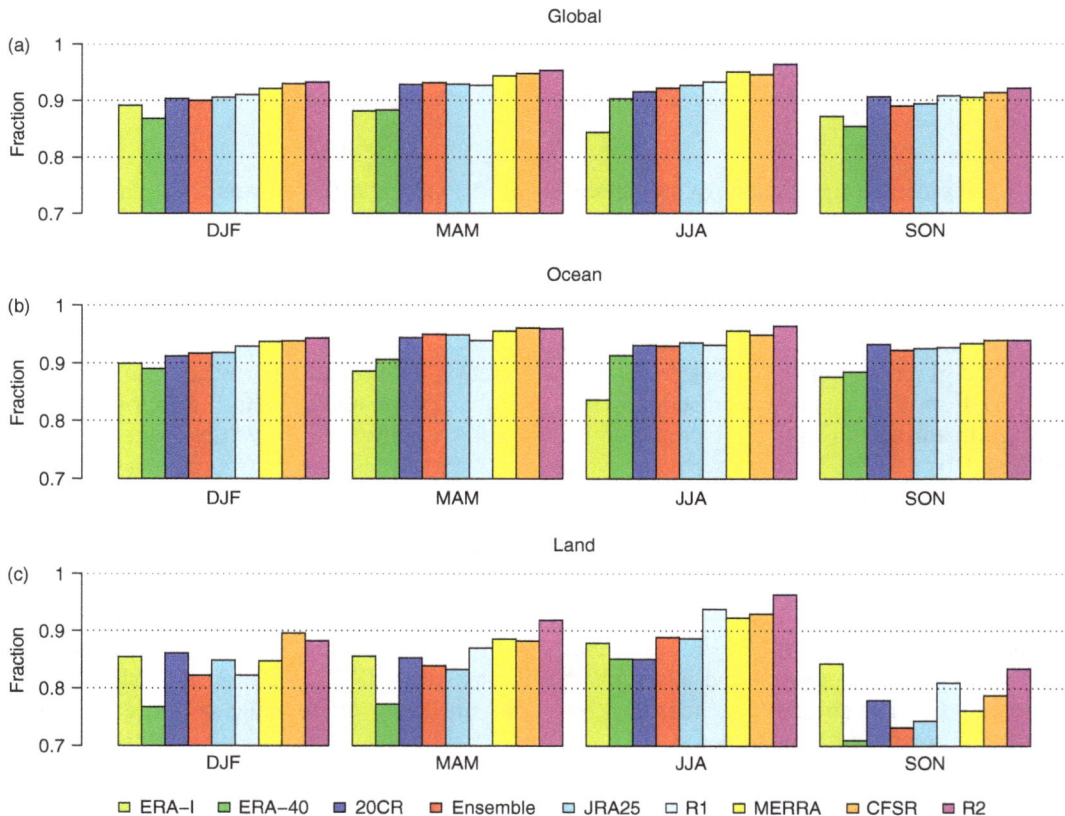

Figure 5. Histograms for fraction of regions where the null hypothesis of no potential predictability is rejected over (a) globe, (b) ocean and (c) land identified by ERA-I, ERA-40, 20CR, ensemble mean, JRA25, R1, MERRA, CFSR and R2 for 2-m temperature in DJF, MAM, JJA and SON over 1979 to 2012.

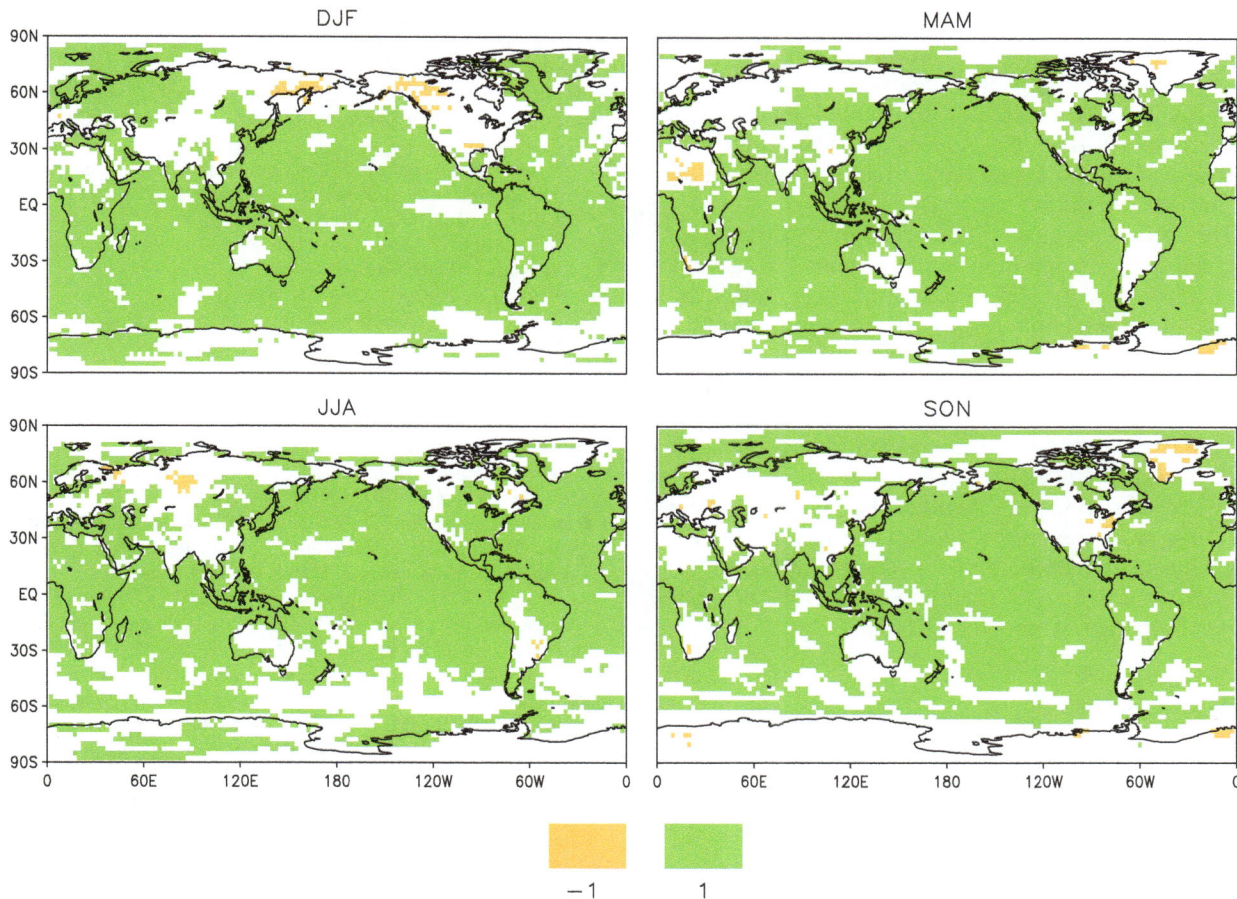

Figure 6. Indicator function of consistency for 2-m temperature from eight reanalyses in DJF, MAM, JJA and SON over 1979 to 2012.

modulated by the sea surface temperature (SST) in the open oceans, which applied in most reanalysis products during our study period.

Agreement between the reanalyses implies that their outputs are not sensitive to the background models and assimilation schemes. A measure of agreement can be gained from the predictability consistency plot in Figure 6. The eight reanalysis datasets consistently reveal more than 69% of the global areas are significantly potentially predictable predominated in the oceans, topical and a small portion of high latitude land areas, while only 0.3% of the globe are insignificant predictable primarily scattered over the northern land region in high latitude. The reanalyses primarily disagree over extratropical land region with the maximum (31%) and minimum (24%) inconsistency occurring in SON and MAM, respectively.

5. Summary and conclusions

Using ANOCOVA, we evaluated and compared potential seasonal predictability in eight reanalyses, including R1, R2, ERA-40, JRA25, ERA-I, MERRA, CFSR and 20CR. For precipitation, the evaluations were conducted between GPCP and seven reanalyses except

20CR, while all reanalyses are compared with the ensemble mean for 2-m temperature.

Assessment of precipitation suggests that both ERA-I and CFSR faithfully reproduce the predictability observed in GPCP, where predictability is high in the tropical Pacific ocean, low in the extratropics and experiences a distinct seasonal variation. In contrast, R1, R2 and MERRA overestimate FPV, and ERA-40 and JRA25 are unrealistic over the tropics. Omitting ERA-40 and JRA25, the remaining five reanalyses are consistent with observations in identifying significance of predictability over a maximum of 38% of the globe in MAM and SON, leaving 62% of the globe with inconsistent predictability.

Predictability of 2-m temperature from all the reanalyses consistently indicates a tropical–extratropical contrast with high FPV in tropical oceans and low values in the extratropical land areas. Additionally, ocean exhibits more persistent predictability pattern than land during all of the seasons. Overall, six of the reanalyses (JRA25, 20CR, R1, MERRA, CFSR and R2) show better agreement with the reanalysis ensemble mean, whereas ERA-40 and ERA-I identify less predictability. There is a higher degree of similarity between the ensemble mean and the reanalyses over ocean than over land. The eight reanalyses identify consistent predictability over a maximum of 69% of the globe with inconsistent estimates occupying the remaining 31% of the globe. It is over the extrotropical land areas that the reanalyses exhibit substantial inconsistency, in particularly over the Asia and North American in DJF and SON, indicating large uncertainty among eight reanalyses over these regions.

This study reveals deficiencies in estimating predictability of precipitation among reanalyses, especially in ERA-40 and JRA25. Such deficiency clearly highlights the model inaccuracy in the representation of the fundamental dynamical and microphysical process that are important to precipitation. Meanwhile, biases in MERRA and JRA-25 are likely linked to the significant shifts caused by the changes in the observing system. The differences in 2-m temperature predictability between reanalyses are manifestations of the different physical processes represented in the background models. Our results are helpful for identifying the areas of needed improvements in future reanalysis as well as directing potential research efforts.

Acknowledgement

This work was sponsored by NASA's Energy and Water Cycle Study (NEWS) program (Grant NNX11AE32G).

References

Adler RF, Huffman GJ, Chang A, Ferraro R, Xie PP, Janowiak J, Rudolf B, Schneider U, Curtis S, Bolvin D, Gruber A, Susskind J, Arkin P, Nelkin E. 2003. The version-2 Global Precipitation Climatology Project (GPCP) monthly precipitation analysis (1979–present). *Journal of Hydrometeorology* 4: 1147–1167.

Andersson E, Bauer P, Beljaars A, Chevallier F, Hólm E, Janisková M, Kållberg P. 2005. Assimilation and modeling of the atmospheric hydrological cycle in the ECMWF forecasting system. *Bulletin of the American Meteorology Society* 86: 387–402.

Bosilovich MG, Chen J, Robertson FR, Adler RF. 2008. Evaluation of global precipitation in reanalyses. *Journal of Applied Meteorology and Climatology* 47: 2279–2299.

Bromwich DH, Nicolas JP, Monaghan AJ. 2011. An assessment of precipitation changes over Antarctica and the Southern ocean since 1989 in contemporary global reanalyses. *Journal of Climate* 24: 4189–4209.

Chaudhuri AH, Ponte RM, Forget G, Heimbach P. 2013. A comparison of atmospheric reanalysis surface products over the ocean and implications for uncertainties in air–sea boundary forcing. *Journal of Climate* 26: 153–170.

Compo GP, Whitaker JS, Sardeshmukh PD, Matsui N, Allan RJ, Yin X, Gleason BE, Vose RS, Rutledge G, Bessemoulin P, Brönnimann S, Brunet M, Crouthamel RI, Grant AN, Groisman PY, Jones PD, Kruk MC, Kruger AC, Marshall GJ, Maugeri M, Mok HY, Nordli Ø, Ross TF, Trigo RM, Wang XL, Woodruff SD, Worley SJ. 2011. The twentieth century reanalysis project. *Quarterly Journal of the Royal Meteorological Society* 137: 1–28, DOI: 10.1002/qj.776.

Dee DP, Uppala SM, Simmons AJ, Berrisford P, Poli P, Kobayashi S, Andrae U, Balmaseda MA, Balsamo G, Bauer P, Bechtold P, Beljaars ACM, van de Berg L, Bidlot J, Bormann N, Delsol C, Dragani R, Fuentes M, Geer AJ, Haimberger L, Healy SB, Hersbach H, Hólm EV, Isaksen L, Kållberg P, Köhler M, Matricardi M, McNally AP, Monge-Sanz BM, Morcrette J-J, Park B-K, Peubey C, de Rosnay P, Tavolato C, Thépaut J-N, Vitart F. 2011. The ERA-Interim reanalysis: Configuration and performance of the data assimilation system. *Quarterly Journal of the Royal Meteorological Society* 137: 553–597, DOI: 10.1002/qj.828.

Feng X, DelSole T, Houser P. 2011. Bootstrap estimated seasonal potential predictability of global temperature and precipitation. *Geophysical Research Letters* 38: L07702, DOI: 10.1029/2010GL046511.

Feng X, DelSole T, Houser P. 2012. Methods for estimating seasonal potential predictability: analysis of Covariance. *Journal of Climate* 25: 5292–5308.

Feng X, DelSole T, Houser P. 2013a. Comparison of statistical estimates of potential seasonal predictability. *Journal of Geophysical Research* 118: 6002–6016, DOI: 10.1002/jgrd.50498.

Feng X, DelSole T, Houser P. 2013b. Comparison of seasonal potential predictability of precipitation. *Journal of Climate* (in press).

Frederiksen CS, Zheng X. 2007. Variability of seasonal-mean fields arising from interaseasonal variability. Part 3: application to SH winter and summer circulations. *Climate Dynamics* 28: 849–866.

Kalnay E, Kanamitsu M, Kistler R, Collins W, Deaven D, Gandin L, Iredell M, Saha S, White G, Woollen J, Zhu Y, Leetmaa A, Reynolds R, Chelliah M, Ebisuzaki W, Higgins W, Janowiak I, Mo KC, Ropelewski C, Wang J, Jenne R, Joseph D. 1996. The NCEP/NCAR 40-year reanalysis project. *Bulletin of the American Meteorological Society* 77: 437–471.

Kanamitsu M, Ebisuzaki W, Woollen J, Yang S-K, Hnilo JJ, Fiorino M, Potter GL. 2002. NCEP-DOE AMIP-II reanalysis (R-2). *Bulletin of the American Meteorological Society* 83: 1631–1648.

Katz RW. 1983. Statistical procedures for making inferences about precipitation changes simulated by an atmospheric general circulation model. *Journal of the Atmospheric Science* 40: 2193–2201.

Kim J-E, Alexander MJ. 2013. Tropical precipitation variability and convectively coupled equatorial waves on submonthly time scales in reanalyses and TRMM. *Journal of Climate* 26: 3013–3030.

Lorenz EN. 1963. Deterministic nonperiodic flow. *Journal of the Atmospheric Science* 20: 130–141.

Madden RA. 1976. Estimates of natural variability of time-averaged seal level pressure. *Monthly Weather Review* 104: 942–952.

Nguyen H, Evans A, Lucas C, Smith I, Timbal B. 2013. The Hadley circulation in reanalyses: climatology, variability, and change. *Journal of Climate* 26: 3357–3376.

Onogi K, Tsutsui J, Koide H, Sakamoto M, Kobayashi S, Hatsushika H, Matsumoto T, Yamazaki N, Kamahori H, Takahashi K, Kadokura S, Wada K, Kato K, Oyama R, Ose T, Mannoji N, Raira R. 2007. The JRA-25 reanalysis. *Journal of the Meteorological Society of Japan* 85: 369–432.

Palmer T, Anderson DLT. 1994. The prospects for seasonal forecasting – a review paper. *Quarterly Journal of the Royal Meteorological Society* 120: 755–793.

Rienecker MM, Suarez MJ, Gelaro R, Todling R, Bacmeister J, Liu E, Bosilovich MG, Schubert SD, Takacs L, Kim G-K, Bloom S, Chen JY, Collins D, Conaty A, Silva AD, Gu W, Joiner J, Koster RD, Lucchesi R, Molod A, Owens T, Pawson S, Pegion P, Redder CR, Reichle R, Robertson FR, Ruddick AG, Sienkiewicz M, Woollen J. 2011. MERRA: NASA's modern-era retrospective analysis for research and applications. *Journal of Climate* 24: 3624–3648.

Robertson FR, Bosilovich MG, Chen J, Miller TL. 2011. The effect of satellite observing system changes on MERRA water and energy fluxes. *Journal of Climate* 24: 5197–5217.

Saha S, Moorthi S, Pan H-L, Wu XR, Wang JD, Nadiga S, Tripp P, Kistler R, Woollen J, Behringer D, Liu HX, Stokes D, Grumbine R, Gayno G, Wang J, Hou Y-T, Chuang H-Y, Juang H-MH, Sela J, Iredell M, Treadon R, Kleist D, Delst PV, Keyser D, Derber J, Ek M, Meng J, Wei HL, Yang RQ, Lord S, Van Den Dool H, Kumar A, Wang WQ, Long C, Chelliah M, Xue Y, Huang BY, Schemm J-K, Ebisuzaki W, Lin R, Xie PP, Chen MY, Zhou ST, Higgins W, Zou C-Z, Liu QH, Chen Y, Han Y, Cucurull L, Reynolds RW, Rutledge G, Goldberg M. 2010. The NCEP climate forecast system reanalysis. *Bulletin of the American Meteorological Society* 91: 1015–1057.

Schenkel BA, Hart RE. 2012. An examination of tropical cyclone position, intensity, and intensity life cycle within atmospheric reanalysis datasets. *Journal of Climate* 25: 3453–3475.

Shukla J, Gutzler DS. 1983. Interannual variability and predictability of 500-mb geopotential heights over the Northern Hemisphere. *Monthly Weather Review* 111: 1273–1279.

10

Estimation of convective precipitation mass from lightning data using a temporal sliding-window for a series of thunderstorms in Southeastern Brazil

João V. C. Garcia,[1]* Stephan Stephany[1] and Augusto B. d'Oliveira[2]

[1] National Institute for Space Research (INPE), Sao Jose dos Campos, Brazil
[2] Center for Monitoring and Warnings of Natural Disasters (CEMADEN-MCTI), Cachoeira Paulista, Brazil

*Correspondence to:
J. V. C. Garcia, Programa de
Pós-graduação Computação
Aplicada (CAP/INPE), Caixa
postal 515, CEP 12245-970,
Sao Jose dos Campos, Brazil.
E-mail: sawamano@gmail.com

Abstract

Some studies have proposed the estimation of convective rainfall from lightning observations by the computation of the rainfall–lightning ratio (RLR). However, as such ratio may depend on season, convective regime and other factors, known approaches failed to provide values of RLR with low variability. An accurate RLR would allow estimating rainfall from lightning data in areas that lack weather radar coverage. This work proposes a straightforward approach for the computation of RLR, based on a temporal sliding-window and a fitting function. It was tested for thunderstorms observed in the Southeastern Brazil with good results.

Keywords: weather radar; lightning; rainfall estimation

1. Introduction

Rainfall estimation is typically performed from weather radar data. However, assuming that convective rainfall can be correlated to lightning, some approaches propose rainfall estimation from lightning data for areas without weather radar coverage, supporting nowcasting. The most common approach is the computation of the rainfall–lightning ratio (RLR), given by the convective rainfall mass per cloud-to-ground (CG) lightning flash. Nevertheless, such ratio may depend heavily on seasonal and geographical factors, local climatology, convective regime, storm type, lightning patterns or intensity, dominant lightning polarity of CG lightning, intracloud to CG ratio and thunderstorm life cycle (Buechler and Goodman, 1990; Soula and Chauzy, 2001; Lang and Rutledge, 2002). Therefore, known approaches may fail to provide values of RLR with low variability (Sist *et al.*, 2010).

A number of studies were performed to estimate the rainfall mass directly from CG lightning observations. Petersen and Rutledge (1998) used the total rainfall mass and the density of CG lightning to examine their relationship on a number of spatial and temporal scales for different parts of the world. The lightning flash incidence is more intense in clouds associated to high-level precipitation, as the electrification increases with altitude as in the case of tall cumulonimbus (Siingh *et al.*, 2010). Tapia *et al.* (1998) computed the RLR by dividing the total convective rainfall mass by the number of CG flashes in a thunderstorm, and proposed a model to reconstruct the spatial and temporal distribution of the rainfall. The summation of the rainfall

distribution of the flashes yields the overall rainfall distribution, which was checked against weather radar data. Kempf and Krider (2003) presented a compilation of RLR values including some obtained from other works, and found values ranging from 38 to 72×10^6 kg per flash for isolated thunderstorms in Florida, Spain and France, and values as high as 5000×10^6 kg per flash for mesoscale thunderstorms in Australia and Central United States. Molinie *et al.* (1999) found values as low as 3×10^6 kg per flash for the Pyrenees, while Williams *et al.* (1992) found values up to 500×10^6 kg per flash for Australia.

The current work proposes a simpler and more accurate approach, the function windowed RLR (WRLR), which employs a temporal sliding-window. This approach is based on the assumption that convective activity is correlated to electrically active cells that correspond to areas with high density of CG strokes. Such density is calculated by the EDDA software that implements standard kernel estimation (Strauss *et al.*, 2010). This software is being evaluated for operational use in order to detect convective precipitation in the recently established Center for Natural Disasters Monitoring and Alert (CEMADEN) in Brazil.

A set of thunderstorms that occurred in 2009 in the Southeastern Brazil was selected from weather radar data to obtain a WRLR function, while another set of January 2010 was employed to test this WRLR function as rainfall estimator. It is expected to include this function as a new module of the EDDA software. This may provide rainfall estimation in parts of Brazil, a huge country that has over 8.5 million km^2, but less than 15% of its area is covered by weather radar. Rainfall estimations can be obtained from meteorological satellites like those of the Tropical Rainfall Measuring Mission (TRMM), National Oceanic Atmospheric Administration (NOAA) or Geostationary Operational Environmental Satellite (GOES) satellite series, but can be imprecise (Ramirez-Beltran et al., 2008; Liao and Meneghini, 2009). On the other hand, Brazil has a ground-based lightning detector network called RINDAT (Brazilian Integrated Lightning Detection Network), one of the largest in the world (Pinto et al., 2006).

2. Data and methodology

2.1. Meteorological data

In this work, meteorological data consists of weather radar and lightning data for the entire year of 2009, plus the first month of 2010. The area of study is around two Brazilian S-band weather radars located in the State of Sao Paulo at the cities of Bauru and Presidente Prudente. This area corresponds to 32 squares with sides of 50 km that are within the range of 150 km of these radars, located at $22°21'30''$S $49°1'42''$W and $22°10'30''$S $51°22'30''$W. Radar data is given by constant altitude plan position indicator (CAPPI) images at 3 km altitude and with 1 km spatial resolution. The energy backscattered by hydrometeors, given by the reflectivity factor Z (in dBZ), is related to the rainfall rate R (in $mm\,h^{-1}$) by the Z–R relationship shown in Equation (1). This work adopted $A = 32$ and $b = 1.65$, according to Calheiros and Gomes (2010).

$$Z = A \times R^b \qquad (1)$$

Lightning data was provided by the Brazilian lightning detection network RINDAT that acquires radio frequency signals emitted by lightning. This network has detection efficiency of 90% and average precision of 500 m in stroke location for the State of Sao Paulo (Naccarato and Pinto, 2009). Lightning stroke data is output in the Universal Lightning ASCII (UALF) format. A lightning flash is composed of one or more strokes. The annual distribution of the number of flashes and strokes, and monthly accumulated rainfall for the year of 2009 for the area of study is shown in Figure 1.

2.2. Original RLR computation

The original RLR computation was proposed by Tapia et al. (1998). In order to estimate a RLR value, partial RLRs are calculated for each one of 22 selected summer storms occurred in Florida in 1992/1993. The resulting RLR value is then given by median of these values. The partial RLRs presented a high variability (from 24 to 365×10^6 kg per flash) that was attributed to different convective regimes. The resulting RLR was 43×10^6 kg per flash.

The model defined by Tapia et al. (1998) allows to estimate rainfall spatial and temporal distribution

using a previously estimated RLR value. A uniform rainfall distribution is assumed in a circle of 5 km radius centered at each CG lightning flash and in a 5-min interval centered at its time of occurrence. The rainfall distribution is then expressed by Equation (2) (x and X_i express geographical coordinates).

$$R(t,x) = C \sum_{t=1}^{N_i} \mathrm{RLR} \times f(t,T_i) \times g(x,X_i), \qquad (2)$$

where $R(t, x)$ = Rainfall rate in mm/h at time t and position x; N_t = Counter for the number of flashes; t_i = Time of occurrence of the i-th flash; x_i = Location of occurrence of i-th flash; RLR = Constant rainfall-lightning ratio (kg/CG flash); C = Unit conversion factor; $f(t,T_i)$ = Dirac delta function for the i-th flash occurred at T_i (checks if $|t-T_i| < 5$ min); $g(x,X_i)$ = Dirac delta function for the i-th flash occurred in X_i (checks if $|x-X_i| < 5$ km).

2.3. Temporal sliding-window based RLR computation

As already stated, the choice of a suitable value for the RLR may be difficult due to its high variability. This seems to be a limitation for the Tapia's model, which employs a constant RLR to estimate convective rainfall from CG lightning data. This work proposes a new approach, based on a precipitation function WRLR of the number of CG lightning strokes (N). Strokes and precipitation refer to a defined window that covers an interval of time (Δt) and a square area Q_j, hence the name WRLR.

This approach is composed of (1) the training phase, in which the WRLR function is derived from known rainfall and stroke data, and (2) the estimation phase that employs the WRLR function to estimate the precipitated mass from stroke data. In the first phase, an area and duration of study is chosen within radar range, comprehending typically many squares Q_j and thousands of Δt's. For each square Q_j, the window advances in time using a sliding-window scheme. Windows without any convective precipitation are discarded. The resulting set of windows provides the data points: data point P_{ij} is given by the pair (m_i, N_i), where m_i is the precipitated mass in the square Q_j and i-th interval of time Δt, and N_i is corresponding number of CG strokes. Outliers corresponding to data points with very high precipitated mass were removed using the Tukey–Kramer method (Tukey, 1977). Finally, a suitable WRLR function is chosen to fit these data points. Once the function must interpolate many data points that may present the same number of strokes, but different values of precipitated mass, average values of the latter quantities were considered. In the second phase, the WRLR function is employed to estimate the precipitated mass in a particular window Q_j and duration Δt that may be placed outside radar range, assuming that climatic characteristics are similar to those of the area of study.

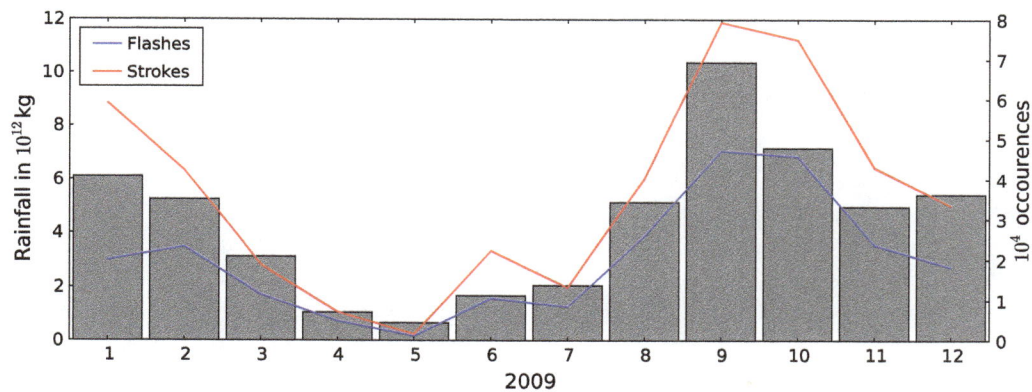

Figure 1. Annual distribution of rainfall and number of lightning flashes and strokes for the area of study in 2009.

According to Sist *et al.* (2010) the correlation between lightning and convective precipitation is more significant than it is to stratiform precipitation. Therefore, this work adopted the criterion proposed in Steiner *et al.* (1995) to filter out nonconvective precipitation. Considering the weather radar grid, this criterion associates to convective precipitation a grid point with reflectivity of at least 40 dBZ or that presents a significant gradient of reflectivity, above a threshold ΔZ for a circle around it, and also all grid points inside the circle. Values of the threshold and the radius of the circle depend on the background intensity. In this work, lightning strokes were adopted for the computation of the WRLR function, as the results were better than those obtained with lightning flashes. Several tests were performed with various values for Q and Δt, resulting in different functions WRLR. Larger and longer windows provide data points with higher precipitated mass and higher number of strokes than smaller and shorter windows. However, the resulting WRLR functions provide precipitation taxes (in $mm\,h^{-1}$) that are equivalent and with similar relative error.

Another point in the proposed methodology is the spatial and temporal rainfall distribution. Tapia *et al.* assumed rainfall distribution as described in Equation (2), based on individual CG lightning flashes. This work assumes rainfall distribution as given by the EDDA software (Strauss *et al.*, 2010) that generates a field of density of occurrence of CG strokes employing kernel density estimation, for the considered area and interval of time. The normalized density of CG strokes is mapped to a density of precipitated mass using its accumulated value for that interval.

3. Analysis of the results

The results refer to the estimation of the precipitated mass from CG lightning stroke data using squares Q with 50 km edge and Δt of 30 min, which were adopted for convenience. The advance of the temporal sliding-window was chosen as 7.5 min providing a significant overlap between consecutive window advances, and thus more data points. The area of study was defined in Section 2.1, composed of 32 squares that were employed for the training and estimation phases. Training data corresponds to the entire year of 2009, while estimation data, to the month of January 2010. The resulting WRLR function is show in Equation (3), while Figure 2 presents the scatterplot of the data points and the function itself.

$$WRLR\,(N) = 941.3 \times N^{0.3878} - 182.1 \qquad (3)$$

The temporal sliding-window with overlap is supposed to smooth the fitting curve (WRLR function)

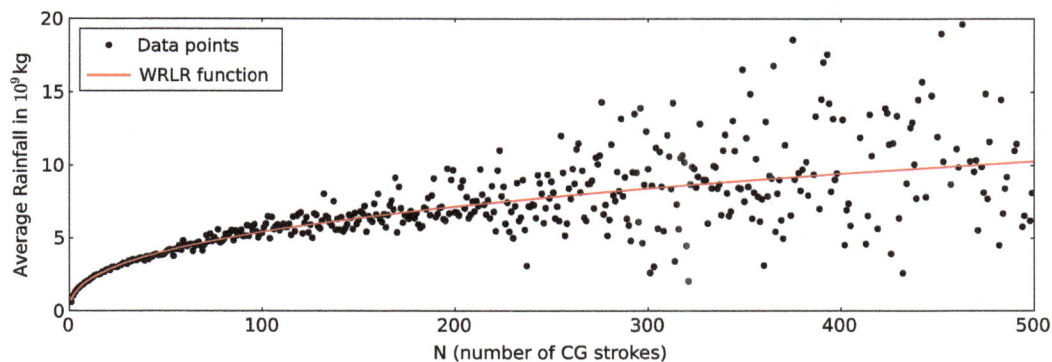

Figure 2. Scatterplot of the data points employed in the training phase and the corresponding WRLR fitting function.

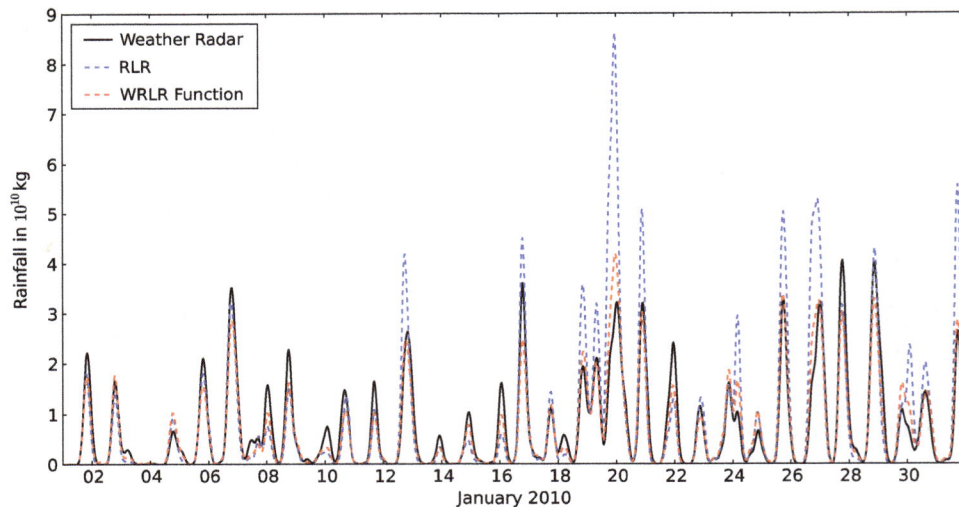

Figure 3. Temporal evolution of the 30-min accumulated rainfall along the month of January 2010 for the considered area given by weather radar, Tapia's model RLR and WRLR function (curves were smoothed by a Gaussian filter).

as it is equivalent to a moving average operator. The scatterplot of Figure 2 shows that low values of N presented a low variability, while the opposite occurred for $N > 150$. However, the latter correspond to less than 1% of the data points. The same figure shows that the function fits well the data points until $N = 100$. The training phase employed 187 735 data points with a total of 235 762 CG flashes or 428 129 CG strokes that correspond to 491 thunderstorms.

The WRLR function obtained in the training phase using data of 2009 was then employed to estimate the precipitated rainfall in January 2010 for the same area. The precipitation inferred from weather radar data was used as reference. The WRLR function was applied for each one of the 32 squares and for each 30-min interval yielding the accumulated precipitation. Tapia's approach RLR was calculated using flashes occurred in the same area along 2009 for the 491 thunderstorms. Partial RLRs were calculated for each thunderstorm, ranging from 0.4 to 1094×10^6 kg per flash and the final RLR was given by the median of these ratios, 219×10^6 kg per flash. The 30-min accumulated values of the precipitated mass were plotted in Figure 3 for the reference (weather radar), and estimations using the Tapia's model RLR (Equation (2)) and the WRLR function (Equation (3)). The curves were smoothed with a one-dimensional Gaussian filter with 2-h width for better visualization. The correlation between Tapia's model RLR estimation and radar precipitation was 0.78, while correlation between the WRLR function estimation and radar was 0.90. These correlations were calculated with unsmoothed data.

The proposed approach does not require selection of thunderstorms, but in order to compare these estimations in a different way, the total precipitated mass for each one of the 47 storms occurred in January 2010 in the same area was computed. Averages, medians and RMSE (root mean square error) values are presented in Table I. The accumulated 30-min

Table I. Some measures of the estimations for January 2010: average precipitated mass per storm and 30-mi accumulated precipitated mass (values in 10^9 kg).

		Weather radar	WRLR function	Tapia's model
Total rainfall per storm	Average	211	197	251
	Median	143	128	117
	RMSE	–	63	186
Accumulated 30-min	Average	10.1	9.5	12.1
	Median	8.4	8.5	5.8
	RMSE	–	2.8	11.8

values for the same thunderstorms for the same area were also computed and a global average for all thunderstorms is presented in the same table, as well as the median and RMSE.

In order to compare a rainfall distribution generated by the Tapia's model and by the proposed approach (WRLR function coupled to the EDDA software), a particular thunderstorm was selected, for the 30-min interval starting at 1 : 08 UTC of 20 January 2010. A squared area with an edge of 300 km was chosen, centered at the weather radar of Bauru. This area has 36 squares with edges of 50 km, differently from the preceding results shown in Figure 2, which included only the 16 inner squares of weather radars of Bauru and Presidente Prudente, in order to comply to the 150 km limit for radar coverage. It is worth to note that the training phase defines a WRLR function using data within this range, but the estimation phase allows to estimate rainfall beyond that range from lightning data, as this was the main goal proposed in this work.

Figure 4 shows the area of that thunderstorm to allow a comparison between the spatial distributions of rainfall given by the weather radar of Bauru, the EDDA/WRLR and the Tapia's model. In the case of the distribution generated by the EDDA/WRLR distribution, the normalized CG stroke density was

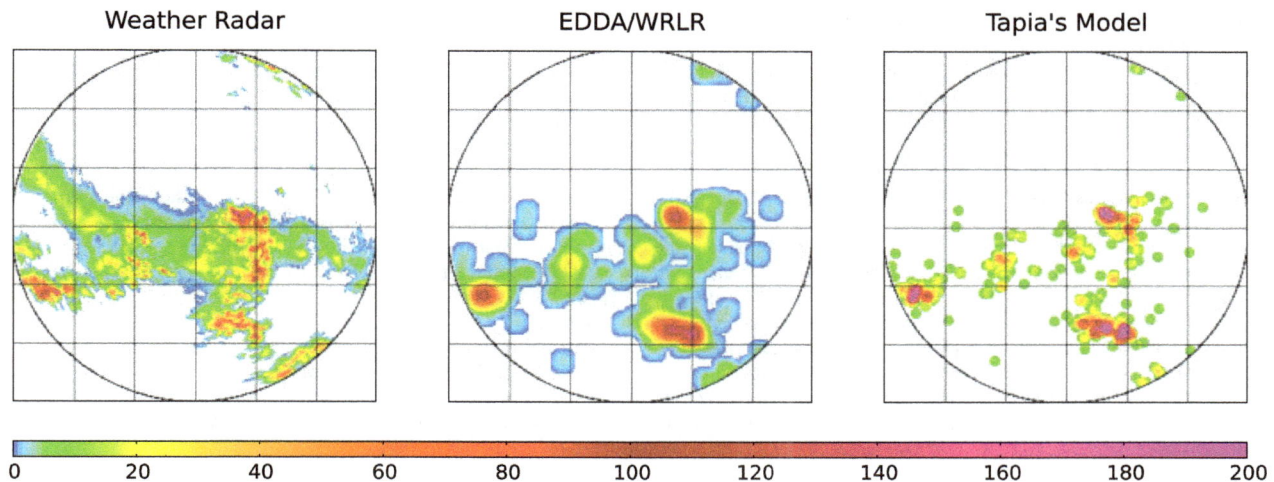

Figure 4. Rainfall distributions generated from radar data (left), the proposed EDDA/WRLR approach (center) and Tapia's model (right) around the weather radar of Bauru for a 30-min interval in 20 January 2010 (color bar shows rainfall rate in $mm\,h^{-1}$).

mapped to rainfall rate. The accumulated values for the 30-min interval, expressed in $10^9\,kg$, were 38.27 (radar), 44.37 (WRLR) and 50.11 (Tapia). Similar comparisons were performed for many other time intervals showing that the spatial distributions given by EDDA/WRLR were compatible with the precipitation observed in radar images.

4. Summary and conclusions

A new approach is proposed to estimate the precipitated mass and rainfall distribution from lightning data. The most common approach, proposed by Tapia *et al.* (1998), is to calculate a constant RLR value given by the median of RLR values obtained for a set of thunderstorms and to estimate rainfall mass assuming a circular distribution around each lightning flash. However, partial RLRs have high variability and may lead to estimation errors. The new approach computes data points composed of the convective precipitated mass and the number of CG strokes for windows with a defined square area and duration. A sliding-window scheme advances each window in time over the same area. Stratiform precipitation was filtered out, as it is usually not related to lightning. The set of data points is then fitted by a suitable WRLR function that can be employed to estimate rainfall beyond weather radar range. The EDDA software, which generates a field of density of occurrence of CG strokes, is also used to estimate rainfall distribution. Assuming weather radar data as reference, the proposed approach yielded a better estimate for the total precipitated mass and for the spatial distribution in a region of Southeast Brazil for the month of January 2010. Further work intends to extend the WRLR function for other regions by defining specific coefficients. In addition, the WRLR will be implemented as a new module of the EDDA software, in order to be evaluated for operational weather monitoring.

Acknowledgements

Authors João Victor Cal Garcia, Stephan Stephany and Augusto B. d'Oliveira thank CNPq (National Council for Scientific and Technological Development of Brazil) for grants 140983/2010-4, PQ 305639/2012-9 and 473053/2010-1, respectively. Authors also thank the Center for Weather Forecasts and Climate Studies (CPTEC/INPE) and the Meteorological Research Institute (IPMet/UNESP) for the meteorological data.

References

Buechler D, Goodman S. 1990. Echo size and asymmetry: impact on NEXRAD storm identification. *Journal of Applied Meteorology* **29**: 962–969, DOI: 10.1175/1520-0450(1990)029<0962:ESAAIO>2.0.CO;2.

Calheiros RV, Gomes, AM. 2010. Flow forecasting in the Corumbataí River basin: radar rainfall stratification and runoff-rainfall relations. ERAD 2010 – The Sixth European Conference on Radar in Meteorology and Hydrology, 1.

Kempf NM, Krider EP. 2003. Cloud-to-ground lightning and surface rainfall during the Great Flood of 1993. *Monthly Weather Review* **131**: 1140–1149, DOI: 10.1175/1520-0493(2003)131<1140:CLASRD>2.0.CO;2.

Lang T, Rutledge S. 2002. Relationship between convective storm kinematics, microphysics and lightning. *Monthly Weather Review* **130**: 2492–2506, DOI: 10.1175/1520-0493(2002)130<2492:RBCSKP>2.0.CO;2.

Liao L, Meneghini R. 2009. Validation of TRMM precipitation radar through comparison of its multiyear measurements with ground-based radar. *Journal of Applied Meteorology and Climatology* **48**: 804–817, DOI: 10.1175/2008JAMC1974.1.

Molinie G, Soula S, Chauzy S. 1999. Cloud-to-ground lightning activity and radar observations of storms in the Pyrénées range. *Quarterly Journal of the Royal Meteorological Society* **125**: 3103–3122, DOI: 10.1256/smsqj.56014.

Naccarato KP, Pinto O. 2009. Improvements in the detection efficiency model for the Brazilian lightning detection network (BrasilDAT). *Atmospheric Research* **91**(2): 546–563, DOI: 10.1016/j.atmosres.2008.06.019.

Petersen W, Rutledge S. 1998. On the relationship between cloud-to-ground lightning and convective rainfall. *Journal of Geophysical Research* **103**: 14025–14040, DOI: 10.1029/97JD02064.

Pinto Jr O, Naccarato K, Saba M, Pinto I, Abdo R, Garcia S, Cazetta Filho A. 2006. Recent upgrades to the Brazilian integrated lightning detection network. 19th International Lightning Detection Conference (ILDC), Tucson, 1.

Ramirez-Beltran ND, Kuligowski RJ, Harmsen WE, Castro JM, Cruz-Pol S, Cardona JM. 2008. Rainfall estimation from convective storms using the hydro-estimator and NEXRAD. *WSEAS Transaction on Systems* **7**(10): 1016–1027.

Siingh D, Kumar S, Singh AK. 2010. Thunderstorms/lightning generated sprite and associated phenomena. *Earth Science India* **3**(II): 24–145.

Sist M, Zauli F, Melfi D, Biron D. 2010. A study about the correlation link between lightning data and meteorological data. 2010 EUMETSAT Meteorological Satellite Conference, Córdoba 1.

Soula S, Chauzy S. 2001. Some aspects of the correlation between lightning and rain activities in thunderstorms. *Atmospheric Research* **56**(1–4): 355–373, DOI: 10.1016/S0169-8095(00)00086-7.

Steiner M, Houze R Jr, Yuter S. 1995. Climatological characterization of three-dimensional storm structure from operational radar and rain gauge data. *Journal of Applied Meteorology* **34**(9): 1978–2007, DOI: 10.1175/1520-0450(1995)034<1978:CCOTDS>2.0.CO;2.

Strauss C, Stephany S, Caetano M. 2010. The EDDA software tool for the generation of density fields of atmospheric discharges for meteorological data mining (in Portuguese). XXXIII Brazilian Congress of Computational and Applied Mathematics (CNMAC-2010), Águas de Lindóia (Brazil), 3: 269–275. ISSN: 1984–8218.

Tapia A, Smith JA, Dixon M. 1998. Estimation of convective rainfall from lightning observations. *Journal of Applied Meteorology* **37**: 1497–1509, DOI: 10.1175/1520-0450(1998)037<1497:EOCRFL>2.0.CO;2.

Tukey J. 1977. *Exploratory Data Analysis*. Addison-Wesley Reading: Boston.

Williams ER, Rutledge SA, Geotis SG, Renno N, Rasmussen E, Rickenbach T. 1992. A radar and electrical study of tropical hot towers. *Journal of the Atmospheric Sciences* **49**(15): 1386–1395, DOI: 10.1175/1520-0469(1992)049<1386:ARAESO>2.0.CO;2.

11

Indian summer monsoon prediction and simulation in CFSv2 coupled model

Gibies George,[1] D. Nagarjuna Rao,[1] C. T. Sabeerali,[1,2] Ankur Srivastava[1] and Suryachandra A. Rao[1,*]

[1] Program for Seasonal and Extended Range Prediction of Monsoon, Indian Institute of Tropical Meteorology, Pune, India
[2] The Center for Prototype Climate Modeling, New York University, Abu Dhabi, UAE.

*Correspondence to:
S. A. Rao, Program for Seasonal and Extended Range Prediction of Monsoon, Indian Institute of Tropical Meteorology, Pune, India.
E-mail: surya@tropmet.res.in

Abstract

Using carefully designed coupled model experiments, we have demonstrated that the prediction skill of the all India summer monsoon rainfall (AISMR) in Climate Forecast System version 2 (CFSv2) model basically comes from the El-Niño Southern Oscillation-Monsoon teleconnection. On the other hand, contrary to observations, the Indian Ocean coupled dynamics do not have a crucial role in controlling the prediction skill of the AISMR in CFSv2. We show that the inadequate representation of the Indian Ocean coupled dynamics in CFSv2 is responsible for this dichotomy. Hence, the improvement of the Indian Ocean coupled dynamics is essential for further improvement of the AISMR prediction skill in CFSv2.

Keywords: Indian summer monsoon; seasonal prediction; ocean–atmosphere coupled dynamics; CFSv2; coupled model; ENSO-monsoon teleconnection

1. Introduction

Early and accurate seasonal prediction of Indian summer monsoon (ISM) during June through September (JJAS) is very important for proper planning and the socioeconomic well-being of India as majority of people in this region depend on rain-fed agriculture for their life and existence. Recently, Gadgil and Srinivasan (2011) have studied the simulation of the all India summer monsoon rainfall (AISMR) in five atmospheric general circulation models (AGCMs) and they have shown that the poor prediction skill in many AGCMs arise due to excessive teleconnection with El-Niño Southern Oscillation (ENSO). Tropical Ocean Global Atmosphere (TOGA), Global Ocean Global Atmosphere, Seasonal Prediction of Indian Monsoon (SPIM) and similar experiments, mainly focus on the response of the atmosphere to the sea surface temperature (SST) forcing (Lau and Nath, 2000; Gadgil and Srinivasan, 2011) based on two-tier modeling strategies where the atmosphere and ocean are treated separately by using the AGCM, which is forced by either observed boundary condition or output from the ocean general circulation models. One of the important limitations of the two-tier experimental design is that in reality a part of the SST especially over the warm pool regions evolves in response to the atmospheric change (Lau and Nath, 2004; Wang et al., 2004; Yu and Lau, 2005).

Recent studies (Kumar et al., 2005; Rajeevan et al., 2012 and the references there in) have reported progress in the multi-model ensemble anomaly correlation coefficient (MME ACC) of AISMR, from 0.28 (DEMETER; Palmer et al., 2004) to 0.45 (ENSEMBLE; Hewitt, 2004) in ocean–atmosphere coupled general circulation models initialised with May initial condition for the period 1960–2005, but the potential predictability of AISMR in coupled models is yet to be achieved. It is important to understand how to further improve the AISMR prediction skill in coupled models. To address this, we have carried out a couple of sensitivity experiments using a state-of-the-art coupled model to quantify the relative importance of the Pacific and Indian Ocean coupled dynamics in modulating the interannual variability of the AISMR. In Section 2, we describe the model and experimental design along with observational datasets used in this study. Section 3 demonstrates the relative importance of the Pacific and Indian Ocean coupled dynamics in simulating the ISM. Section 4 summarizes the results.

2. Model and experiment design

To understand the relative role of active ocean dynamics of different basins in forcing the interannual variability of the ISM, we have carried out a set of sensitivity experiments along with control (CTL) run using CFSv2 coupled model. The National Centers for Environmental Prediction (NCEP) Climate Forecast System version 2 (CFSv2; Saha et al., 2014b) is a state-of-the-art coupled climate model developed by the NCEP, USA. The atmospheric component of the CFSv2 is Global Forecast System with a spectral resolution of T126 and 64 hybrid vertical levels and the ocean component is Geophysical Fluid Dynamics Laboratory Flexible Modeling System Modular Ocean Model version 4p0d. In addition, it has a four-layer NOAH land surface model and a three-layer dynamical sea ice model coupled together with the atmosphere and ocean components in Earth

System Modeling Framework. In this study, both CTL and sensitivity runs are made in hindcast mode (initialised every year) and integrated for 9 months lead time, for the period 1982–2009. Both CTL and sensitivity runs are an ensemble mean of five realizations of CFSv2 T126 model runs initialized with the February initial conditions (00z05Feb, 00z10Feb, 00z15Feb, 00z20Feb and 00z25Feb). The CFSv2 reanalysis (Saha et al., 2010) obtained from NCEP are used as the initial conditions for the model run.

The CTL run (same as original CFSv2) has ocean dynamics and ocean-atmospheric coupling all over the globe while in the Indian Ocean slab (ISLAB) run, active ocean dynamics are removed from the tropical Indian Ocean, instead a uniform slab (50 m) over the tropical Indian Ocean (30°S to 30°N; 45°E to 120°E) provides the flux-driven SST. We have carried out separate sensitivity experiments by prescribing Mixed Layer Depth (MLD) as 40, 50 and 60 m, but modest changes to this fixed MLD do not alter the major conclusions of this study. Hence, the results from those sensitivity experiments which have a uniform 50 m slab MLD are discussed in this study. Some of the earlier studies used prescribed climatological SST forcing to remove the influence of SST variations in some basin (Lau and Nath, 2004; Yu and Lau, 2005; Achuthavarier et al., 2012). There is another set of sensitivity studies (Yokoi et al., 2012) using prescribed climatological wind stress fluxes to the ocean in order to remove atmospheric coupling while keeping ocean dynamics. Both of the above-mentioned strategies use some kind of climatological forcing field into the coupled system. Hence, the ocean and atmosphere over the same basin exist in two different states (one is in climatological and the other is in dynamical state). This resembles the two-tier modeling strategy of standalone models. Spatial and temporal variations in the MLD [e.g. E2 experiment of Krishnan et al. (2011)] are not prescribed in the sensitivity experiment because it is a well-known fact that MLD variations are also partly due to coupled ocean-atmosphere dynamics. The coupled sensitivity experiment designed for this study uses a strategy by which the thermodynamical forcing is maintained while completely removing ocean dynamics in a particular basin (This indirectly removes the dynamical component of ocean-atmosphere coupling as well). The 50 m uniform regional slab ocean in this experiment resembles Tropical Ocean Global Atmosphere with Mixed Layer Ocean experiment (TOGA-ML) experiment of Lau and Nath (2000), Slab ocean experiment of Dommenget (2010) and E1 experiment of Krishnan et al. (2011), except for the differences in the domain. In ISLAB run, the atmosphere is coupled with fully dynamical Ocean elsewhere outside the tropical Indian Ocean. Hence, the difference between ISLAB and CTL runs isolates the role of Indian Ocean coupled dynamics. Similarly, the Pacific Ocean slab (PSLAB) run is identical to the ISLAB run except that coupled dynamics are removed only from the tropical Pacific Ocean (30°S to 30°N; 120°E to 75°W).

Therefore, the comparison of the PSLAB run and CTL run isolates the role of Pacific Ocean coupled dynamics. As the model integration is for 9 months, we have not used any flux correction in any of the sensitivity experiments as model-simulated SST does not drift significantly.

In order to evaluate the model-simulated rainfall, we use the pentad precipitation from Global Precipitation Climatology Project (Xie et al., 2003). We use the monthly Extended Reconstructed SST version 3 (Smith et al., 2008) to verify the model-simulated SST.

3. Results and discussion

3.1. Model biases in simulating SST and precipitation

Figure 1(a) shows the seasonal mean SST in tropics, which clearly displays warm SSTs (>28 °C) over the Indo-Pacific region covering the eastern Arabian Sea, the entire Bay of Bengal, the eastern equatorial Indian Ocean and the western tropical Pacific, including the South Pacific Convergence Zone, coinciding with well-known warm pool regions. The CFSv2 CTL run is able to capture the observed large-scale spatial pattern of SST (figure not shown). However, it exhibits a cold SST bias over the entire Indian Ocean, northwest and southwest Pacific Ocean and a narrow region of cold bias over the central equatorial Pacific (Figure 1(c)). Warm SST bias over the south/north eastern tropical Pacific Ocean (Figure 1(c)) is also observed. Previous study by Pokhrel et al. (2012) argued that the cold SST bias over the Indian Ocean in CFSv2 CTL run is due to dry surface atmosphere and associated increase in the latent heat flux. Furthermore, we have noticed that the model (CTL run) overestimates the net heat loss (Unit: Peta Watt here after PW) from the northern Indian Ocean (observed −0.61 PW; CTL run −0.98 PW) and underestimates heat gain over the southern tropical Indian Ocean (observed 1.34 PW; CTL run −0.30 PW). As a result, the model underestimates (cold bias) the SSTs over the Indian Ocean. Detailed analysis of the heat budget will be reported elsewhere. The warm SST bias over the north/south eastern Pacific may be a result of misrepresentation of stratus cloud decks in the eastern Pacific and the resulting penetration of more shortwave radiation to the surface, as reported by Zheng et al. (2011).

The seasonal mean precipitation over south Asian summer monsoon region show three zones of maximum rainfall, namely the head Bay of Bengal, the eastern equatorial Indian Ocean and the western coast of India (Figure 1(b)). The CFSv2 is able to capture these zones of maximum rainfall over ISM domain (Saha et al., 2014a). In spite of cold SST bias (Figure 1(c)), the model overestimates rainfall over oceanic regions (Figure 1(d)), probably due to the fact that the SSTs over these regions are still above the critical SST (27.5 °C) for convection to occur. Contrary to this, the rainfall

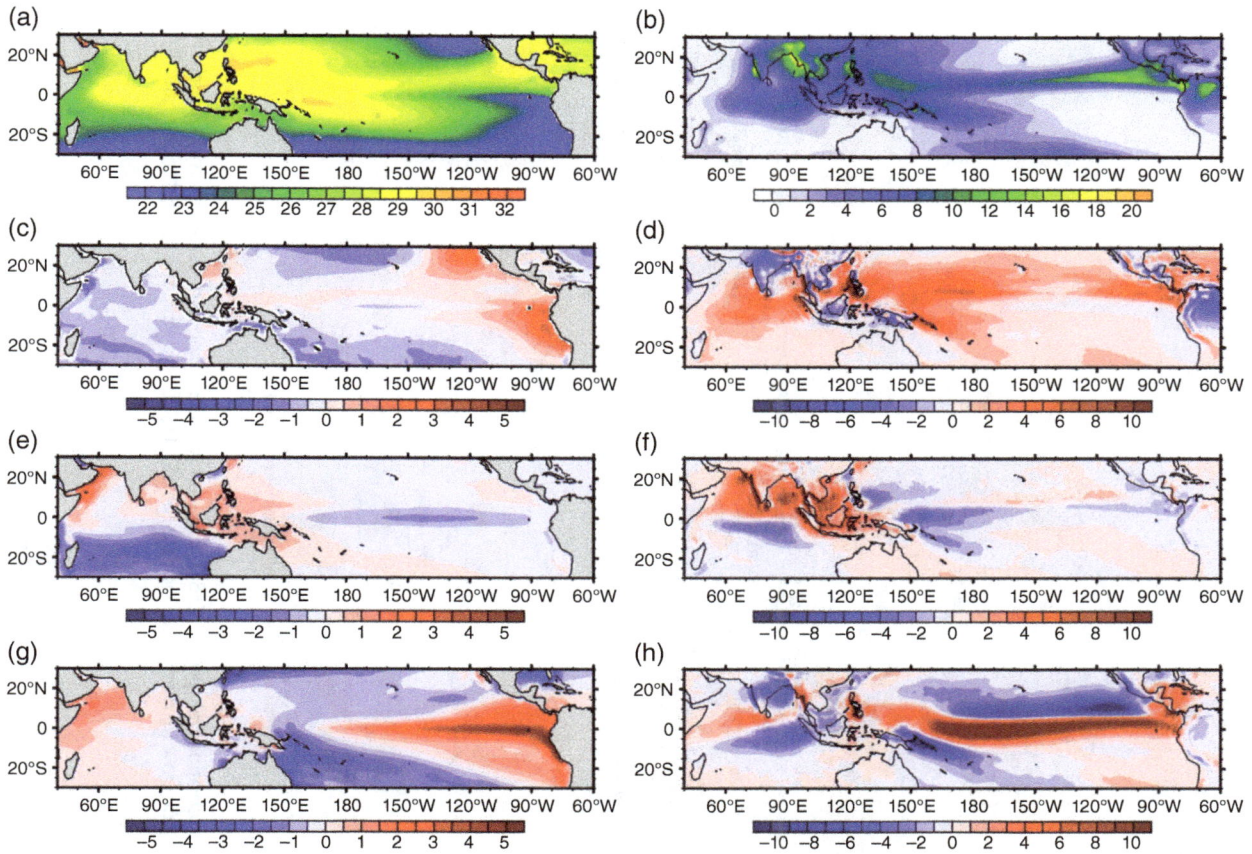

Figure 1. (a) Observed seasonal (JJAS) climatological SST, (b) observed seasonal (JJAS) climatological precipitation, (c) seasonal SST bias of CFSv2 control (CTL) run, (d) seasonal precipitation bias of CFSv2 control (CTL) run, (e) seasonal SST difference between ISLAB and CTL runs (ISLAB − CTL run), (f) seasonal precipitation difference between ISLAB and CTL runs (ISLAB − CTL run), (g) seasonal SST difference between PSLAB and CTL runs (PSLAB − CTL run), (h) seasonal precipitation difference between PSLAB and CTL runs (PSLAB − CTL run).

over land regions is underestimated (Figure 1(d)). The dry bias over the Indian land mass is not unique to the CFSv2 model, but many CMIP5 models have also shown a similar bias in the precipitation simulation (Sabeerali *et al.*, 2013). Recent study by Saha *et al.* (2012) has reported that the dry bias over the Indian land region can be reduced by correcting the biases of Eurasian snow cover. In order to understand how coupled dynamics in each basin control model biases, two sensitivity experiments are carried out. The details of the sensitivity experiments are already described in Section 2. Here, we discuss results from these sensitivity experiments.

The CFSv2 ISLAB run, compared with CTL run, shows that absence of Indian Ocean dynamics results in colder (warmer) SST in the southern (northern) tropical Indian Ocean (Figure 1(e)), indicating the lack of southward heat transport. The cold SST bias over the equatorial central Pacific has increased in the ISLAB run (Figure 1(e)) and cold biases over the northwest and southwest Pacific remain unchanged (Figure 1(e)). These results indicate that the Indian Ocean coupled dynamics have a crucial role in determining the SSTs over the southern Indian Ocean and the equatorial central Pacific, whereas the Indian Ocean coupled

dynamics have no significant role in deciding the SSTs over the northwest and southwest Pacific. In response to the reduced SST bias over the northern Indian Ocean, the dry bias over the Indian landmass has decreased in the ISLAB run (Figure 1(f)). By prescribing the slab in the Indian Ocean and calculating the SST simply from the net heat flux, it is clear that in the absence of active coupled dynamics, the SST in the northern tropical Indian Ocean exhibits slight warm bias (due to the absence of southward heat transport) in contrast to the cold bias (due to overestimation of southward heat transport) in the CTL run. This study confirms that the SSTs in the northern Indian Ocean are primarily determined by the surface heat fluxes, as suggested by Shenoi *et al.* (2002), whereas the realistic active coupled dynamics in the Indian Ocean are important in determining the correct SSTs over the southern Indian Ocean in CFSv2 because the southward (westward) heat transport through the northern (eastern) boundary dominantly determines the SSTs over southern tropical Indian Ocean. The SST–precipitation lead–lag relationship (Figure 2) in the warm pool regions (Bay of Bengal, Eastern Equatorial Indian Ocean and Northwest Pacific) are not captured in the model due to strong cold bias (up to −1 °C). However, in ISLAB

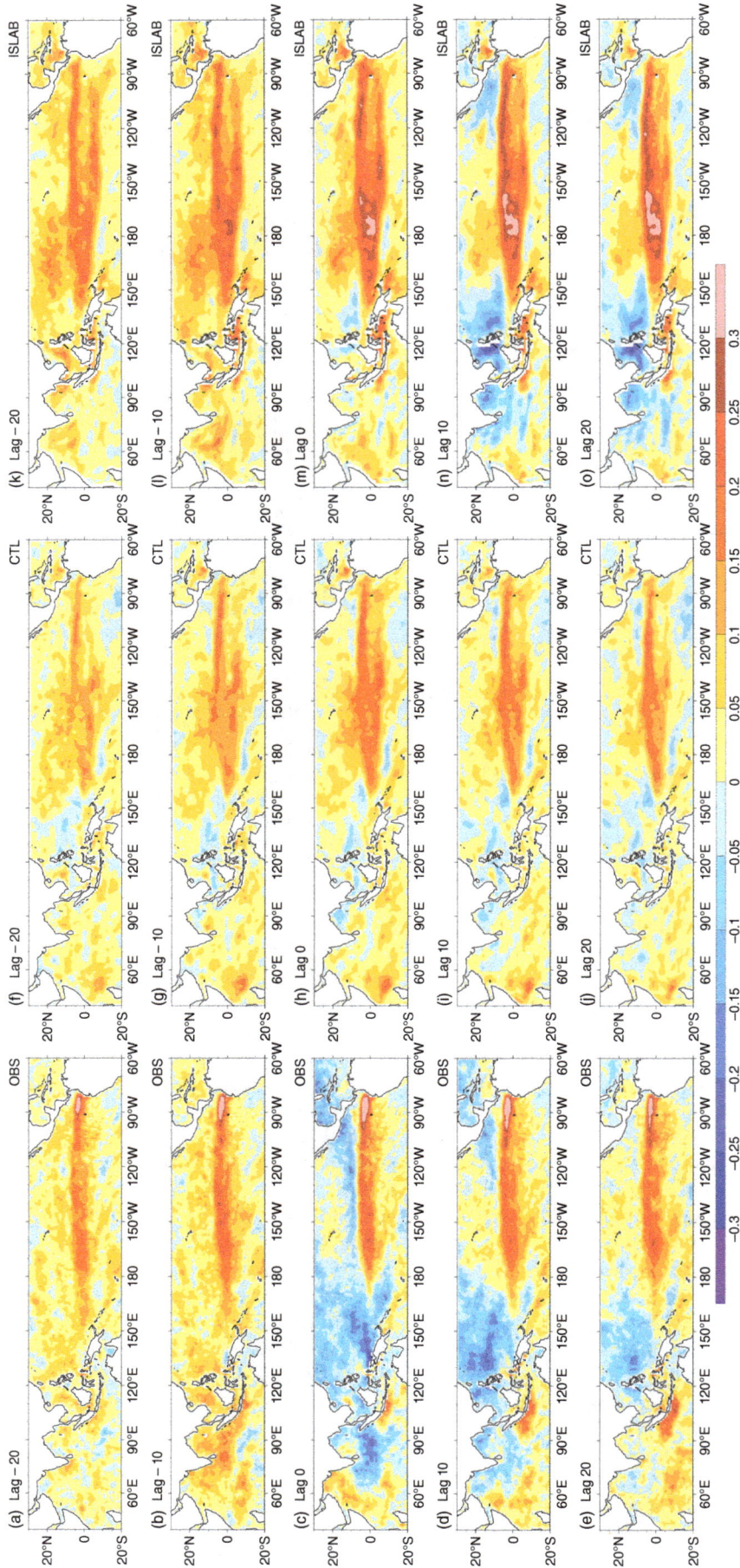

Figure 2. Spatial map of lead–lag relationship between SST and precipitation at each grid point for observation (left panel), CFSv2 CTL run (middle panel) and ISLAB run (right panel). First row marked as lag −20 indicate that the SST lead 20 days before the precipitation peak and similarly bottom row marked lag 20 indicate the SST lag 20 days after the precipitation peak.

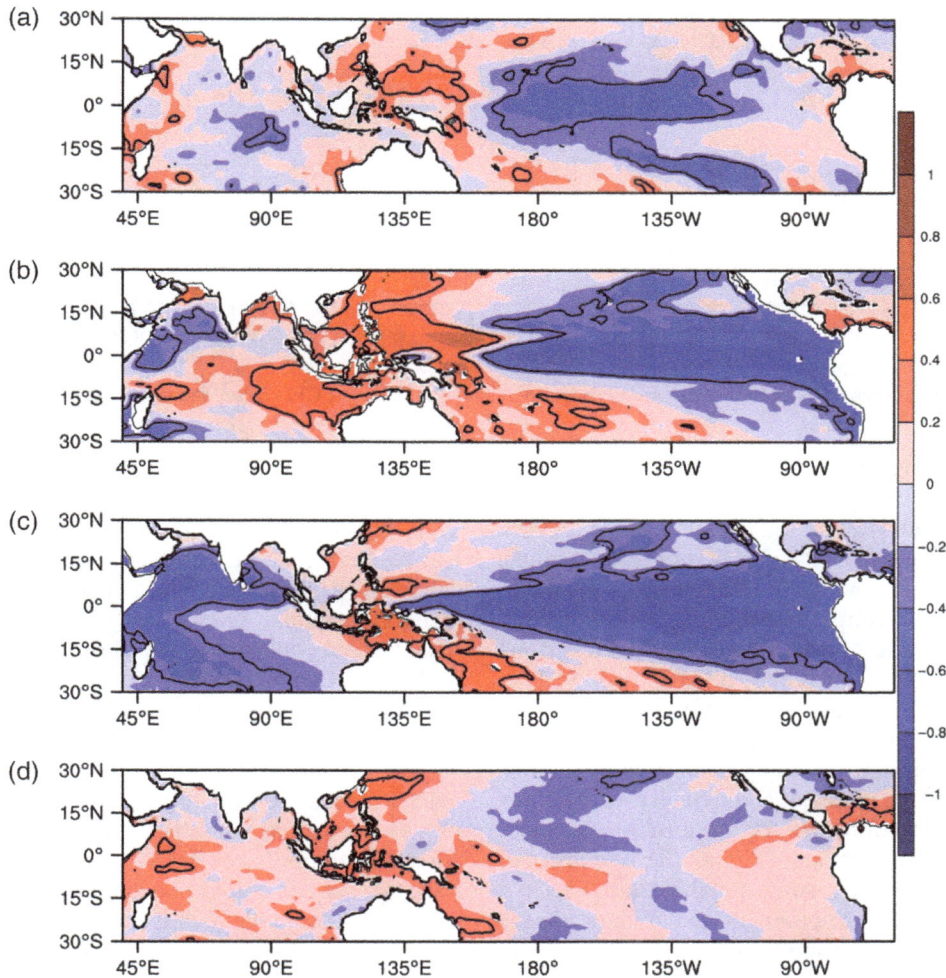

Figure 3. Spatial map of seasonal (JJAS) SST correlated with AISMR index for (a) observation, (b) CFSv2 CTL run, (c) ISLAB run, (d) PSLAB run. Statistically significant values (95% confidence level) are contoured.

run, due to better simulation of SSTs in the warm pool regions, the lead–lag relationship of air–sea interaction is reasonably simulated.

In the absence of the Pacific Ocean coupled dynamics (Dommenget, 2010), the warm bias over the eastern equatorial Pacific Ocean has strengthened in magnitude and spatial extent (Figure 1(g)), which resembles a perennial El-Niño condition. This is due to the absence of upwelling along the coast of Peru and eastern equatorial Pacific in the PSLAB run. During an El-Niño event, the associated teleconnection forces a warming over the western Indian Ocean and Arabian Sea (Murtugudde and Busalacchi, 1999; Venzke et al., 2000). Similar warming associated with perennial El-Niño type bias is noticed in PSLAB run. Similarly, the strengthening of cold bias over the northwest and southwest Pacific Ocean (Figure 1(g)) is due to absence of the dynamics associated with subtropical gyre and western boundary current. The perennial El-Niño type bias in the PSLAB run further enhances the dry bias over the Indian land region and there exists a wet bias all along tropical oceans (Figure 1(h)). This response is due to the well-known ENSO–monsoon

teleconnection relation, wherein El-Niño condition over the Pacific forces subsidence over the Indian land region (Rajeevan and Pai, 2007, and the references therein).

3.2. Teleconnections associated with AISMR

ENSO and IOD are the two dominant modes of climate variability over tropical oceans and the interannual variability of the AISMR is mainly related to these two modes (Kumar et al., 2006, and the references therein). The spatial pattern of correlation between the observed AISMR and global SST shows a negative correlation over the central/eastern tropical Pacific and major portions of the central/eastern Indian Ocean (Figure 3(a)), while positive correlation is observed over the western Pacific warm pool region and along the Somali–Oman coast (Figure 3(a)). The CFSv2 CTL run is able to capture this large-scale spatial pattern of observed correlation over the Pacific Ocean (Figure 3(b)). However, compared with observations the Pacific ENSO is strongly coupled to AISMR in CFSv2 CTL run (Figure 3(b)). As evident in Figure 3(a), several studies

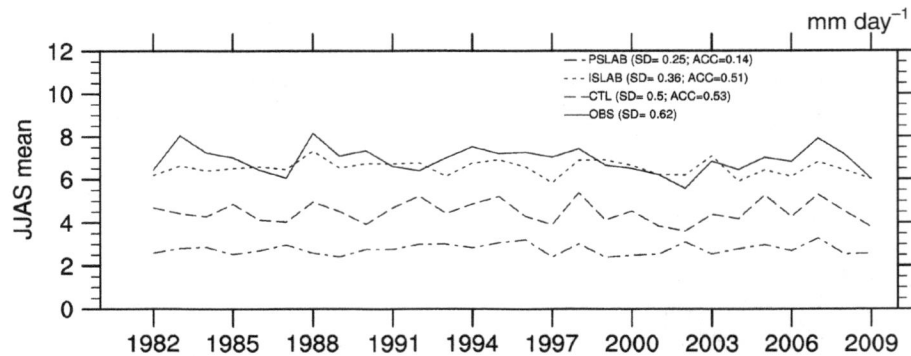

Figure 4. Interannual variations of Indian summer monsoon rainfall for different CFSv2 run and observation.

in recent times have highlighted that the central Pacific warming (El-Niño Modoki) is more conducive to force drought condition over India (Kumar *et al.*, 2006) compared with eastern Pacific warming (canonical El-Niño). However, many coupled models failed to capture this relationship (Wang *et al.*, 2015). CFSv2 also suffers from the same limitation by which strong negative correlation between AISMR and SST is concentrated over the eastern Pacific. In contrast to observations, the CTL run shows positive correlations all over central/eastern Indian Ocean and a negative correlation along the monsoon wind track (Figure 3(b)). The positive correlation over the western Pacific warm pool region is very strong (Figure 3(b)) compared with observations. This suggests that in CFSv2, a negative dipole-like structure in the Indian Ocean and La-Niña-like condition in the tropical Pacific will enhance precipitation over India. On the other hand, in observations, a positive dipole-like structure and a La-Niña-like structure are associated with good monsoon condition. This indicates that in CFSv2 the monsoon teleconnections over the tropical Indian Ocean are exactly opposite to the observed teleconnection. This suggests that further improvement in AISMR skill is possible by improving the simulation of Indian Ocean teleconnections.

In the ISLAB run, the teleconnection pattern over the Pacific Ocean is almost identical to the CTL run, although the influence of the eastern/central Pacific SST on the AISMR has increased and the negative correlation has further extended to the western Pacific (Figure 3(c)). The extension of negative correlation to western Pacific is due to extension of easterlies into the western Pacific in ISLAB run (figure not shown). Contrary to the CTL run, strong negative correlations are noticed in the ISLAB run along the path of monsoon cross-equatorial flow (Figure 3(c)), which suggests that the strong monsoon strengthens the cross-equatorial flow, and thereby, enhancing the latent heat loss from the ocean and cooling SST along its path (Shukla and Misra, 1977).

However, in the PSLAB run, the teleconnection between the Indian Ocean SST and AISMR is marginally improved compared with observations in the western tropical Indian Ocean wherein positive

correlation between the western Indian Ocean SST and AISMR is faithfully captured (Figure 3(d)). However, the negative correlation pattern over the central and eastern equatorial Indian Ocean is not captured (Figure 3(d)), demonstrating that central and eastern equatorial Indian Ocean coupled dynamics are not represented properly in PSLAB run also. In the absence of Pacific Ocean coupled dynamics in PSLAB run, teleconnections associated with ENSO have completely disappeared from the Pacific Ocean as expected (Figure 3(d)).

The CTL run reasonably captures the spatial patterns of SST (as well as rainfall) in the tropical Indian Ocean correlated with Niño 3.4 (figure not shown). These indicate that the ENSO teleconnections are reasonably captured in the model, while the AISMR teleconnection with the Indian Ocean SSTs are misrepresented in the model.

3.3. Interannual variations of AISMR

The evolution of the AISMR over years (Figure 4) shows that the magnitude of rainfall is underestimated in the CFSv2 CTL run compared with observations. The magnitude of AISMR in ISLAB run is comparable with observations (Figure 4). In contrast, in the PSLAB run, the magnitude of the AISMR is exceptionally weak (Figure 4). This result could be interpreted in two different ways: (1) in the CTL run, Indian Ocean coupled dynamics are not simulated reasonably, hence the convection over the equatorial Indian Ocean is overestimated and resulted in subsidence over the Indian landmass through modulation of local Hadley cell; (2) in ISLAB run, the northern (southern) Indian Ocean exhibits warm (cold) bias, due to the absence of meridional heat transport and resulting in enhanced (suppressed) convection over the northern (southern) Indian Ocean. Both of the above interpretations suggest that proper representation of air–sea heat fluxes and ocean dynamics in the tropical Indian Ocean can improve the AISMR simulation. In order to make better predictions, the ENSO–monsoon teleconnection and Indian Ocean SST–monsoon teleconnection should be better represented in the model. The CFSv2 T126L64 in general underestimates the interannual variance of

AISMR in all runs, particularly in ISLAB and PSLAB runs compared with observations. The observed inter-annual variance of the AISMR anomaly is about $0.36\,\mathrm{mm^2\,day^{-2}}$, while the same is $0.25\,\mathrm{mm^2\,day^{-2}}$ in CTL run. In ISLAB run, the interannual variance drops to $0.12\,\mathrm{mm^2\,day^{-2}}$ (50% of CTL), which suggests that 50% of interannual variance of AISMR in CFSv2 is related to the Indian Ocean dynamics. Similarly, in PSLAB run, the interannual variance drops to $0.06\,\mathrm{mm^2\,day^{-2}}$ (20% of CTL), indicating that 80% of AISMR variance in CFSv2 is related to the teleconnection from the Pacific Ocean dynamics. On the other hand, the root mean square error (RMSE) is maximum in the PSLAB run (0.7), while it is less for the CTL and ISLAB (0.5) run. ACC of AISMR drops significantly in the PSLAB run (0.14) compared with CTL (0.53) and ISLAB runs (0.51). Large RMSE and small ACC in the PSLAB run indicate that interannual variations and phase of the AISMR anomaly in CFSv2 are mainly driven by ENSO–monsoon teleconnections in the model. It should be noted that the Indian Ocean dynamics contribute significantly to the interannual variance of AISMR.

4. Summary

The CFSv2 model has a dry bias over Indian land region, while the misrepresentation of Indian Ocean dynamics leads to improper ocean–atmosphere interactions and overestimated oceanic rainfall coexisting with cold SST biases. Better simulation of ocean–atmosphere interactions and reduced dry bias over Indian land along with simulation of warmer SST in northern Indian Ocean in the ISLAB run confirms that the dry bias over the Indian landmass is primarily due to cold SST simulated in the tropical Indian Ocean. Similarly, significant drop in AISMR variance in ISLAB run suggests that ocean dynamics in the Indian Ocean are important for the proper simulation of the interannual variance of AISMR. The study reveals the relative importance of the Indian Ocean and Pacific Ocean coupled dynamics in determining the seasonal mean rainfall over the Indian land region and its hindcast skill. The major portion of the prediction skill of AISMR basically comes from the Pacific Ocean teleconnections, and it is reasonably captured in the CFSv2. These results suggest that the Indian Ocean SST bias should be minimized to reduce the seasonal mean dry bias over the Indian land region, whereas the Indian Ocean SST and AISMR teleconnections should be reasonably captured to improve the AISMR prediction skill. These findings highlight the need to improve the Indian Ocean coupled dynamics in CFSv2 for the further improvement of the prediction skill of AISMR. Even though this study is based on CFSv2 model, similar biases are reported in other models from leading climate centers and hence the findings from this study will be useful for addressing the biases in other models also.

Acknowledgements

The Indian Institute of Tropical Meteorology is funded by the Ministry of Earth Sciences, Government of India. Gibies George and D. Nagarjuna Rao acknowledge the Council of Scientific and Industrial Research for their research fellowship. C. T. Sabeerali acknowledges the support from the Monsoon Mission project. We thank National Center for Atmospheric Research for making available the NCL software. All the data sources are duly acknowledged. We acknowledge the National Centers for Environmental Prediction Climate Forecast System team and Monsoon Desk.

References

Achuthavarier D, Krishnamurthy V, Kirthman BP, Huang B. 2012. Role of the Indian Ocean in the ENSO-Indian summer monsoon teleconnection in NCEP climate forecast system. *Journal of Climate* **25**: 2490–2508.

Dommenget D. 2010. The slab ocean El Niño. *Geophysical Research Letters* **37**: L20701, doi: 10.1029/2010GL044888.

Gadgil S, Srinivasan J. 2011. Seasonal prediction of the Indian monsoon. *Current Science* **3**: 343–353.

Hewitt CD. 2004. Ensembles-based predictions of climate changes and their impacts. *Transactions of the American Geophysical Union* **85**(52): 566, doi: 10.1029/2004EO520005.

Krishnan R, Suchithra S, Swapna P, Kumar V, Ayantika DC, Mujumdar M. 2011. The crucial role of ocean–atmosphere coupling on the Indian monsoonanomalous response during dipole events. *Climate Dynamics* **37**(1–2): 1–17, doi: 10.1007/s00382-010-0830-2.

Kumar KK, Hoerling M, Balaji R. 2005. Advancing dynamical prediction of Indian monsoon rainfall. *Geophysical Research Letters* **32**: 2–5, doi: 10.1029/2004GL021979.

Kumar KK, Balaji R, Hoerling M, Gary TB, Cane M. 2006. Unraveling the mystery of Indian monsoon failure during El Ni no. *Science* **314**(5796), doi: 10.1126/science.1131152.

Lau NC, Nath MJ. 2000. Impact of ENSO on the variability of the Asian-Australian monsoons as simulated in GCM experiments. *Journal of Climate* **13**(24): 4287–4309, doi: 10.1175/1520-0442(2000)013<4287.

Lau NC, Nath MJ. 2004. Coupled GCM simulation of atmosphere–ocean variability associated with zonally asymmetric SST changes in the tropical Indian Ocean. *Journal of Climate* **17**(2): 245–265, doi: 10.1175/15200442(2004)017<0245.

Murtugudde R, Busalacchi AJ. 1999. Interannual variability of the dynamics and thermodynamics of the tropical Indian ocean. *Journal of Climate* **12**: 2300–2326.

Palmer TN, Doblas-Reyes FJ, Hagedorn R, Alessandri A, Gualdi S, Andersen U, Feddersen H, Cantelaube P, Terres J-M, Davey M, Graham R, Délécluse P, Lazar A, Déqué M, Guérémy J-F, Díez E, Orfila B, Hoshen M, Morse AP, Keenlyside N, Latif M, Maisonnave E, Rogel P, Marletto V, Thomson MC. 2004. Development of a European multi-model ensemble system for seasonal to interannual prediction (DEMETER). *Bulletin of American Meteorological Society* **85**: 853–872.

Pokhrel S, Rahaman H, Parekh A, Saha SK, Dhakate A, Hemantkumar S, Chaudhari HS, Rakesh MG. 2012. Evaporation-precipitation variability over Indian Ocean and its assessment in NCEP Climate Forecast System (CFSv2). *Climate Dynamics* **39**(9–10): 2585–2608, doi: 10.1007/s00382-012-1542-6.

Rajeevan M, Pai DS. 2007. On the El Niño-Indian monsoon predictive relationships. *Geophysical Research Letters* **34**(4): 1–4, L04704, doi: 10.1029/2006GL028916.

Rajeevan M, Unnikrishnan CK, Preethi B. 2012. Evaluation of the ENSEMBLES multi-modelseasonal forecasts of Indian summer monsoon variability. *Climate Dynamics* **38**(11–12): 2257–2274, doi: 10.1007/s00382-011-1061-x.

Sabeerali CT, Ramu DA, Dhakate A, Salunke K, Mahapatra S, SuryachandraRao A. 2013. Simulation of boreal summer intraseasonal oscillations in the latest CMIP5coupled GCMs. *Journal of Geophysical Research Atmosphere* **118**(10): 4401–4420, doi: 10.1002/jgrd.50403.

Saha S, Shrinivas M, Hua-Lu P, Xingren W, Wang J, Nadiga S, Tripp P, Kistler R, Woollen J, Behringer D, Liu H, Stokes D, Grumbine R, Gayno G, Wang J, Hou Y-T, Chuang H-Y, Juang H-MH, Sela J, Iredell M, Treadon R, Kleist D, Van Delst P, Keyser D, Derber J, Ek M, Meng J, Wei H, Yang R, Lord S, Van Den Dool H, Kumar A, Wang W, Long C, Chelliah M, Xue Y, Huang B, Schemm J-K, Ebisuzaki W, Lin R, Xie P, Chen M, Zhou S, Higgins W, Zou C-Z, Liu Q, Chen Y, Han Y, Cucurull L, Reynolds RW, Rutledge G, Goldberg M. 2010. The NCEP climate forecast system reanalysis. *Bulletin of American Meteorological Society* **91**: 1015–1057, doi: 10.1175/2010BAMS3001.1.

Saha SK, Pokhrel S, Chaudhari HS. 2012. Influence of Eurasian snow on Indian summer monsoon in NCEP CFSv2 freerun. *Climate Dynamics* **41**(7–8): 1801–1815, doi: 10.1007/s00382-012-1617-4.

Saha SK, Pokhrel S, Chaudhari HS, Dhakate A, Shewale S, Sabeerali CT, Salunke K, Hazra A, Mahapatra S, Rao AS. 2014a. Improved simulation of Indian summer monsoon in latest NCEP climate forecast system free run. *International Journal of Climatology* **34**(5): 1628–1641, doi: 10.1002/joc.3791.

Saha S, Shrinivas M, Wu X, Wang J, Nadiga S, Tripp P, Behringer D, Hou Y-T, Chuang H-Y, Iredell M, Ek M, Meng J, Yang R, Mendez MP, Van Den Dool H, Zhang Q, Wang W, Chen M, Becker E. 2014b. The NCEP Climate Forecast System version 2. *Journal of Climate* **27**: 2185–2208, doi: 10.1175/JCLI-D-12-00823.1.

Shenoi SC, Shankar D, Shetye SR. 2002. Diffrences in heat budgets of the near-surface Arabian Sea and Bay of Bengal: implications for the summer monsoon. *Journal of Geophysical Research* **107**(C6): 5.1–5.14, doi: 10.1029/2000JC000679.

Shukla J, Misra BM. 1977. Relationships between sea surface temperature and wind speedover the Central Arabia Sea, and monsoon rainfall over India. *Monthly Weather Review* **105**(8): 998–1002, doi: 10.1175/1520-0493(1977)105<0998.

Smith TM, Reynolds RW, Peterson TC, Lawrimore J. 2008. Improvements to NOAA's historical merged land–ocean surface temperature analysis (1880–2006). *Journal of Climate* **21**(10): 2283–2296, doi: 10.1175/2007JCLI2100.1.

Venzke S, Latif M, Villwock A. 2000. The coupled GCM ECHO-2. Part II: Indian Ocean response to ENSO. *Journal of Climate* **13**(8): 1371–1383, doi: 10.1175/1520-0442(2000)013<1371.

Wang B, Kang IS, Lee JY. 2004. Ensemble simulations of Asian–Australian monsoon variability by 11 AGCMs. *Journal of Climate* **17**(4): 803–818, doi: 10.1175/1520-0442(2004)017<0803.

Wang B, Xiang B, Li J, Webster PJ, Rajeevan MN, Liu J, Ha K-J. 2015. Rethinking Indian monsoon rainfall prediction in the context of recent global warming. *Nature Communications* **6**: 7154, doi: 10.1038/ncomms8154.

Xie P, Janowiak JE, Arkin PA, Adler R, Gruber A, Ferraro R, Huffman GJ, Curtis S. 2003. GPCP pentad precipitation analyses: an experimental dataset based on gauge observations and satellite estimates. *Journal of Climate* **16**(13): 2197–2214.

Yokoi T, Tozuka T, Yamagata T. 2012. Seasonal and interannual variations of the SST above the Seychelles Dome. *Journal of Climate* **25**(2): 800–814, doi: 10.1175/JCLI-D-10-05001.1.

Yu JY, Lau KM. 2005. Contrasting Indian Ocean SST variability with and without ENSO influence: a coupled atmosphere–ocean GCM study. *Meteorology and Atmospheric Physics* **90**(3–4): 179–191, doi: 10.1007/s00703-004-0094-7.

Zheng Y, Shinoda T, Lin JL, Kiladis GN. 2011. Sea surface temperature biases under the stratus cloud deck in the southeast Pacific Ocean in 19 IPCC AR4 coupled general circulation models. *Journal of Climate* **24**: 4139–4164, doi: 10.1175/2011JCLI4172.1.

12

Decadal trends of global precipitation in the recent 30 years

Xiaofan Li,[1]* Guoqing Zhai,[1] Shouting Gao[2] and Xinyong Shen[3]

[1]*Department of Earth Science, Zhejiang University, Hangzhou, Zhejiang 310027, China*
[2]*Laboratory of Cloud-Precipitation Physics and Severe Storms (LACS), Institute of Atmospheric Physics, Chinese Academy of Sciences, Beijing 100029, China*
[3]*Key Laboratory of Meteorological Disaster of Ministry of Education, Nanjing University of Information Science and Technology, Jiangsu 210044, China*

Correspondence to:
X. Li, Department of Earth Science, Zhejiang University, 38 Zheda Road, Hangzhou, Zhejiang 310027, China.
E-mail: xiaofanli@zju.edu.cn

Abstract

Decadal trends of global precipitation are examined using the Climate Prediction Center (CPC) Merged Analysis of Precipitation (CMAP), Global Precipitation Climatology Project (GPCP), and National Centers for Environmental Prediction (NCEP)/National Center for Atmospheric Research (NCAR) reanalysis data. The decadal trends of global precipitation average diverge a decreasing trend for the CMAP data, a flat trend for the GPCP data, and an increasing trend for the reanalysis data. The decreasing trend for the CMAP data is associated with the reduction in high precipitation. The flat trend for the GPCP data is related to the offset between the increase in high precipitation and the decrease in low precipitation. The increasing trend for the reanalysis data corresponds to the increase in high precipitation.

Keywords: decadal trends; global precipitation; precipitation statistics

1. Introduction

Precipitation is an important component in regulating global water cycle at multi-temporal and spatial scales. There are no rain gauge observations over the vast open ocean until the satellite retrievals are available. The studies of climate-scale variability of precipitation have been limited at regional scales (e.g. Dore, 2005; Pauling *et al.*, 2006). The rain gauge and satellite retrievals lead to the generation of long-term global precipitation data such as the CMAP [Climate Prediction Center (CPC) Merged Analysis of Precipitation] data (e.g. Xie and Arkin, 1997) and the GPCP (Global Precipitation Climatology Project) data (e.g. Adler *et al.*, 2003; Huffman *et al.*, 2009). Gu *et al.* (2007) analyzed the GPCP data and showed an upward change over the tropical ocean and a downward change over the tropical land during the period 1979–2005.

In this study, the CMAP and GPCP data and the reanalysis data developed by the National Centers for Environmental Prediction (NCEP)/National Center for Atmospheric Research (NCAR) are compared to highlight the similarities and differences in decadal variability of observed and simulated global precipitation. The differences could come from the quality of observational data for decadal analysis and model physics. The data are briefly discussed in Section 2. The decadal trends of global precipitation are analyzed using the annually averaged precipitation data in Section 3. The precipitation amounts are accumulated and precipitation rate are zonally averaged to explain the trends of global precipitation. A summary is given in Section 4.

2. Data

The monthly mean versions of the following three data sets are used in this study. The observational precipitation data include the CMAP (Xie and Arkin, 1997) and the GPCP data (Huffman *et al.*, 2009). The CMAP data were constructed by merging gauge and satellite estimates. The satellite retrievals include Geostationary Operational Environmental Satellite (GOES) precipitation index (GPI), outgoing longwave radiation (OLR) precipitation index (OPI), special sensor microwave/imager (SSM/I) scattering and SSM/I emission, and microwave sounding unit (MSU). The data cover the latitudes from 88.75°S northward to 88.75°N, and longitudes from 1.25°E eastward. The horizontal resolution is 2.5° latitude × 2.5° longitude.

The GPCP is one of the global energy and water experiment (GEWEX) analyses of the water and energy cycle organized by the GEWEX radiation panel. The GPCP data were developed from the SSM/I retrievals, merged geosynchronous- and low-Earch-orbit infrared data, the OPI data, the estimate from Television Infrared Observation Satellite Operational Vertical Sounder (TOVS) and Advanced Infrared Sounders (AIRS), and the global precipitation-gauge data. The version 2.1 (Huffman *et al.*, 2009) of the GPCP data used in this study also has horizontal resolution of 2.5° latitude × 2.5° longitude.

The precipitation data were developed by a joint project between the NCEP and the NCAR, which involves the recovery of land surface, ship, rawinsonde, pibal, aircraft, satellite, and other data (Kalnay *et al.*, 1996; Kistler *et al.*, 2001). These data were then quality controlled and assimilated with a data assimilation system. The project produces new atmospheric analyses using historical data (1948 onwards) and analyses of the current atmospheric state (Climate Data Assimilation System, CDAS). The data have the horizontal resolution of 1.905° latitude × 1.875° longitude.

3. Results

The analysis of annual averages from monthly and globally averaged precipitation data shows a decreasing trend for CMAP data (Figure 1(a)), a flat trend for GPCP data (Figure 1(b)), and an increasing trend for reanalysis data. The linear trends of globally averaged precipitation rate are −0.0032 mm day^{-1} year^{-1} for CMAP data, 0.0003 mm day^{-1} year^{-1} for GPCP data, and 0.0119 mm day^{-1} year^{-1} for the reanalysis data. To examine the differences in decadal variation of annually averaged precipitation data, we take the horizontal distribution of difference in precipitation rate between temporal averages from the periods 1994–2008 to 1979–1993. The negative differences largely cover ocean areas, in particular, the Intertropical Convergence Zone (ITCZ) and the South Pacific Convergence Zone (SPCZ) (Figure 2(a)) and are consistent with the decreasing trend of globally averaged precipitation rate for the CMAP data (Figure 1(a)). The reduction of precipitation from 1979–1993 to 1994–2008 is weaker, and the negative differences occupy fewer areas and are weaker over the ITCZ and SPCZ in the GPCP data (Figure 2(b)) than in the CMAP data, which causes the flat trend of globally averaged precipitation rate for the GPCP data (Figure 1(b)). The negative differences only occur over the equatorial areas, and are surrounded by the large positive differences in the calculation of the reanalysis data (Figure 2(c)). Thus, the globally averaged precipitation rate in the reanalysis data shows a strong increasing trend (Figure 1(c)). The divergent trends between the observed and modeled global precipitation imply different precipitation statistics. Thus, we analyze the contributions of different rain intensities to the changes in globally averaged precipitation rates.

Precipitation amount is accumulated from each grid for precipitation-rate (P_S) bins with an interval of 0.3 mm day^{-1} during the periods 1979–1993 (CP_{S1}) and 1994–2008 (CP_{S2}), respectively. Figure 3 shows accumulated precipitation amount during the period 1979–1993 and the difference in accumulated precipitation amount between 1994–2008 and 1979–2008 $(CP_{S2}-CP_{S1})$ as the function of precipitation rate (P_S) for the three precipitation data sets. The CMAP data show that the maximum cumulative precipitation amount in the period 1979–1993 occurs when precipitation rates are 1.8–2.1 mm day^{-1} (Figure 3(a)). The difference in cumulative precipitation amount between the periods 1994–2008 and 1979–1993 is positive when the precipitation rate is lower than 2.4 mm day^{-1} and the maximum positive difference appears in the

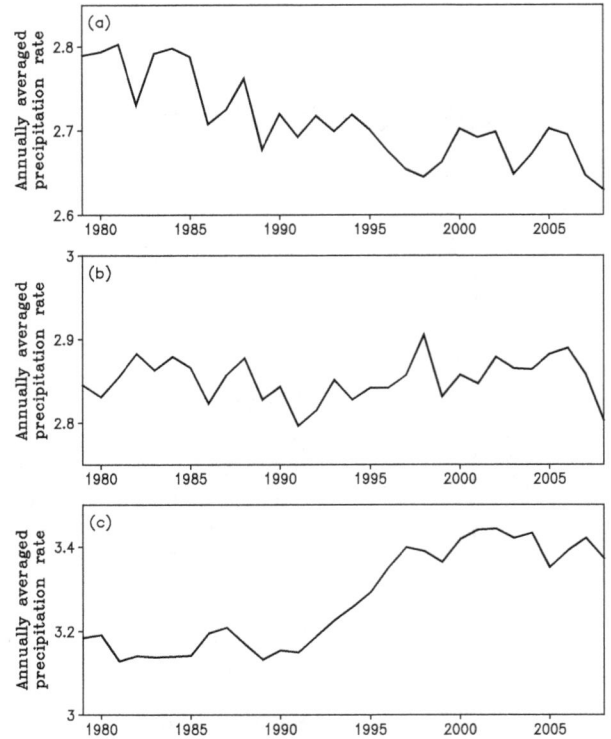

Figure 1. Time series of precipitation rate (mm day^{-1}) annually and globally averaged using (a) CMAP, (b) GPCP, and (c) NCEP/reanalysis data from 1979 to 2008.

bin of 1.2–1.5 mm day^{-1} (Figure 3(b)). The difference generally is negative when precipitation rate is higher than 2.4 mm day^{-1} and the maximum negative difference occurs in the bin of 3–3.3 mm day^{-1}. The weak precipitation in CMAP data has an increasing trend, whereas their high precipitation data has a decreasing trend. The decreasing trend in globally averaged CMAP precipitation rate is associated with the decrease in high precipitation. The decrease in high precipitation over oceans could be an artifact of input data change and atoll sampling error (Yin et al., 2004). Lau and Wu (2007) analyzed the CMAP data over the Tropics and found a decreasing trend of precipitation rate, which is attributable to the decreasing trend in moderate (25–75% by precipitation amount) precipitation events in their analysis of probability distribution functions of tropical precipitation.

The GPCP data reveal that the cumulative precipitation peak appears around the precipitation rates of 3.0–3.3 mm day^{-1} (Figure 3(c)). The maximum reductions in cumulative precipitation amount from 1979–1993 to 1994–2008 occur around the cumulative precipitation peak (Figure 3(d)). The cumulative precipitation amount increases from 1979–1993 to 1994–2008 when the precipitation rate is higher than 5 mm day^{-1}. Opposite to CMAP precipitation data, the high precipitation shows an increasing trend, but the low precipitation reveals a reducing trend, and the flat trend in the global GPCP precipitation data is related to the offset.

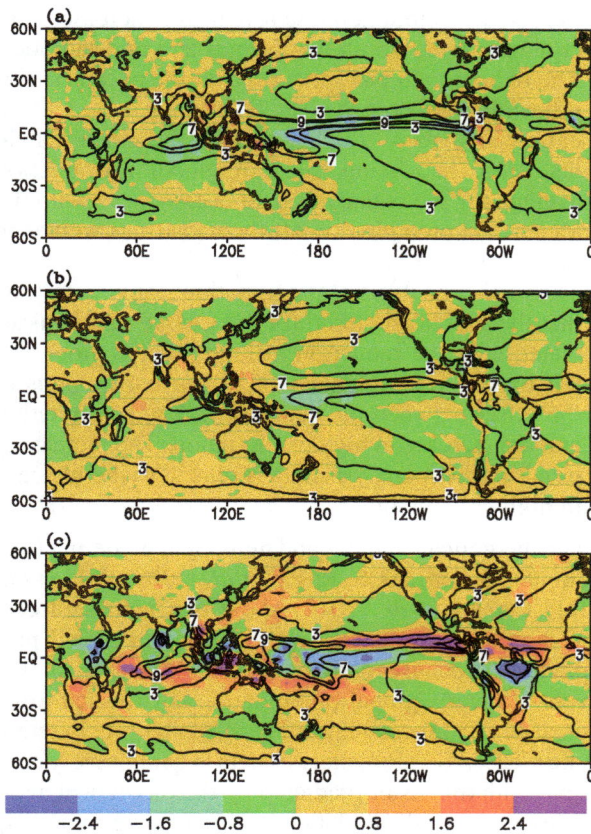

Figure 2. Horizontal distributions of precipitation rates (mm day^{-1}) averaged in the period of 1979–1993 (contour) and differences (color shaded) between time averages from 1994 to 2008 and from 1979 to 1993 calculated from (a) CMAP data, (b) GPCP, and (c) NCEP/NCAR reanalysis data.

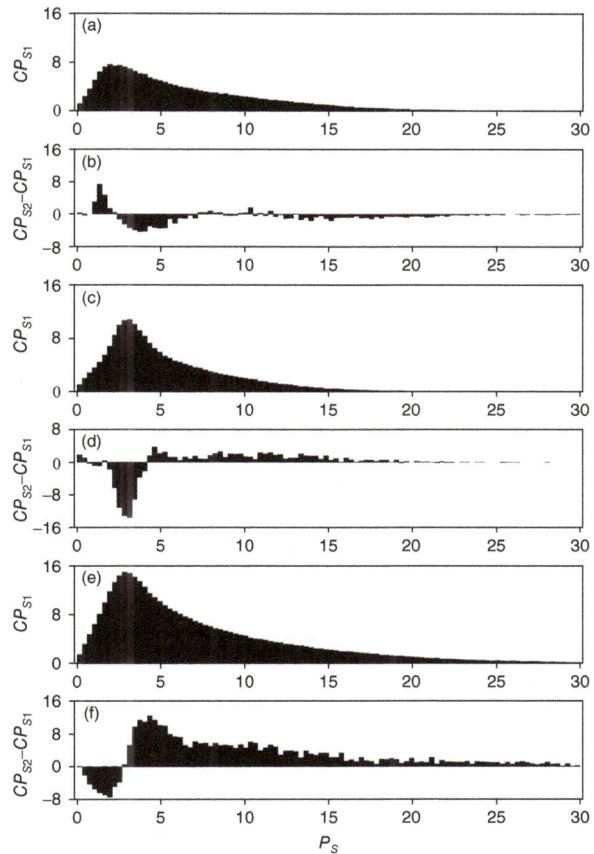

Figure 3. Cumulative precipitation amount (CP_{S1}) from the period 1979–1994 in (a) CMAP, (c) GPCP, and (e) reanalysis data and difference in cumulative precipitation amount ($CP_{S2}-CP_{S1}$) for (b) CMAP, (d) GPCP, and (f) reanalysis data between the periods 1995–2009 (CP_{S2}) and 1979–1994 (CP_{S1}) in y-axis as a function of precipitation rate (P_S) in x-axis. Plotting scales are 10^8 mm for CP_{S1}, 10^7 mm for $CP_{S2}-CP_{S1}$, and mm day^{-1} for P_S.

The reanalysis data show that the cumulative reanalysis precipitation amount reaches its maximum when precipitation rates are 2.7–3 mm day^{-1} (Figure 3(e)), which is similar to the GPCP precipitation data. The difference for precipitation rates of lower than 2.7 mm day^{-1} is negative, whereas the difference for precipitation rates of higher than 2.7 mm day^{-1} is positive (Figure 3(f)). The maximum positive and negative differences appear in the bins of 1.8–2.1 mm day^{-1} and 4.2–4.5 mm day^{-1}, respectively. Like the GPCP precipitation data, the high precipitation in the reanalysis data has an increasing trend, determining the increasing trend in globally averaged precipitation rate.

The analysis of zonally averaged precipitation rate during 1979–1993 shows that the precipitation rate for the reanalysis data is slightly higher than that for the CMAP data, but it is significantly higher than that for the GPCP data in the latitudes of the equator – 10°N (Figure 4(a)). The precipitation rate for the reanalysis data is significantly higher than those for the GPCP and CMAP data but the precipitation rate for the CMAP data is slightly higher than that for the GPCP data in the latitudes of the equator – 10°S. The precipitation rate for the CMAP data is much lower than those for the reanalysis and GPCP data in the latitudes of 40–60°S. The precipitation rates for the three data sets are

similar in the other latitudes. Thus, the zonally averaged precipitation rates are higher for the reanalysis data than for the CMAP data, which is consistent with those found in the study by Latham *et al.* (2012). The zonally averaged precipitation rate for the reanalysis data is generally increased from the periods 1979–1993 to 1994–2008, whereas it is decreased around the equator (Figure 4(b)). The maximum increases for the reanalysis data around 15°S and 10°N (Figure 4(b)) are collocated with the maximum zonally averaged precipitation rates (Figure 4(a)), which is consistent with the precipitation increase for the precipitation rates of higher than 3 mm day^{-1}(Figure 3(f)). The zonally averaged precipitation rate for the GPCP data is barely changed from the periods 1979–1993 to 1994–2008, which is consistent with the offset between the precipitation decrease around 3 mm day^{-1} and precipitation increase for the precipitation rates of higher than 5 mm day^{-1} (Figure 3(d)). The zonally averaged precipitation rate for the CMAP data is generally is reduced from the periods 1979–1993 to 1994–2008. The maximum reduction for the CMAP data around 10°N, 10°S, and 45°S are consistent with the precipitation decrease

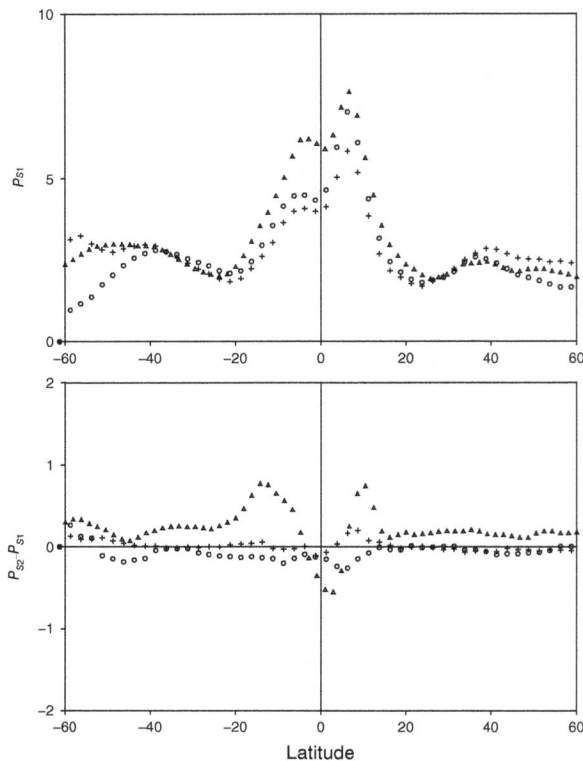

Figure 4. Meridional distribution of (a) rain rate (P_{S1}) averaged zonally and from the period 1979–1993 and (b) differences ($P_{S2} - P_{S1}$) in zonally averaged rain rate between 2008–1994 (P_{S1}) and 1979–1993 (P_{S2}) for CMAP (open circle), GPCP (cross), and reanalysis (triangle). Unit is mm day^{-1}.

for the precipitation rates of higher than 3 mm day^{-1} (Figure 3(b)). Although El Niño/La Niña has significant effects on tropical precipitation, our analysis here reveals that the increased precipitation rate for the reanalysis data and the decreased precipitation rate for the CMAP data occur not only in the Tropics, but also in the mid-latitudes, supporting the trends of globally averaged precipitation rates shown in Figure 1.

4. Summary and discussions

In this study, the decadal trends of global precipitation are calculated and compared using the CMAP, GPCP, and NCEP/NCAR reanalysis monthly precipitation data over the past 30 years from 1979 to 2008. The major results include:

- The decadal trend of annually and globally averaged precipitation depends on a decreasing trend for the CMAP data, a flat trend for GPCP data, and an increasing trend for the reanalysis data.
- The analysis of horizontal distributions of differences in temporally averaged precipitation between the second (1993–2008) and the first (1979–1993) 15 years shows that the decreasing trend in the CMAP data is associated with the reduction in precipitation over the oceans. The further analysis of difference in zonally averaged precipitation rate reveals the increased precipitation rate in both the Tropics and mid-latitudes. The reduction in precipitation over the oceans is significantly weaker in the GPCP data than in the CMAP data, which shows the flat trend in the global GPCP data. The increasing trend of global precipitation average for the reanalysis data is associated with the increase in precipitation off the equator as well as in the mid-latitudes.

- The further analysis of precipitation statistics reveals that the decreasing trend for the CMAP data is associated with the reduction in high precipitation. The flat trend for the global GPCP data corresponds to the offset between the decrease in low precipitation and the increase in high precipitation. The increasing trend for the reanalysis data is related to the increase in high precipitation.

The theoretical analysis and numerical modeling (e.g. Held and Soden, 2006; Rasch *et al.*, 2009; Bala *et al.*, 2011; Latham *et al.*, 2012) show that enhanced carbon dioxide could lead to the increase in global precipitation. Because carbon dioxide increases in the recent 30 years, global precipitation is expected to be increased. The increasing trend in global precipitation in the NCEP/NCAR reanalysis data is qualitatively consistent with the result predicted by the theoretical analysis as the reanalysis data are largely determined by the model physics. The analysis of precipitation statistics shows that the increase in carbon dioxide may lead to the increase in high precipitation rate. But the increase in high precipitation rate may come mainly from the precipitation area off the equator, and the magnitudes of the difference in precipitation between the second and the first 15-year average (Figures 2(c) and 4(b)) are similar to the 30-year global average (Figure 1(c)); is such strong off-equator precipitation difference model-dependent? Since all the three data show the reducing trend along the equatorial Pacific (Figures 2 and 4(b)), does the increasing trend in global precipitation is the reanalysis data reflect the decadal trend in the real atmosphere? Unlike the reanalysis data, the high precipitation in the CMAP data shows a weak decreasing trend for both globally and zonally averaged precipitation rates, which is associated with the reduction in strong precipitation. The zonally averaged precipitation rate for the GPCP data is barely changed meridionally from the first 15 years to the second 15 years. The diverged trends of globally averaged precipitation rate from the three data sets in this study cannot provide a clear and realistic climate signal. Therefore, it is necessary to conduct in-depth analysis for these data sets in order to obtain physically consistent decadal trends for global precipitation when the quality of observational data and model physics is improved.

Acknowledgements

The authors thank Dr P.-P. Xie at Climate Precipitation Center, NOAA, for the CMAP precipitation data and the two anonymous reviewers for their constructive comments. X. Li was supported by 985 funding of Zhejiang University. G. Zhai was supported by the National Natural Science Foundation of China (Grant No. 41175047), and the National Key Basic Research and Development Project of China (Grant No. 2013CB430100). S. Gao was supported by the National Key Basic Research and Development Project of China under Grant No. 2012CB417201, and the National Natural Sciences Foundation of China under Grant Nos. 40930950 and 41075043. X. Shen was supported by the National Key Basic Research and Development Project of China under Grant Nos. 2013CB430103 and 2011CB403405, the National Natural Science Foundation of China under Grant Nos. 41075058 and 41175065, and the Priority Academic Program Development of Jiangsu Higher Education Institutions, China (Grant No. PAPD2014).

References

Adler RF, Huffman GJ, Chang A, Ferraro R, Xie PP, Janowiak J, Rudolf B, Schneider U, Curtis S, Bolvin D, Gruber A, Susskind J, Arkin P, Nelkin E. 2003. The version 2 Global Precipitation Climatology Project (GPCP) monthly precipitation analysis (1979–present). *Journal of Hydrometeorology* 4: 1147–1167.

Bala G, Caldeira K, Nemani R, Cao L, Ban-Weiss G, Ho-Jeong S. 2011. Albedo enhancement of marine clouds to counteract global warming: impacts on the hydrological cycle. *Climate Dynamics* 37: 915–931, DOI: 10.1007/s00382-010-0868-1.

Dore MHI. 2005. Climate change and changes in global precipitation patterns: what do we know? *Environment International* 31: 1167–1181.

Gu G, Adler RF, Huffman GJ, Curtis S. 2007. Tropical precipitation variability on interannual-to-interdecadal and longer time scales derived from the GPCP monthly project. *Journal of Climate* 20: 4033–4046.

Held IM, Soden BJ. 2006. Robust responses of the hydrological cycle to global warming. *Journal of Climate* 19: 5686–5699.

Huffman GJ, Adler RF, Bolvin DT, Gu G. 2009. Improving the global precipitation record: GPCP Version 2.1. *Geophysical Research Letters* 36: L17808, DOI: 10.1029/2009 GL040000.

Kalnay E, Kanamitsu M, Kistler R, Collins W, Deaven D, Gandin L, Iredell M, Saha S, White G, Woollen J, Zhu Y, Leetmaa A, Reynolds R, Chelliah M, Ebisuzaki W, Higgins W, Janowiak J, Mo KC, Ropelewski C, Wang J, Jenne R, Joseph D. 1996. The NCEP/NCAR 40-Year Reanalysis Project. *Bulletin of the American Meteorological Society* 77: 437–471.

Kistler R, Kalnay E, Collins W, Saha S, White G, Woollen J, Chelliah M, Ebisuzaki W, Kanamitsu M, Kousky V, van den Dool H, Jenne R, Fiorino M. 2001. The NCEP-NCAR 50-year reanalysis: monthly means CD-ROM and documentation. *Bulletin of the American Meteorological Society* 82: 247–268.

Latham J, Bower K, Choularton T, Coe H, Connolly P, Cooper G, Craft T, Foster J, Gadian A, Galbraith L, Iacovides H, Johnston D, Launder B, Leslie B, Meyer J, Neukermans A, Ormond B, Parkes B, Rasch PJ, Rush J, Salter S, Wang H, Wang Q, Wood R. 2012. Marine cloud brightening. *Philosophical Transactions of the Royal Society A: Mathematical, Physical & Engineering Sciences* 370: 4217–4262, DOI: 10.1098/rsta.2012.0086.

Lau K-M, Wu H-T. 2007. Detecting trends in tropical precipitation characteristics, 1979-2003. *International Journal of Climatology* 27: 979–988.

Pauling A, Luterbacher J, Casty C, Wanner H. 2006. 500 years of gridded high resolution precipitation reconstructions over Europe and the connection to large scale circulation. *Climate Dynamics* 26: 387–405.

Rasch PJ, Latham J, Chen CC. 2009. Geoengineering by cloud seeding: Influence on sea ice and climate system. *Environmental Research Letters* 4: 045112–045119, DOI: 10.1088/1748-9326/4/4/045112.

Xie P, Arkin PA. 1997. Global precipitation: a 17-year monthly analysis based on gauge observations, satellite estimates and numerical model outputs. *Bulletin of the American Meteorological Society* 78: 2539–2558.

Yin X, Gruber A, Arkin P. 2004. Comparison of the GPCP and CMAP merged gauge–satellite monthly precipitation products for the period 1979–2001. *Journal of Hydrometeorology* 5: 1207–1222.

13

Hurricane simulation using different representations of atmosphere–ocean interaction: the case of Irene (2011)

P. A. Mooney,[1]* D. O. Gill,[1] F. J. Mulligan[2] and C. L. Bruyère[1]

[1] Mesoscale and Microscale Meteorology Laboratory, National Center for Atmospheric Research (NCAR), Boulder, CO, USA
[2] Department of Experimental Physics, Maynooth University, Kildare, Ireland

*Correspondence to:
P. A. Mooney, Mesoscale and Microscale Meteorology Laboratory, National Center for Atmospheric Research, P.O. Box 3000, Boulder, CO 80307-3000, USA.
E-mail: pmooney@ucar.edu

Abstract

Three approaches to represent sea surface temperatures (SSTs) in atmospheric models have been investigated using the Weather Research and Forecasting model: (1) prescribing SSTs every 6 h from reanalysis, (2) a one-dimensional ocean mixed-layer model and (3) a fully coupled regional ocean model. Hurricane Irene (2011) was chosen as the test case. All three options produced results comparable to observations immediately after storm passage but only options (1) and (3) captured recovery to pre-storm conditions which suggests both are feasible approaches for long-term simulations of tropical cyclones. Option (2) merits further investigation because of its greater computational efficiency and reduced complexity.

Keywords: Weather Research and Forecasting (WRF) model; sea surface temperatures; ocean mixed-layer model; coupled atmosphere–ocean model; WRF–ROMS; hurricanes

1. Introduction

Sea surface temperature plays an important role in the life cycle of a tropical cyclone (TC) and is one of the main factors for cyclogenesis (e.g. Bruyère *et al.*, 2012). As a TC passes over the warm ocean surface, it reduces the temperature of the sea surface leaving a cold wake that continues to cool for up to 2 days afterwards and may extend for hundreds of kilometres adjacent to the storm track (Dare and McBride, 2011). The magnitude of this cooling can be up to 9 °C as shown by Lin *et al.* (2003) for the case of Kai-Tak (2000) in the South China Sea. This cooling depends strongly on the TC intensity, its translational speed and the depth of the ocean mixed layer (Dare and McBride, 2011).

The time required for SSTs to return to their climatological values varies widely between TCs with recovery periods ranging from 1 to 60 days (Hart *et al.*, 2007) with the majority recovering within 30 days after a TC has passed (Dare and McBride, 2011). These lingering cold wakes can impact seasonal TC activity as later storms may interact with them; the probability for cyclones to encounter a cold wake is ~10% on average (Balaguru *et al.*, 2014). This additional mixing may also be important on longer time scales through its impact on the large-scale slowly varying ocean overturning circulation, and may impact the long-term climatology of TCs (Dare and McBride, 2011). Clearly, it is important to include the ocean response to hurricanes in atmospheric models used for long-term studies of TCs. Since most atmospheric models represent the oceanic response solely through changes in SSTs, it is essential to accurately represent them in models.

When the focus of hurricane studies moves from weather forecasting to regional climate studies, computational efficiency becomes increasingly important and the benefits of more realistic representations of physical processes in the model must be critically assessed. This study examines three different methods of representing SSTs in the Weather Research and Forecasting (WRF) model to determine the impact of trading computational efficiency for model configuration, to inform the representation of SSTs in atmospheric models for long-term studies of TCs. Hurricane Irene (2011) was chosen as a case study because most forecast models failed to correctly predict the cyclone intensity due to underestimated storm-induced upper-ocean cooling (Glenn *et al.*, 2016). Simulations are evaluated against the best track, satellite and buoy data.

2. Model domains and details

2.1. The Weather Research and Forecasting (WRF) model

All simulations use the WRF model (Skamarock *et al.*, 2008) over the two domains shown in Figure 1 with two-way nesting, 51 model levels and a model top at 10 hPa. The outer domain in Figure 1 has a grid spacing of 36 km with 340 × 260 (east–west × north–south) grid points, while the inner domain has a 12-km spacing with 802 × 511 grid points. Each simulation covers the 14-day period beginning at 0000 UTC on the 23 August 2011 with initial conditions and 6-h boundary conditions derived from ERA-Interim (Dee *et al.*, 2011).

The WRF parameterization schemes used are the Community Atmosphere Model (Collins *et al.*, 2004)

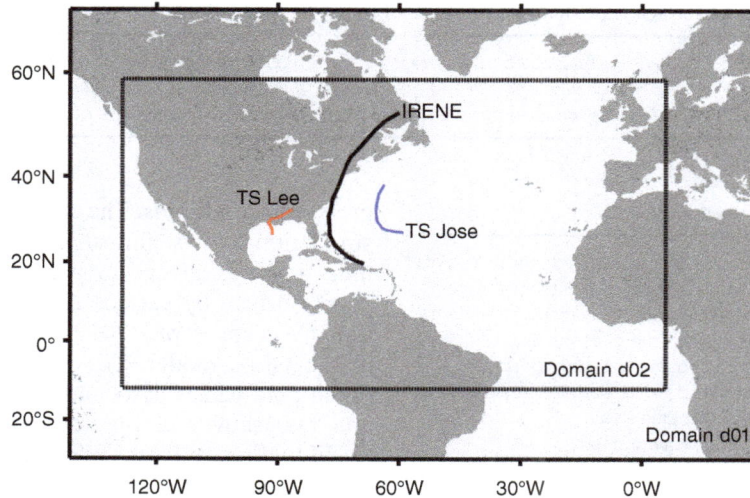

Figure 1. Outer 36-km WRF domain d01 and inner 12-km WRF domain d02. Also shown are the tracks of hurricanes Irene (23 August 2011 to 30 August 2011) and Tropical Storms (TS) Lee (2 September 2011 to 6 September 2011) and Jose (26 August 2011 to 29 August 2011).

longwave and shortwave radiation schemes, the Kain-Fritsch cumulus scheme (Kain and Fritsch, 1990; Kain, 2004), the Yonsei University planetary boundary layer scheme (Hong *et al.*, 2006), the WRF single moment six-class scheme (Hong and Lim, 2006) and the Noah land surface model (Chen and Dudhia, 2001). Analysis of 16 WRF simulations using different combinations of physical parameterization schemes has shown that this combination accurately simulates the track and minimum pressures of Irene (not shown). Three different approaches of updating SSTs are used with this WRF configuration (see Table 1) to simulate the passage of Irene.

2.2. Representation of the ocean surface

2.2.1. ERA-Interim SSTs

In this simulation, WRF uses daily averaged SSTs from ERA-Interim, which are derived from the Operational Sea Surface Temperature and Sea Ice Analysis (OSTIA; Stark *et al.*, 2007). WRF uses the scheme described in Zeng and Beljaars (2005) to modify these daily averages in response to surface winds and changes in the radiative fluxes (e.g. diurnal variations in shortwave radiation).

2.2.2. One-dimensional (1-D) ocean mixed-layer (OML) model

This is a simple 1-D OML model (Davis *et al.*, 2008) based on Pollard *et al.* (1972). Neither horizontal advection nor the pressure gradient are accounted for in this model which simply requires specification of the initial mixed-layer depth (climatological values obtained from www.ifremer.fr/cerweb/deboyer/mld; de Boyer Montégut *et al.*, 2004), the deep layer temperature lapse rate (default 0.14 K/m) and the wind stress at the ocean surface which is provided by the WRF model.

2.2.3. Regional Ocean Modelling System (ROMS)

ROMS (Shchepetkin and McWilliams, 2005) is a three-dimensional (3-D) regional ocean model with terrain following coordinates that solves the free surface, hydrostatic, primitive equations. In this simulation, ROMS is configured using a single domain which covers the same area as the outer WRF domain. The ROMS domain uses a grid spacing of 12 km and 30 stretched vertical levels. Values for the initial conditions and open boundaries are generated from the global HYbrid Coordinate Ocean Model with Naval Research Lab Coupled Ocean Data Assimilation (HYCOM/NCODA; http://tds.hycom.org/thredds/dodsC/GLBa0.08/expt_90 .9; Cummings, 2005).

A 10-day spin-up period is used for ROMS and HYCOM/NCODA tracer and velocity fields are provided using Orlanski-type radiation conditions in conjunction with relaxation. The Flather method was used to obtain boundary values for the free-surface and depth-averaged velocity from HYCOM/NCODA. The Generic Length Scale vertical mixing scheme (Warner *et al.*, 2005) was used to calculate the vertical turbulent mixing and specify the quadratic drag formulation for the bottom friction. Other parameters for ROMS are shown in Table 2.

The coupled WRF–ROMS modelling system used in this study is the Coupled Ocean–atmosphere–Wave–Sediment Transport (Warner *et al.*, 2010) modelling system. These models exchange data once per hour: WRF receives SSTs from ROMS every hour while providing ROMS with wind stress, sea level pressure and surface heat fluxes.

2.3. Data sets

ERA-Interim was obtained on a global 0.75° grid from the European Centre for Medium Range Weather Forecasting data server: http://data.ecmwf.int/data.

Table 1. Summary of the ocean surface representation in each of the WRF simulations.

WRF–ERAI	WRF with updated daily averaged SSTs from ERA-Interim; a diurnal cycle is imposed on the input SST data.
WRF–OML	WRF run with a simple 1-D ocean mixed layer model
WRF–ROMS	WRF fully coupled to a 3-D hydrostatic, primitive equation regional ocean model system (ROMS).

Table 2. Model parameters used in ROMS.

L	1015	Number of I-direction interior rho-points
M	775	Number of J-direction interior rho-points
h_{max}	5000 m	Maximum depth of computational domain
h_{min}	50 m	Minimum depth of computational domain
θ_s	5	Sigma coordinate stretching factor
θ_b	0.4	Sigma coordinate bottom stretching factor
dt (baroclinic)	30 s	Baroclinic time step
dt (barotropic)	1 s	Barotropic time step
Outflow	10 days	
Inflow	0.5 days	

The OSTIA is provided by GHRSST, UKMO and MyOcean and obtained from http://podaac.jpl.nasa. gov/dataset/UKMO-L4HRfnd-GLOB-OSTIA. This is a Level 4 SST analysis produced daily on a global 0.054° grid by the UK Met Office using optimal interpolation. Best track data for observed hurricanes (Figure 1) were obtained from the International Best Track Archive for Climate Stewardship (Knapp et al., 2010). Buoy data (locations shown in Figure 2) were obtained from the US National Data Buoy Center: http://www.ndbc.noaa.gov/to_station.shtml.

3. Results

Figure 2 shows the difference between pre-storm SSTs (23 August 2011) and post-storm SSTs (27 August 2011; 1 September 2011; 5 September 2011) for the OSTIA satellite data (row one) and the three simulations. This sequence of SST differences shows the cooling generated by Irene and the recovery to pre-storm conditions. Figure 2(a) shows that Irene's passage caused SST cooling of approximately 2–3 °C by the 27 August 2011, while the observed recovery to pre-storm conditions is shown in Figure 2(b) and (c). A wide wake with greater cooling on the right side of the track is clearly evident.

Corresponding results from WRF–ERAI (Figure 2(d)–(f)) are in excellent agreement with the OSTIA results. Small differences arise from the imposed diurnal cycle (described above) and the interpolation of OSTIA's SSTs (12 km) to the coarser ERA-Interim grid (~80 km). Figure 2(e) shows that WRF–ERAI SSTs have begun recovery to pre-storm conditions five days after the passage of Irene. Four days later, the SSTs have almost returned to pre-storm conditions (Figure 2(f)).

The behaviour of SSTs in the WRF–OML simulation is shown in Figure 2(g)–(i). Irene's passage generates a slightly colder wake than observations. Of interest to this study is the absence of a recovery to

pre-storm conditions. The cold wake in the WRF–OML simulation shows almost no deterioration even after nine days. Recovery to pre-storm conditions is primarily driven by surface fluxes and horizontal advection (Vincent et al., 2012) which is not represented in the OML model. The WRF–OML has two additional cold wakes associated with tropical storms Jose and Lee shown in Figure 1. Lee is also present in OSTIA (Figure 2(c)), WRF–ERAI (Figure 2(f)) and WRF–ROMS (Figure 2(k)–(l)).

The characteristics of Irene's cold wake in the WRF–ROMS simulation are similar to those observed in the satellite SSTs – wide with greater cooling on the right side of the track. Similar to the WRF–ERAI simulation and satellite SSTs, the cold wake in WRF–ROMS shows some recovery after five days with SSTs partially returning to pre-storm conditions after nine days. However, the WRF–ROMS simulation generates a colder wake than both WRF–ERAI and OSTIA.

This is also evident in Figure 3 which shows the mean SSTs over a square area, whose width is four times the radius of maximum winds and centred on the location of maximum cyclone intensity for each simulation (see Figure 2). The evolution of the SSTs in Figure 3 shows the rapid cooling as Irene passes and the subsequent recovery or the absence of a recovery in the case of WRF–OML. SSTs in WRF–ROMS and WRF–OML show very similar initial cooling, and both show greater cooling than either WRF–ERAI or OSTIA.

While the lower SSTs in WRF–ROMS and WRF–OML may be caused by a slower than observed translational speed, it must be noted that satellite SST measurements are based on a very thin layer of the ocean surface while simulated SSTs can represent a layer several centimetres deep (Costa et al., 2012). For this reason, the simulated SSTs are also compared to measurements from the three buoys (m44014, m44065 and m44013) shown in Figure 2.

A comparison of simulated and observed SSTs, 10-m wind speeds and mean sea level pressure (MSLP) at each buoy is shown in Figure 4. At buoy m44014, the WRF–ERAI simulation shows greater cooling than the other simulations (Figure 4(a)) but it still fails to reach the cold SSTs observed at the buoy. Both WRF–ERAI and WRF–ROMS SSTs show a slow recovery to pre-storm conditions which agrees with the observed rate of warming at the buoy. WRF–ROMS and WRF–OML temperatures show little initial cooling (approximately 1 °C) with no return to pre-storm conditions in the case of WRF–OML. At buoy m44065, SSTs cool by approximately 4 °C as the hurricane passes and continue to cool at a much slower rate for the next day

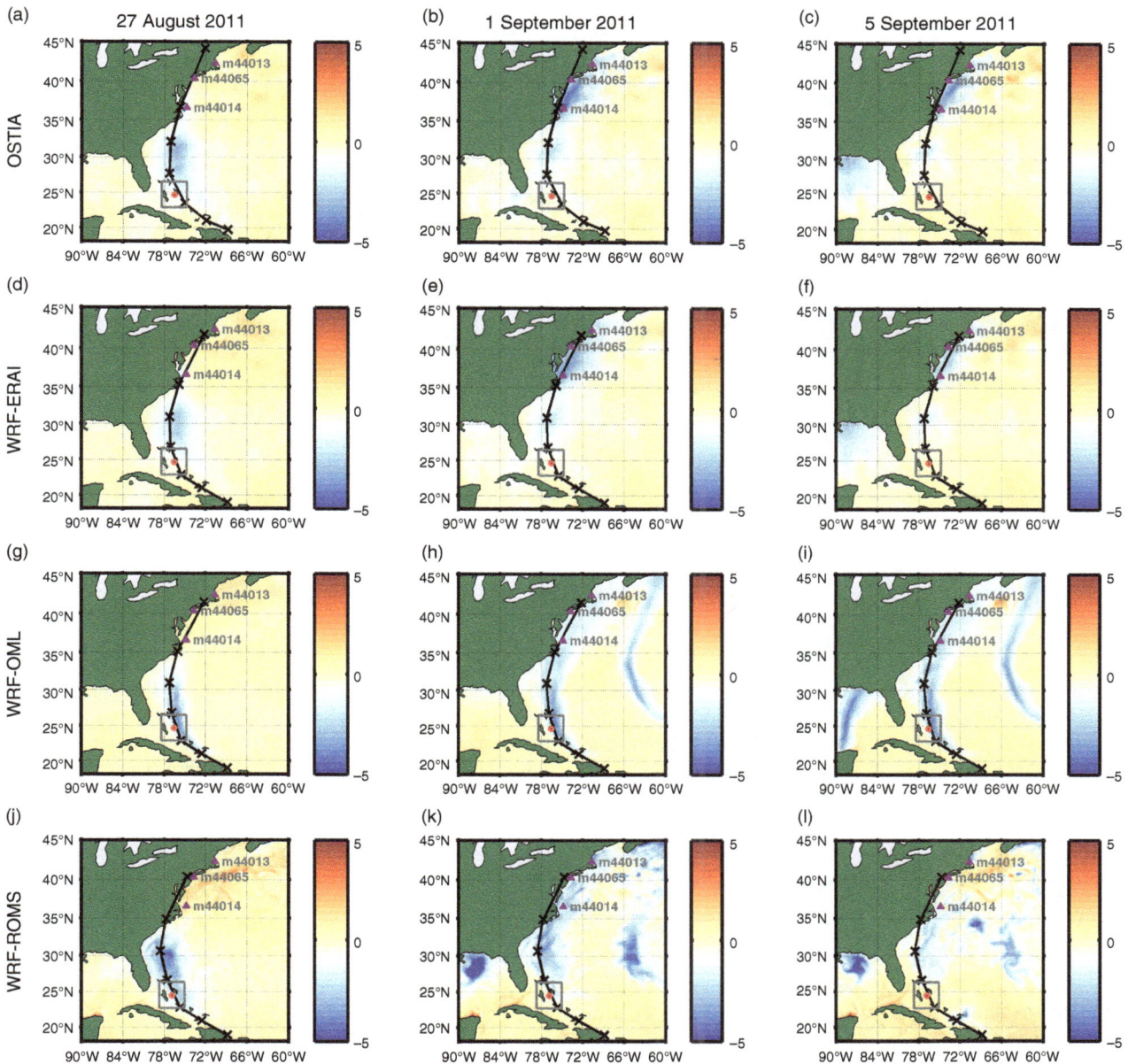

Figure 2. Post-storm cooling in OSTIA sea surface temperatures (a) difference between the 27 August 2011 and the 23 August 2011, (b) difference between the 1 September 2011 and the 23 August 2011, (c) difference between the 4 September 2011 and the 23 August 2011. (d)–(f) Same as (a)–(c) except for WRF–ERAI. (g)–(i) Same as (a)–(c) except for WRF–OML. (j)–(l) Same as (a)–(c) except for WRF–ROMS. SSTs in the square box centred on the red dots are averaged at each time step and the results are displayed in Figure 3.

or so (see Figure 4(b)). This is followed by rapid warming which quickly reduces to a rate of approximately 0.25 °C/day. A week after Irene passed, SSTs at buoy m44065 are still 2 °C colder than pre-storm temperatures. WRF–ERAI SSTs display most of this observed behaviour with the exception of the rapid warming immediately following the passage of Irene. Although WRF–ROMS does simulate the correct temperature three days after the hurricane passed, its SSTs do not show the observed 4 °C drop in temperature immediately following Irene's passage nor the rapid warming subsequent to it. WRF–OML shows cooling that is intermediate between WRF–ERAI and WRF–ROMS

but again shows no evidence of recovery to pre-storm conditions.

Although the SSTs at buoy m44013 experience greater cooling (almost 8 °C) than buoy m44065, they behave in a similar way – rapid cooling, followed by a brief period of rapid warming and then a slow oscillatory recovery. This behaviour is well captured by WRF–ROMS that outperforms the other simulations at this buoy, and shows almost 6 °C of cooling. WRF–ERAI SSTs show a similar rate of recovery in the week following Irene's passage, however, it shows only 2 °C of cooling as the hurricane passes and no rapid recovery. The WRF–OML shows very

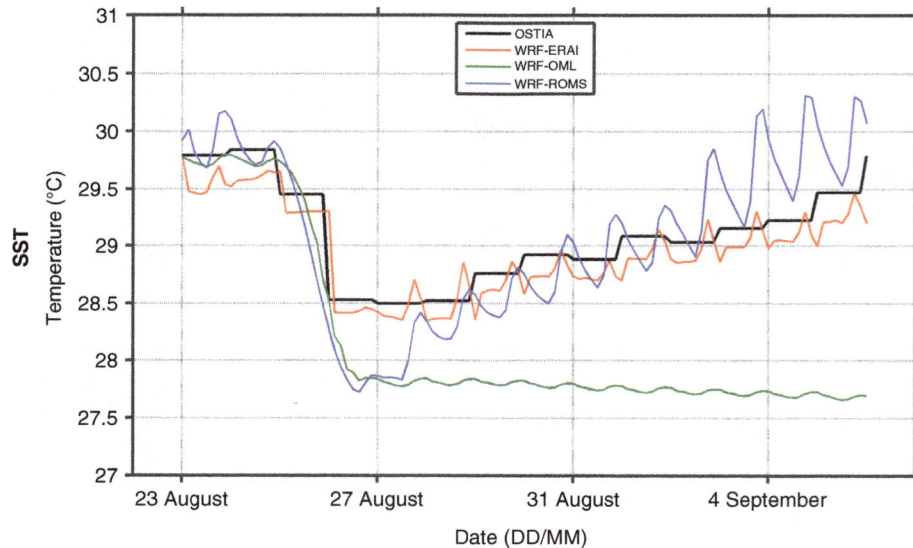

Figure 3. Sea surface temperatures from each simulation and OSTIA averaged over a square box whose width is four times the radius of maximum winds at maximum storm intensity (~4 × 90 km) centred on the location where the hurricane reaches maximum intensity (see red dot near 25°N on Figures 2(a), (d), (g) and (j)).

little cooling (approximately 1 °C) as the hurricane passes.

Figure 4(d)–(f) and Figure 4(g)–(i) show the measured and simulated 10-m wind speeds and MSLP at each buoy. The MSLPs from WRF–ERAI and WRF–OML are in good agreement with those measured at the buoy, except the simulated minimum occurs slightly later than the observed minimum. This is due to the slower than observed translational speed of the simulated cyclones. Similarly, WRF–ROMS shows a delay in the timing of the minimum pressure, which is also 10–15 hPa weaker than the minimum measured at the buoys.

Table 3 shows the root mean squared error (RMSE) of each simulated track (determined from the location of minimum pressure) from the best-observed track. Somewhat surprisingly WRF–ROMS has the greatest departure from the best-observed track. This is partially due to the slower translational speed of the WRF–ROMS simulation.

RMSEs for minimum sea level pressures are also included in Table 3. Clearly the hurricane intensity is well captured by all three simulations, which is in good agreement with the behaviour in pressure noted earlier at the buoys.

4. Discussion and Conclusions

The processes that cause cooling of the upper ocean mixed layer can be divided into those that are responsible for cooling the ocean surface within hours of the TC arrival and those that continue the cooling for up to two days after the TC has passed (Hart et al., 2007). The dominant mechanisms for cooling the ocean surface are transient upwelling and wind-driven oceanic turbulence that causes vertical mixing and entrainment

of colder water from below the thermocline into the overlying mixed layer (Vincent et al., 2012). Other processes include enhanced surface sensible and latent heat fluxes from the ocean to the atmosphere driven primarily by the winds near the radius of maximum wind (Price, 1981), horizontal transport of warm water away from the storm centre (Leipper, 1967), precipitation falling into the ocean surface and radiative losses (Brand, 1971). Those processes which cool the sea surface temperature in the vicinity of the storm have the greatest impact as they reduce the amount of heat and moisture available to the tropical storm which in turn limits its intensity (Yablonsky et al., 2015). In addition to influencing the individual TC, the cold wake may also impact other storms that interact with it at a later stage (Balaguru et al., 2014).

In WRF, the coupled ocean model (WRF–ROMS) is the most physically realistic simulation of the air–sea interaction and it produces SSTs similar to observed values, but it is computationally expensive. For example, WRF–ROMS used 1.5 times as many processors and 60% more computational time than WRF–ERAI.

The WFR–ERAI is a good, less-expensive alternative to WRF–ROMS and as a result it continues to be in widespread use by the regional climate modelling community. However, it cannot represent the feedback between the atmosphere and ocean. This could have implications for long-term climate simulations of TCs, where TCs in regional models do not coincide with TC tracks in SSTs from the parent model. This can impact TC genesis, track and intensity in regional climate simulations.

WRF with the one-dimensional ocean mixed-layer model (WRF–OML) is capable of simulating the cold wake in the SSTs caused by the passage of hurricane Irene which makes it useful for short-term studies.

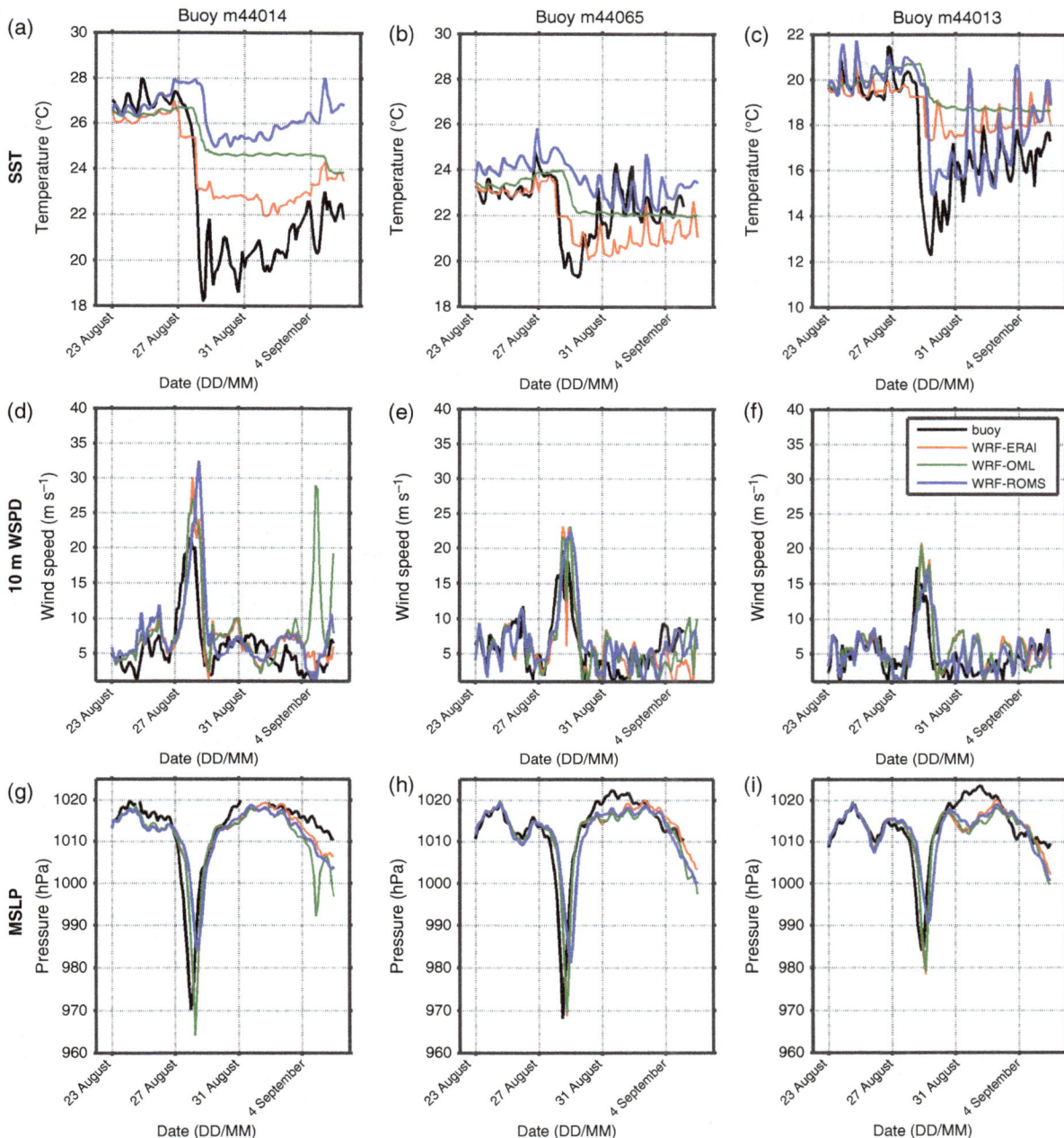

Figure 4. (a)–(c) Simulated and measured sea surface temperatures at buoys m44014, m44065 and m44013 covering the period 23 August 2011 to 6 September 2011. (d)–(f) Same as (a)–(c) except for 10-m wind speeds. (g)–(i) Same as (a)–(c) except for mean sea level pressure.

Recovery to climatological values is primarily driven by surface fluxes and horizontal advection (Vincent *et al.*, 2012). As the OML model does not include horizontal advection, it is foreseeable that its SSTs fail to recover to pre-storm conditions. While this suggests that the OML model is unsuitable for longer term studies, its computational efficiency and physically based representation of oceanic cooling make it an attractive option for future work. One approach which would retain computational efficiency will focus on representing the recovery of SSTs by adding a relaxation term based on empirical data.

Table 3. Root mean squared errors (RMSEs) for the tracks and intensity of hurricane Irene simulated by the three simulations listed in Table 1.

	WRF–ERAI	WRF–OML	WRF–ROMS
Tracks (km)	56	58	113
Intensity (hPa)	9	9	8

Acknowledgements

The authors are grateful to Dr. John Warner (USGS, Woods Hole) for providing access to the COAWST modelling system. NCAR is sponsored by the National Science Foundation

(NSF). This work was partially supported by NSF EASM Grant AGS-1048829 and the Research Partnership to Secure Energy for America. The authors acknowledge high-performance computing support from Yellowstone (ark:/85065/d7wd3xhc) provided by NCAR's Computational and Information Systems Laboratory, sponsored by the NSF. Part of the data analysis in this work was undertaken with climate data operators (CDO).

References

Balaguru K, Taraphdar S, Leung LR, Foltz GR, Knaff JA. 2014. Cyclone-cyclone interactions through the ocean pathway. *Geophysical Research Letters* **41**: 6855–6862, doi: 10.1002/2014GL061489.

de Boyer Montégut C, Madec G, Fischer AS, Lazar A, Iudicone D. 2004. Mixed layer depth over the global ocean: an examination of profile data and a profile-based climatology. *Journal of Geophysical Research* **109**: C12003, doi: 10.1029/2004JC002378.

Brand S. 1971. The effects on a tropical cyclone of cooler surface waters due to upwelling and mixing produced by a prior tropical cyclone. *Journal of Applied Meteorology* **10**(5): 865–874.

Bruyère CL, Holland GJ, Towler E. 2012. Investigating the use of a genesis potential index for Tropical Cyclones in the North Atlantic Basin. *Journal of Climate* **25**: 8611–8626, doi: 10.1175/JCLI-D-11-00619.1.

Chen F, Dudhia J. 2001. Coupling an advanced land surface-hydrology model with the Penn State-NCAR MM5 modeling system Part I: model implementation and sensitivity. *Monthly Weather Review* **129**: 569–585.

Collins WD, Rasch PJ, Boville BA, Hack JJ, McCaa JR, Williamson DL, Kiehl JT, Briegleb B, Bitz C, Lin S-J, Zhang M, Dai Y. 2004. Description of the NCAR Community Atmosphere Model (CAM3.0). NCAR Technical Note, TN-464+STR, NCAR, Boulder, CO, USA, 102–143.

Costa P, Gómez B, Venâncio A, Pérez E, Pérez-Munuzuri V. 2012. Using the Regional Ocean Modeling System (ROMS) to improve the sea surface temperature predictions of the MERCATOR Ocean System. In *Advances in Spanish Physical Oceanography*, Espino M, Font J, Pelegrí JL, Sánchez-Arcilla A (eds). Scientia Marina 76S1: Barcelona, Spain; 165–175. ISSN: 0214-8358; doi: 10.3989/scimar.03614.19E.

Cummings JA. 2005. Operational multivariate ocean data assimilation. *Quarterly Journal of the Royal Meteorological Society* **131**: 3583–3604, doi: 10.1256/qj.05.105.

Dare RA, McBride JL. 2011. Sea surface temperature response to tropical cyclones. *Monthly Weather Review* **139**(12): 3798–3808.

Davis CAW, Wang W, Chen SS, Chen Y, Corbosiero K, DeMaria M, Dudhia J, Holland G, Klemp J, Michalakes J, Reeves H, Rotunno R, Snyder C, Xiao Q. 2008. Prediction of landfalling hurricanes with the advanced hurricane WRF model. *Monthly Weather Review* **136**: 1990–2005, doi: 10.1175/2007MWR2085.1.

Dee DP, Uppala SM, Simmons AJ, Berrisford P, Poli P, Kobayashi S, Andrae U, Balmaseda MA, Balsamo G, Bauer P, Bechtold P, Beljaars ACM, van de Berg L, Bidlot J, Bormann N, Dlesol C, Dragani R, Fuentes M, Geer AJ, Haimberger L, Healy SB, Hersbach H, Hólm EV, Isaken L, Källberg P, Kohler M, Matricardi M, McNally AP, Monge-Sanz BM, Morcrette J-J, Park BK, Peubey C, de Rosnay P, Tavoloato C, Thépaut J-N, Vitart F. 2011. The ERA-Interim reanalysis: configuration and performance of the data assimilation system. *Quarterly Journal of the Royal Meteorological Society* **137**: 553–597, doi: 10.1002/qj.828.

Glenn SM, Miles TN, Seroka GN, Xu Y, Forney RK, Yu F, Roarty H, Schofield O, Kohut J. 2016. Stratified coastal ocean interactions with tropical cyclones. *Nature Communications* **7**: 10887, doi: 10.1038/ncomms10887.

Hart RE, Maue RN, Watson MC. 2007. Estimating local memory of tropical cyclones through MPI anomaly evolution. *Monthly Weather Review* **135**(12): 3990–4005.

Hong S-Y, Lim J-OJ. 2006. The WRF single moment six class scheme (WSM6). *Journal of the Korean Meteorological Society* **42**: 129–151.

Hong S-Y, Noh Y, Dudhia J. 2006. A new vertical diffusion package with an explicit treatment of entrainment processes. *Monthly Weather Review* **134**: 2318–2341.

Kain JS. 2004. The Kain-Fritsch convective parameterization: an update. *Journal of Applied Meteorology* **43**: 170–181.

Kain JS, Fritsch JM. 1990. A one-dimensional entraining/detraining plume model and its application in convective parameterization. *Journal of the Atmospheric Sciences* **47**: 2784–2802.

Knapp KR, Kruk MC, Levinson DH, Diamond HJ, Neumann CJ. 2010. The International Best Track Archive for Climate Stewardship (IBTrACS): unifying tropical cyclone data. *Bulletin of the American Meteorological Society* **91**: 363–376.

Leipper DF. 1967. Observed ocean conditions and Hurricane Hilda. *Journal of the Atmospheric Sciences* **24**: 182–196.

Lin I, Liu WT, Wu C-C, Wong GTF, Hu C, Chen Z, Liang W-D, Yang Y, Liu KK. 2003. New evidence for enhanced ocean primary production triggered by tropical cyclone. *Geophysical Research Letters* **30**: 1718, doi: 10.1029/2003GL017141.

Pollard RT, Rhines PB, Thompson RORY. 1972. The deepening of the wind-mixed layer. *Geophysical Fluid Dynamics* **4**(1): 381–404, doi: 10.1080/03091927208236105.

Price JF. 1981. Upper ocean response to a hurricane. *Journal of Physical Oceanography* **11**(2): 153–175.

Shchepetkin AF, McWilliams JC. 2005. The regional ocean modeling system: a split-explicit free-surface, topography-following coordinates ocean model. *Ocean Model* **9**: 347–404, doi: 10.1016/jocemod.2004.08.002.

Skamarock WC, Klemp JB, Dudhia J, Gill DO, Barker DM, Duda MG, Huang X-Y, Wang W, Powers JG. 2008. A description of the advanced research WRF Version 3. NCAR Technical Note, NCAR, Boulder, CO, USA.

Stark JD, Donlon CJ, Martin MJ, McCulloch ME. 2007. OSTIA: an operational, high resolution, real time, global sea surface temperature analysis system, Oceans 07 IEEE Aberdeen, conference proceedings. In *Marine Challenges: Coastline to Deep Sea*. IEEE: Aberdeen, Scotland.

Vincent EM, Lengaigue M, Madec G, Vialard J, Samson G, Jourdain NC, Menkes CE, Julien S. 2012. Processes setting the characteristics of sea surface cooling induced by tropical cyclones. *Journal of Geophysical Research* **117**: C02020, doi: 10.1029/2011JC007396.

Warner JC, Sherwood CR, Arango HG, Signell RP. 2005. Performance of four turbulence closure methods implemented using a generic length scale method. *Ocean Modelling* **8**: 81–113.

Warner JC, Armstrong B, He R, Zambon JB. 2010. Development of a Coupled Ocean–Atmosphere–Wave–Sediment Transport (COAWST) modeling system. *Ocean Modelling* **35**(3): 230–244.

Yablonsky RM, Ginis I, Thomas B, Tallapragada V, Sheinin D, Bernardet L. 2015. Description and analysis of the ocean component of NOAA's Operational Hurricane Weather Research and Forecasting Model (HWRF). *Journal of Atmospheric and Oceanic Technology* **32**(1): 144–163, doi: 10.1175/JTECH-D-14-00063.1.

Zeng X, Beljaars A. 2005. A prognostic scheme of sea surface skin temperature for modeling and data assimilation. *Geophysical Research Letters* **32**: L14605, doi: 10.1029/2005GL023030.

14

Sea surface temperature influence on a winter cold front position and propagation: air–sea interactions of the 'Nortes' winds in the Gulf of Mexico

G. A. Passalacqua,[1,2,]* J. Sheinbaum[2] and J. A. Martinez[1]

[1]*Facultad de Ciencias Marinas, UABC, Ensenada, Mexico*
[2]*Departamento de Oceanografía Física, CICESE, Ensenada, Mexico*

Correspondence to:
G. A. Passalacqua, Facultad de Ciencias Marinas, UABC, Apdo. #76, Ensenada, Baja California, C.P. 22800, Mexico.
E-mail:
gino.passalacqua@uabc.edu.mx

Abstract

A high-resolution, regional atmospheric model with different sea surface temperature (SST) boundary conditions (BC) is used to examine the air–sea interactions of the winter cold fronts (CF) advancing over the Gulf of Mexico (GoM). Comparison with oceanic-buoy 10-m wind, 2 m air temperature (AIR.2 m), sea level pressure (SLP) and SST reveals good agreement with observations. The CF propagation speed was significantly affected by the SST: higher SST produced faster CF traveling speeds. Using a 1-D ocean mixed layer model as BC reduced the air–sea fluxes and the CF propagated slower but in accordance with the reanalysis data; representing an improvement in the numerical modeling of the CF propagation and airmass modification over the GoM.

Keywords: air–sea interactions; cold front; SST; WRF

1. Introduction

During winter, synoptic-scale cold and dry polar air masses propagate southward along the eastern side of the Rocky Mountains into GoM, resulting in fast moving CF popularly known as 'Nortes' winds. As the CF advances over the GoM warm waters, the wind accelerates producing substantial air-mass modification and intensive exchanges of heat, moisture and momentum between the ocean and atmosphere (Nowlin and Parker, 1974; Garreaud, 2001). The cold airflow trailing the CF has strong lower-tropospheric northerly winds; regularly surpassing $15 \, \mathrm{m \, s^{-1}}$, cooling the air temperature by $10-15\,°C$ in a period of 24 h, and increasing SLP by $15-30$ mb (Merrill, 1992; Schultz *et al.*, 1997).

The evolution of a CF over the GoM waters is mainly affected by two non-exclusive governing mechanisms: large surface heat flux (HF) and cold air damming (CAD) dynamics. Before arriving to the GoM, the northerly flow is dominated by CAD dynamics (Colle and Mass, 1995, hereafter CM95). On the central and Eastern GoM, the CF is mainly affected by large surface HF that generates vertical mixing. This mechanism appears to be dominant in the planetary boundary layer (PBL) during the CF progression (Mailhot, 1992; Merrill, 1992; Thompson and Burk, 1993).

Short-term forecasts or simulations (7–10 days) generally assume that SST changes do not propagate into atmospheric adjustments. However, studies in other regions with cold–dry air outbreaks (e.g. Gulf of Lyon, Lebeaupin Brossier *et al.*, 2009 and Lebeaupin Brossier *et al.*, 2013, Gulf Stream, Booth *et al.*, 2012), have shown the importance of changing mixed layer heat content and the higher thermodynamic ocean memory which considerably impact the air–sea interactions and the evolution of CFs.

Because significant air–sea interactions have being observed immediately after the passage of a CF over the GoM (Nowlin and Parker, 1974), our main purpose is to evaluate albeit in the simplest manner the impact of air–sea processes in the evolution of a 'Norte' event under different SST BCs setups. Three numerical experiments using the Weather Research and Forecasting (WRF) model version 3.4.1 (Skamarock and Klemp, 2008) are carried out representing different ocean BCs: (1) a constant in time, but spatially varying SST; (2) a time and space varying SST field taken from daily reanalysis, with a superimposed predictive SST diurnal variation scheme (Zeng and Beljaars, 2005); and (3) a 1-D ocean mixed layer that models the effect of wind-driven vertical mixing (Pollard *et al.*, 1973), usually employed in hurricane simulations (Davis *et al.*, 2008).

2. Model description and experimental setup

In this study, we use the WRF with the Advanced Research dynamic solver, WRF-ARW (Skamarock and Klemp 2008), which integrates the fully compressible, non-hydrostatic equations of motion on a terrain-following vertical coordinate system. The physics packages used in this study were implemented after an exhaustive review of literature applying WRF for similar weather events (see Table S1, Supporting

Information for details). To validate the spatial patterns, we compared the model results against Climate Forecast System Reanalysis (CFSR; Saha *et al.*, 2010).

The model includes three one way nested domains with 108, 36 and 12 km grid spacing. The coarser domain encompasses most of North America, and part of Central America, from 60° N to 16°S and 164°W to 29°W. Our analysis in centered on the highest resolution domain, which includes the whole GoM and a substantial area of the North American Cordillera and the Sierra Madre mountains in order to have a realistic depiction of their topographic impact. All domains have 74 vertical levels from sea level to 50 mb, with the lowest level at 10 m over the ocean surface and 36 levels below 750 mb. Each simulation runs from the 1 to 15 January 2010, to simulate conditions before and after the main 'Norte' event that occur on 7 January.

The initial and BCs were obtained from the NCEP-FNL, with a resolution of 1° × 1° (available at http://rda.ucar.edu/datasets/ds083.2/). For SST surface initial and BCs (when SSTs vary in time), the NCEP Real Time Global (RTG) 0.5° × 0.5° SST analysis (Thiebaux *et al.* 2003) was used (available at http://polar.ncep.noaa.gov/sst/rtg_low_res/). All the experiment use RTG SST data from 1 January 2010 as initial condition, and for fixed SST experiment (CTE) it was used for the entire run. In the evolving SST experiment (DAY), the SSTs are updated every 24 h. For the ocean mixed layer experiment (OML), the initial mixed layer depth was set to 20 m; although the initial depth is on the shallower end, we decided to use this initial value throughout the domain to force a rapid response. A value of gamma (deep ocean stratification parameter) of 0.14 was used for the simulation.

3. Results

3.1. Synoptic description

Time series of the three WRF experiments as well as observational data from NDBC buoy 42001, located 330 km South of Southwest Pass, Louisiana (Figure 1), show the arrival of a CF on 1 January at 1200 (UTC). With the CF, SLP (Figure 1(b)) and both AIR.2 m and SST (Figure 1(c) and (d) respectively) decrease afterwards. On 7 January, a wind shift produced by a 'Return Flow' (Crisp and Lewis, 1992) generates a drop of roughly 8 mb in SLP, and AIR.2 m increases roughly 10 °C. The main 'Norte' arrives on 8 January at 0650 h with a dramatic change in 10 m wind speed magnitude (WSPD) and direction, producing a drastic drop in the AIR.2 m, and a consistent increase in SLP. The wind veers rapidly from southerly to northerly, and accelerates from 7 m s⁻¹ to persistent winds above 12 m s⁻¹ for more than 48 hours. The AIR.2 m gradually rises on the early hours of 8 January until the end of the simulation. We will focus our discussion using results from this buoy which is located in the middle of the GoM; although we also did comparisons with

other NDBC buoys (42035, 42047, 42002 and 42055) to validate the model.

3.2. Discussion

The observed SST at buoy 42001 (Figure 1(d)) has a slight cooling trend. The ocean cools about 2 °C after the first CF event (1 January) but then warms up 1 °C maintaining a nearly constant value. After the arrival of the main CF on 7 January, the ocean cools down slowly to 21 °C, warming up again after 13 January. The WRF experiments with variable SSTs, (DAY and OML), also show a cooling trend with DAY producing a more noticeable cooling only after the main CF of 7 January, but almost no change before that (Figure 1(d)). OML is somewhat closer to observations during the first week of the simulation but clearly it has a cooling SST bias (Figure 1(d)). Mallard *et al.* (2013) use a method whereby the mixed-layer routine is reinitialized every day from observed SSTs to reduce the bias in month-long hurricane simulations. We decided to use the mixed layer model with no restarts to investigate its impact. Experiment CTE has the warmest (unchanging) SST, OML has the lowest, and DAY run is in-between, but both DAY and OML improve the simulation of the atmospheric variables (Figure 1(a)–(c)).

Although not shown, all the compared NDBC buoys and their correspondent simulation time series present a similar behavior as NDBC 42001. There are clear biases because all simulations underestimate SLP, and AIR.2 m is mostly warmer than observations throughout the entire run. The intensity of the warmer/weaker AIR.2 m/SLP bias is buoy dependant. Biases are related to the CF air–sea interaction, as OML with coldest SST and lower HF (Figure 2) consistently produces better results.

The 10 m winds are less affected by the changes in surface BCs and relatively similar in all experiments; they all show good agreement with CFSR both in magnitude (not shown) and direction (Figure 2). OML has the lowest winds root mean square errors when compared with the buoys.

To better understand the impact of the changing BCs on a larger scale, we now look at differences in the spatial structure and time evolution of some key aspects of its synoptic evolution. In particular we investigate the structure of the CF and identify its leading edge (CF_LE) by analyzing the magnitude of AIR.2 m and WSPD gradients. We define the CF_LE as the region where both quantities have maximum values across the GoM, in accordance with the concepts of CM95 and Schultz and Steenburgh (1999). Both criteria are consistent with each other and appear to be in good agreement with the CFSR data over the ocean. The procedure is deemed sufficient to locate the front position and is consistent with the methods used for detecting weather fronts (Hope *et al.*, 2014).

Figure 2 displays a snap shot of the CF progression over the GoM on 8 January 2010 at 0600 UTC including 10 m wind vectors, SLP, latent heat flux (LHF) and

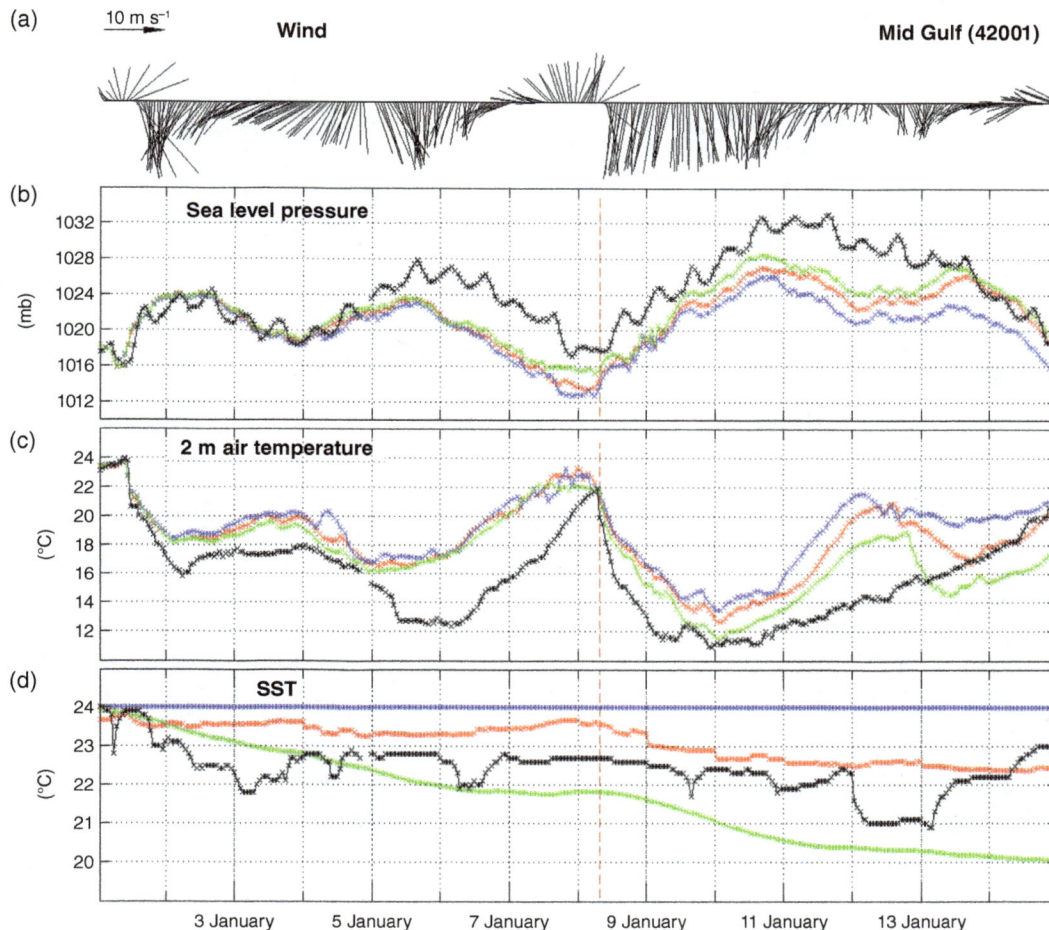

Figure 1. Time series of wind vectors (a) sea level pressure (b) 2 m air temperature (c) and, SST (d) at NDBC Buoy Mid-Gulf (42001). The black line represents the observations at the buoy. Blue, red and green represent the WRF data from the CTE, DAY and OML model experiments, correspondingly. The red dashed vertical line signals the time of arrival of the main CF at the buoy location.

the CF_LE. A sharp LHF contrast trails the CF_LE in CFSR and in all the WRF experiments (Fig. 2). CFSR and OML show magnitudes of $100-300$ W m^{-2} before the CF_LE and values higher than 500 W m^{-2} behind the front, related to the wide-spread invasion of cold air. CFSR has weaker LHF than all the WRF experiments with OML (due to its cooling bias) being the closest. LHF patterns are also different: CTE and DAY produce similar and very high LHF (up to 600 W m^{-2}) on a band along the Louisiana, Texas and western Mexican GoM coast with particularly high values over Texas, whereas CFSR has maximum values off-shore (on the Louisiana, Texas region). It is not clear whether the large LHF values near the coast in the northern and western GoM produced by WRF are realistic.

It is necessary to emphasize that the values of LHF of OML are more in accordance with CFSR values, even though that experiment has higher LHF especially in the western GoM and in northern coastal areas where the CF enters in contact with the ocean (Figure 2). CFSR relative humidity (not shown) exhibits to be wetter than any WRF experiment. Furthermore, post-frontal values for CFSR LHF tend to have a negative bias in the

GoM for the winter months (Xue *et al.*, 2011) and for the annual mean (Wang *et al.*, 2011). Therefore, care should be exercised in the above comparisons that use CFSR as a reference. Nevertheless, our results indicate that the OML latent (Figure 2(b)) and sensible (not shown) heat fluxes are more in agreement with CFSR (Figure 2(d)) than any of the other experiment.

We noticed that the traveling speed of the CF in each simulation was different. This is a very relevant factor for the amount of precipitation delivered by CFs in the GoM, as less winter precipitation is produced by faster moving fronts (Pérez *et al.*, 2014). To confirm this premise, CF_LE traveling speeds were calculated from observed NDBC buoys along the central GoM and each WRF experiment at the same locations (Table 1). All the WRF simulation experience an increase of the CF_LE velocity as the CF enters the GoM waters. CTE has the most dramatic acceleration as the front progresses southward. OML accelerates at the lowest rate, reaching CF_LE traveling speeds comparable with NDBC speeds.

To emphasize these results we present the position of the CF_LE for CFSR and the WRF experiments at four

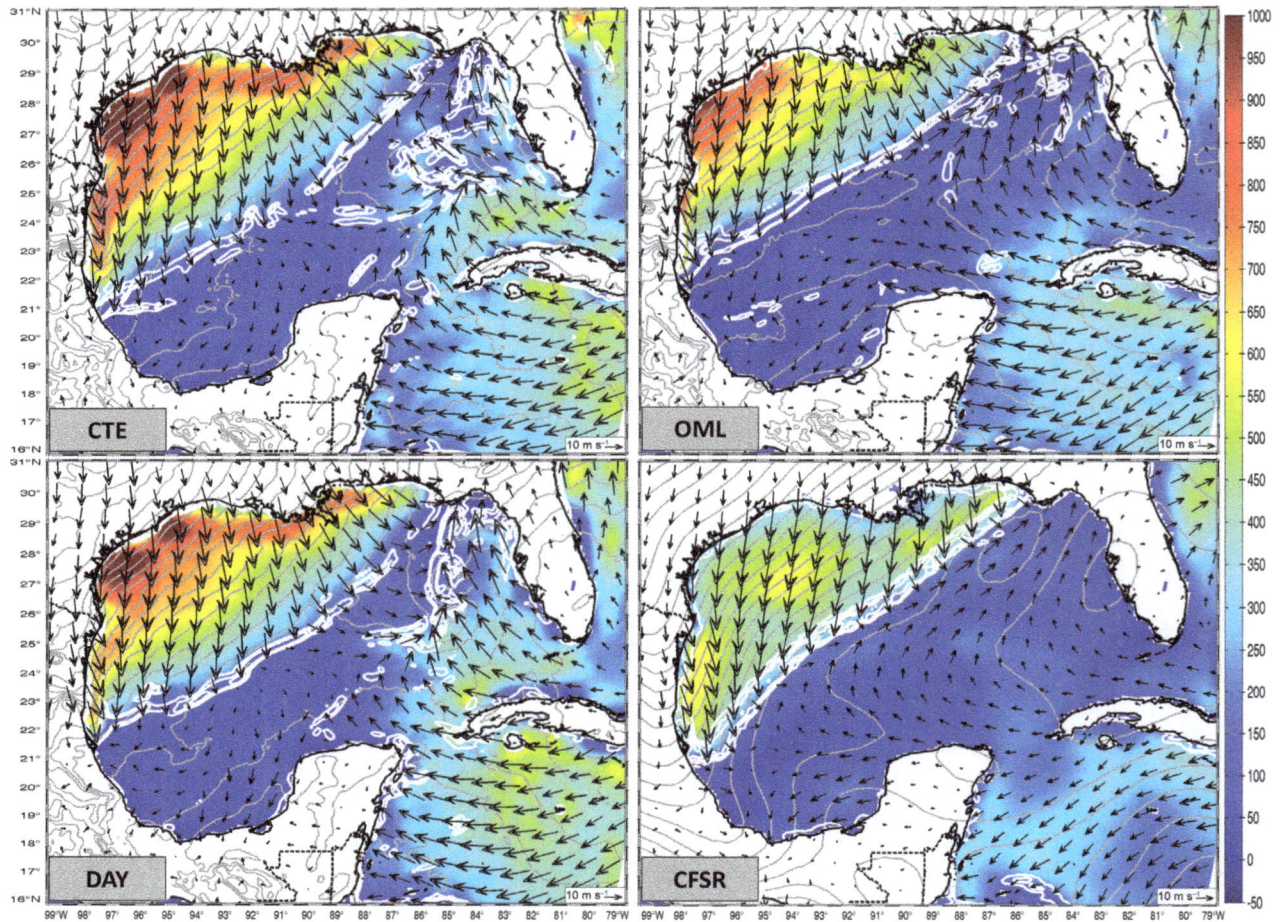

Figure 2. CFSR reanalysis data and WRF CTE, OML and DAY experiments for latent heat (W m^{-2}, colors), 10 m wind vectors (black vectors), sea level pressure (hpa, gray contours) and 10-m wind speed gradient (s^{-1}, white contours) at 0600 UTC on 8 January 2010. Only higher values of wind speed gradient are presented, ranging from 0.0001 to 0.0002 s^{-1} with contours every 0.00005 s^{-1}. Wind speed gradient contours represent the positions of the CF. The black dashed line represents country borders.

Table I. Cold front leading edge travelling speeds (m s^{-1}) calculated based on the arrival time of the front at each buoy and the distance between them. The percentages indicate the WRF experiments travelling speed in comparison to the observed NDBC buoy.

	Cold front propagation speed						
	CTE		**DAY**		**OML**		**NDBC**
Buoy	**Speed**	**%**	**Speed**	**%**	**Speed**	**%**	**Speed**
GALV–TABSV	6.6952	−23.81	11.7170	+33.34	11.7170	+33.34	8.7875
TABSV–WGULF	16.2770	+58.34	13.0220	+26.67	10.8510	+5.55	10.2800
WGULF–CMPCH	19.5080	+83.35	13.0050	+22.23	11.7050	+10.01	10.6400

different times, displaying the progression of the CF under the different BCs (Figure 3). We found differences in the CF_LE position between simulations as the CF crosses the GoM. In general, the CTE front, with highest HF, is more frontolytically and is always ahead than the other simulations and CFSR. Contrastively, OML with the lowest HF, has a well defined CF_LE

structure and is always farther north in the simulations. A small wedge on the CF_LE is noticeable along the Mexican coast particularly under CFSR, suggesting an effect of CAD dynamics on the difference in CF_LE position. In CTE and DAY, pre-frontal wind shifts influencing the CF progression (Schultz, 2005) are located 200 to 300 km ahead of the CF_LE (Figure 2(a) and(c)), probably as a consequence of stronger easterly winds (Figure 3(b) and (c)). Only OML, with the coldest SST, maintains almost an unbroken and well defined CF_LE when crossing the GoM, and generally collocates well with CFSR.

4. Conclusions

We implemented the high-resolution, regional atmospheric WRF model to represent the progression of winter CF in the GoM with the interest of investigating its response to three different SST BCs: constant, daily observed and an ocean mixed layer model. Each simulation lasted 14 days, focusing our results on the event entering the GoM on 7 January 2010. In general, the model is in good agreement with NDBC buoy SLP,

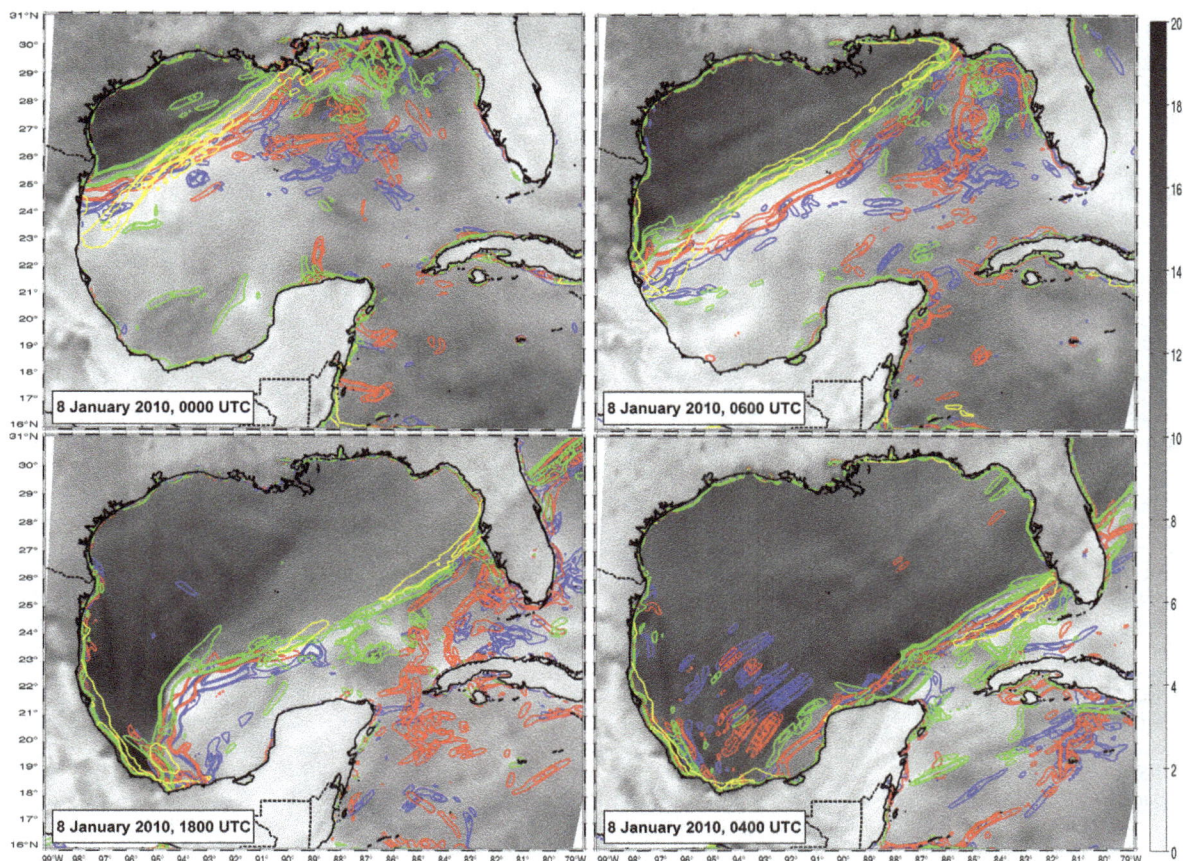

Figure 3. Cold front leading edge (CF_LE) progression in the GoM for CFSR (yellow), CTE (blue), DAY (red) and OML (green). Contours of WSPD gradient, representing the CF_LE, are shown with values ranging from 0.0001 to 0.0002 s^{-1} and contoured every 0.00005 s^{-1}. As reference, the WSPD (m s^{-1}) from the OML experiment is shown as background (gray scale).

AIR.2 M and surface winds. The biases obtained are strongly influenced by the SST BCs, as CTE produced higher biases and OML is the closest to observation values.

Based on the type of weather event, we found that analysis of AIR.2 m and gradient magnitude of the WSPD magnitude are a good method to locate the CF_LE position and its progression. Different CF propagation speeds were found for the different WRF experiments (Figure 3), and were corroborated by calculating them from observed NDBC buoys along the central GoM, revealing sensitivity to the SST specification and consequently the HF, resulting in higher speeds under CTE and the lowest in OML.

The air–sea interactions are strongly modified by SST BCs that significantly influence the corresponding HF; and have an effect on the structure and organization of the CF. Higher HF have a more frontolytic effect over the front (Burk and Thompson, 1992; CM95). As OML has lower SST and consequently the lowest HF, the CF_LE has a well define and organized structure across the GoM (Figures 2(b) and 3). In the CTE and DAY experiments, the front organization and structure are considerably affected.

Our results suggest that the use of 1-D ocean mixed layer model with the WRF system improves the representation of the CAO in the GoM and its corresponding CF, in accordance with Nicholls and Decker, in press results. We based our statement in the better representation of the CF_LE, and a well-organized and defined front without a pre-frontal wind shift, trailed by a sharp gradient of LHF with similar magnitudes as in the CFSR data. In addition, OML has a closer to buoy-observed SLP and AIR.2 m. Although, OML trails the CFSR front on the western Gulf and over Florida, it has a similar overall position, and OML has closer to buoy-observed propagation speeds (Table 1). In general terms, the OML front has a better performance on the progression, structure and organization of the CAO over the GoM. Details about how the SST BC affect the mechanisms involved on the CF progression speed, should be addressed by studying different events during the Nortes season, in order to take account of the intra-seasonal variability and obtain a more robust analysis. Nevertheless, this short paper highlights the importance of SST BC and the resulting air–sea interaction processes for short and medium term weather forecasting. We have shown that these interactions have considerable implications for the CF winds, heat, and humidity fluxes that affect the CF organization and propagation speed; consequently, could impact the amount of precipitation each storm produces.

References

Booth JF, Thompson L, Patoux J, Kelly KA. 2012. Sensitivity of midlatitude storm intensification to perturbations in the sea surface temperature near the Gulf Stream. *Monthly Weather Review* **140**(4): 1241–1256, doi: 10.1175/mwr-d-11-00195.1.

Burk SD, Thompson WT. 1992. Airmass modification over the Gulf of Mexico: mesoscale model and airmass transformation model forecasts. *Journal of Applied Meteorology* **31**(8): 925–937, doi: 10.1175/1520-0450(1992)031<0925:amotgo>2.0.co;2.

Colle BA, Mass CF. 1995. The structure and evolution of cold surges east of the Rocky Mountains. *Monthly Weather Review* **123**(9): 2577–2610, doi: 10.1175/1520-0493(1995)123<2577:TSAEOC >2.0.CO;2.

Crisp CA, Lewis JM. 1992. Return flow in the Gulf of Mexico. Part I. A classificatory approach with a global historical perspective. *Applied Meteorology* **31**(8): 868–881, doi: 10.1175/1520-0450 (1992)031<0868:RFITGO>2.0.CO;2.

Davis C, Wang W, Chen SS, Chen YS, Corbosiero K, DeMaria M, Dudhia J, Holland G, Klemp J, Michalakes J, Reeves H, Rotunno R, Snyder C, Xiao QN. 2008. Prediction of landfalling hurricanes with the Advanced Hurricane WRF model. *Monthly Weather Review* **136**(6): 1990–2005, doi: 10.1175/2007mwr2085.1.

Garreaud RD. 2001. Subtropical cold surges: regional aspects and global distribution. *International Journal of Climatology* **21**(10): 1181–1197, doi: 10.1002/joc.687.

Hope P, Keay K, Pook M, Catto J, Simmonds I, Mills G, McIntosh P, Risbey J, Berry G. 2014. A comparison of automated methods of front recognition for climate studies: a case study in Southwest Western Australia. *Monthly Weather Review* **142**(1): 343–363, doi: 10.1175/MWR-D-12-00252.1.

Lebeaupin Brossier C, Drobinski P. 2009. Numerical high-resolution air-sea coupling over the Gulf of Lions during two Tramontane/Mistral events. *Journal of Geophysical Research, [Atmospheres]* **114**: 21, doi: 10.1029/2008JD011601.

Lebeaupin Brossier C, Drobinski P, Beranger K, Bastin S, Orain F. 2013. Ocean memory effect on the dynamics of coastal heavy precipitation preceded by a mistral event in the northwestern Mediterranean. *Quarterly Journal of the Royal Meteorological Society* **139**(675): 1583–1597, doi: 10.1002/qj.2049.

Mailhot J. 1992. Numerical simulation of airmass transformation over the Gulf of Mexico. *Journal of Applied Meteorology* **31**(8): 946–963, doi: 10.1175/1520-0450(1992)031<0946:NSOATO>2.0. CO;2.

Mallard MS, Lackmann GM, Aiyyer A. 2013. Atlantic hurricanes and climate change. Part I: experimental design and isolation of thermodynamic effects. *Journal of Climate* **26**(13): 4876–4893, doi: 10.1175/JCLI-D-12-00182.1.

Merrill RT. 1992. Synoptic analysis of the GUFMEX return-flow event of 10–12 March 1988. *Journal of Applied Meteorology* **31**(8): 849–867, doi: 10.1175/1520-0450(1992)031<0849: SAOTGR>2.0.CO;2.

Nicholls SD, Decker SG. 2015. Impact of coupling an ocean model to WRF nor'easter simulations. *Monthly Weather Review* **140**(12): 4997–5016, doi: 10.1175/MWR-D-15-0017.1.

Nowlin WD, Parker CA. 1974. Effects of a cold-air outbreak on shelf waters of Gulf-of Mexico. *Journal of Physical Oceanography* **4**(3): 467–486, doi: 10.1175/1520-0485(1974)004<0467:EOACAO >2.0.CO;2.

Pérez EP, Magaña V, Caetano E, Kusunoki S. 2014. Cold surge activity over the Gulf of Mexico in a warmer climate. *Frontiers Earth Science* **2**: 19, doi: 10.3389/feart.2014.00019.

Pollard RT, Rhines PB, Thompson RORY. 1973. The deepening of the wind-mixed layer. *Geophysical Fluid Dynamics* **4**(4): 381–404.

Saha S, Moorthi S, Pan H-L, Wu X, Wang J, Nadiga S, Tripp P, Kistler R, Woollen J, Behringer D, Liu H, Stokes D, Grumbine R, Gayno G, Wang J, Hou Y-T, Chuang H-Y, H. Juang H-M, Sela J, Iredell M, Treadon R, Kleist D, Van Delst P, Keyser D, Derber J, Ek M, Meng J, Wei H, Yang R, Lord S, van den Dool H, Kumar A, Wang W, Long C, Chelliah M, Xue Y, Huang B, Schemm J-K, Ebisuzaki W, Lin R, Xie P, Chen M, Zhou S, Higgins W, Zou C-Z, Liu Q, Chen Y, Han Y, Cucurull L, Reynolds RW, Rutledge G, Goldberg M. 2010. The NCEP climate forecast system reanalysis. *Bulletin of the American Meteorological Society* **91**(8): 1015–1057, doi: 10.1175/2010BAMS3001.1.

Schultz DM. 2005. Review of cold fronts with prefrontal troughs and wind shifts. *Monthly Weather Review* **133**(8): 2449–2472, doi: 10.1175/MWR2987.1.

Schultz DM, Steenburgh WJ. 1999. The formation of a forward-tilting cold front with multiple cloud bands during Superstorm 1993. *Monthly Weather Review* **127**(6): 1108–1124, doi: 10.1175/ 1520-0493(1999)127<1108:TFOAFT>2.0.CO;2.

Schultz DM, Bracken WE, Bosart LF, Hakim GJ, Bedrick MA, Dickinson MJ, Tyle KR. 1997. The 1993 Superstorm cold surge: frontal structure, gap flow, and tropical impact. *Monthly Weather Review* **125**(1): 5–39, doi: 10.1175/1520-0493(1997)125<0005: TSCSFS>2.0.CO;2.

Skamarock WC, Klemp JB. 2008. A time-split nonhydrostatic atmospheric model for weather research and forecasting applications. *Journal of Computational Physics* **227**(7): 3465–3485, doi: 10.1016/j.jcp.2007.01.037.

Thiebaux J, Rogers E, Wang WQ, Katz B. 2003. A new high-resolution blended real-time global sea surface temperature analysis. *Bulletin of the American Meteorological Society* **84**(5): 645, doi: 10.1175/BAMS-84-5-645.

Thompson WT, Burk SD. 1993. Postfrontal boundary-layer modification over the western Gulf of Mexico during GUFMEX. *Journal of Applied Meteorology* **32**(9): 1521–1537, doi: 10.1175/1520-0450(1993)032<1521:PBLMOT>2.0.CO;2.

Wang WQ, Xie PP, Yoo SH, Xue Y, Kumar A, Wu XR. 2011. An assessment of the surface climate in the NCEP climate forecast system reanalysis. *Climate Dynamics* **37**(7–8): 1601–1620, doi: 10.1007/s00382-010-0935-7.

Xue Y, Huang BY, Hu ZZ, Kumar A, Wen CH, Behringer D, Nadiga S. 2011. An assessment of oceanic variability in the NCEP climate forecast system reanalysis. *Climate Dynamics* **37**(11–12): 2511–2539, doi: 10.1007/s00382-010-0954-4.

Zeng XB, Beljaars A. 2005. A prognostic scheme of sea surface skin temperature for modeling and data assimilation. *Geophysical Research Letters* **32**(14): 4, doi: 10.1029/2005GL023030.

15

A hybrid model based on latest version of NCEP CFS coupled model for Indian monsoon rainfall forecast

D. R. Pattanaik[1]* and Arun Kumar[2]

[1] India Meteorological Department, Pune, India
[2] NOAA/NWS/NCEP, Climate Prediction Centre, College Park, MD, USA

*Correspondence to:
 D. R. Pattanaik, India
Meteorological Department,
Pune, India.
E-mail: pattanaik_dr@yahoo.co.in

Abstract

The forecast skill of all India summer monsoon rainfall (AISMR) during June to September (JJAS) in the new version of the National Centers for Environmental Prediction (NCEP)'s Climate Forecast System version 2 (CFSv2) is analyzed by considering 28 years (1982–2009) retrospective forecasts. The spatial patterns of JJAS mean rainfall and its interannual variability is more realistic over the Indian monsoon region in CFSv2 as compared to previous version of NCEP's CFS. A hybrid (dynamical–empirical) model based on the forecast variables of CFSv2 is developed for AISMR, which shows correlation that is highly significant with observed AISMR. The hybrid model correctly predicted the observed AISMR departure of 2013.

Keywords: all India summer monsoon rainfall; NCEP CFSv2; seasonal forecast skill; hybrid model

I. Introduction

The variability spectrum of all India summer monsoon rainfall (AISMR) over India during June to September (JJAS) in last 113 years (1901–2013) ranges from a departure of about −25% in 1918 to about +23% in 1917 from its long period average (LPA) of ≈89 cm. The droughts (deficient rainfall) and floods (excess rainfall) are two extremes of the interannual variability of monsoon rainfall. This variability has profound impact on all sectors of national economy including agriculture, water resource and others. An overwhelming majority of cropped area in India (around 57%) is rainfed and falls within the medium- and low range of monsoonal rainfall (MoAg, 2009). Large areas are therefore affected vagaries of southwest monsoon. Most parts of peninsular, central and northwest regions of India are most prone to periodic drought (Mamoria, 1978).

The prediction of AISMR during JJAS is vital for the policy planning and national economy for the agro-economic country like India. The monsoon prediction in this timescale is mainly done by using statistical and dynamical models. Sir H. F. Blanford, the founder Head of India Meteorological Department (IMD), made the first attempt for estimating the prospective rains by utilizing the indications provided by the preceding winter and spring snowfall over the Himalayas (Blanford, 1884). In the last few decades, many statistical models have been developed for the prediction of AISMR (Shukla and Mooley, 1987; Gowariker et al., 1991; Rajeevan et al., 2003; Sahai et al., 2003; Pattanaik et al., 2005). The statistical forecasting method employed by IMD has met with varying degrees of

success but with no significant improvements in the forecast skill over a long period, particularly in forecasting an extreme rainfall season. On the other hand, the dynamical prediction has evolved over the years to a stage where coupled general circulation models (CGCMs) are now employed for routine, seasonal climate prediction by some operational forecasting centers.

In the last few decades, many dynamical models have been tested for predicting the summer monsoon rainfall (Palmer et al., 1992; Chen and Yen, 1994; Sperber and Palmer, 1996; Shukla et al., 2000; Saha et al., 2006; Krishnan et al., 2011; Kumar and Krishnamurti 2012; Saha et al., under review). Until 2005, the skill of monsoon prediction by dynamical models was too poor (Gadgil et al., 2005; Pattanaik and Kumar, 2010; Preethi et al., 2010). However, since then, there have been considerable improvements in the skill of the dynamical models (Delsole and Shukla, 2012; Krishnamurti and Kumar, 2012; Rajeevan et al., 2012). Delsole and Shukla (2012) demonstrated that the skill of predicting AISMR with coupled atmosphere–ocean models initialized in May is statistically significant and is much higher than empirical prediction methods. Although the current generation of coupled models are improving, still the prediction of Indian monsoon remains a challenging task. A point to note is that the coupled models have shown better skill in predicting large-scale variables compared to that of Indian monsoon rainfall. As the large-scale features are better predicted in Climate Forecast System version 1 (CFSv1), Pattanaik and Kumar (2010) and Pattanaik et al. (2012) used the large-scale variables, and their forecasts from CFSv1, as possible predictors for the hybrid model for downscaling of real-time CFSv1 forecast for the predictions of AISMR.

The present study evaluated the skill of AISMR forecast in the latest version of the National Centers for Environmental Prediction (NCEP)'s operational coupled model (Saha *et al.*, 2014) known as the Climate Forecast System version 2 (CFSv2) and compared it with the corresponding forecast skill in the earlier (CFSv1) version of the model (Saha *et al.*, 2006). The question we address is whether there was significant improvement in the forecast skill of AISMR in CFSv2 compared to CFSv1? Can we use the current-generation-coupled model for the real-time forecast of AISMR? How the skill of the coupled model for the forecast of AISMR can be further improved by using the hybrid concept (dynamical–empirical model) developed based on the forecast variables from coupled models having significant relationship with AISMR? The reason for exploring hybrid forecast of Indian monsoon rainfall is that rainfall being a localized process is much harder to predict and large-scale circulation can be relatively easier to predict. There have been some attempts earlier to statistically downscale monsoon rainfall from the model forecasts. Zhu *et al.* (2008) used the variability of the 500-hPa geopotential height (GPH) for the downscaling of summer monsoon precipitation anomaly. They used 500-hPa GPH because it is predicted well by the models in contrast to precipitation anomalies. They also found that the downscaled precipitation has higher forecast skill in the South China Sea, western North Pacific and the East Asia Pacific regions, where the anomaly correlation coefficient (ACC) has been improved by 0.14, corresponding to the conventional multi-model ensemble (MME) forecast. In the present study, an attempt was made to develop a new hybrid model for the real-time forecast of AISMR based on the CFSv2 hindcasts period of 28 years (1982 to 2009) and was tested for the real-time forecast of AISMR during 2013.

2. Model hindcast data and observed data

2.1. Model hindcast

As stated earlier, the NCEP coupled model has been upgraded in many respects from version 1 (CFSv1) to version 2 (CFSv2). The details of the modeling components of both the versions of NCEP CFS models are given in the following sections.

2.1.1. NCEP's Climate Forecast System version 1 (CFSv1)

The atmospheric component of the operational CFSv1 is the NCEP atmospheric GFS model. The oceanic component is the GFDL Modular Ocean Model version 3 (MOM3). The atmospheric component of the CFSv1 has spectral triangular truncation of 62 waves (T62) in the horizontal (equivalent to nearly a 200 km Gaussian grid) and a finite difference in the vertical with 64 sigma layers (Saha *et al.*, 2006). The atmospheric initial

conditions were from the NCEP/DOE Atmospheric Model Inter-comparison Project (AMIP) II Reanalysis (R2) data (Kanamitsu *et al.*, 2002), and the ocean initial conditions were from the NCEP Global Ocean Data Assimilation (GODAS) (Behringer and Xue, 2004), which was made operational at NCEP in September 2003. An extensive set of retrospective forecasts ('hindcasts') was generated and covered a 25-year period (1981–2005) in order to assess forecast performance. Each forecast run is full 9-month integration with 15 initial conditions that span each month. In the present analysis, the hindcast analysis obtained with 15 initial conditions of the months March, April and May are used for the skill analysis of CFSv1 forecast for the simulation of Indian monsoon rainfall during JJAS.

2.1.2. NCEP's Climate Forecast System version 2 (CFSv2)

The second version of the NCEP CFS (CFSv2) was made operational at NCEP in March 2011. The atmospheric model has a spectral triangular truncation of 126 waves (T126) in the horizontal (equivalent to nearly a 100 km grid resolution) and a finite difference in the vertical with 64 sigma-pressure hybrid layers (Saha *et al.* under review). In CFSv2, the ocean model changed from MOM version 3 to MOM version 4.0 (MOM4p0). The domain and resolution of MOM4p0 changed from a quasi-global domain (75°S–65°N) to a fully global domain. Increasing resolution from 1°×1° (1/3° within 10° of the equator) to 1/2° × 1/2° (1/4° within 10° of the equator). The vertical grid of 40 Z-levels with variable resolution (23 levels in the top 230 m) is retained.

One major component of CFSv2 was to have a coupled atmosphere-ocean-sea ice-land Reanalysis from 1979 to 2011 with the new system (resulting in the Climate Forecast System Reanalysis, CFSR) at NCEP for the purpose of creating initial conditions for CFSv2 retrospective forecasts for 28-year period (1982–2009). The vertical coordinate for the CFSR is the same as that in the operational CDAS.

The sea-ice model in CFSv2 is an interactive three layer (two layer of sea ice and one layer of snow) sea ice model with five categories of sea-ice thickness representing different types of sea ice. The NOAH land-surface model (Ek *et al.*, 2003) used in CFSv2 was first implemented in the GFS for operational medium-range weather forecast (Mitchell *et al.*, 2005) and then in the CFSR (Saha *et al.*, 2010). Within CFSv2, NOAH is employed in both the coupled land-atmosphere-ocean model to provide land-surface prediction of surface fluxes (surface boundary conditions), and in the Global Land Data Assimilation System (GLDAS) to provide the land-surface analysis and evolving land states. NOAH has four soil layers (10, 30, 60 and 100 cm) with frozen soil physics included. Above all, CFSv2 was designed to improve consistency between the model states and the initial states produced by the data assimilation system. Thus, the land-surface model in CFSv2 is a new modeling package. In the present analysis, the CFSv2 hindcast

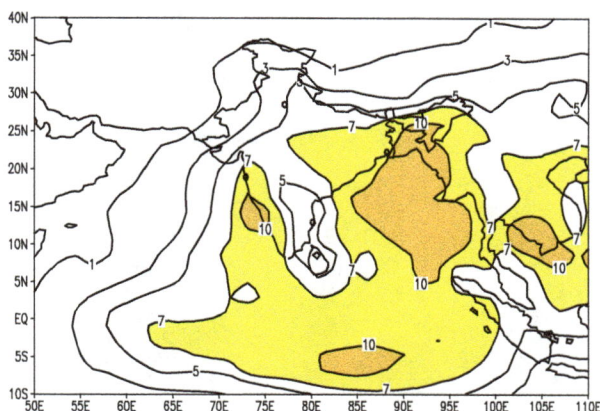

Figure 1. Spatial climatological rainfall (mm/day) for 25-year period from 1981 to 2005 valid for JJAS from verification analysis (Xie–Arkin). Rainfall with more than 7 mm/day is shaded.

analysis obtained with 24 initial conditions of the months March, April and May are used during the 28 years (1982–2009) for the skill analysis of forecast of AISMR in CFSv2 during JJAS.

2.2. Observed data used for verification

The rainfall analysis for verification is obtained from the global monthly precipitation using gauge observations, satellite estimates and numerical model outputs (Xie and Arkin, 1996). The other observed variables such as the wind vectors, GPH, etc. used for the development of hybrid model are obtained from the NCEP reanalysis (Kalnay et al., 1996).

3. Forecast skill of AISMR in CFSv2 and CFSv1

3.1. Simulation of mean monsoon rainfall

The interannual variability of monsoon in a coupled model depends on how correctly it simulates the teleconnection patterns, which also depends on the realistic simulation of mean features (Shukla, 1984; Fennessy et al, 1994). Xie and Arkin (1996) had made an attempt to merge and reproduce rainfall distributions over land and ocean with the use of surface observations, satellite data, buoys data and outputs from numerical models (GCM). This has been used for the comparison with the forecast rainfall from CFSv2 and CFSv1. The observed climatology of JJAS rainfall is shown in Figure 1. The observed climatology shows two rainfall maxima, one over the west coast region and the other over the head Bay of Bengal. Along with these two maxima the observed climatology has a zone of less rainfall over the northwestern parts of the country and the rain shadow region of Tamil Nadu situated on the southeastern coastal state of India (Figure 1). The corresponding seasonal mean monsoon forecast rainfall during JJAS along with the biases of models with different lead times from CFSv2 and CFSv1 during the respective

hindcast periods is shown in Figures 2 and 3, respectively. As seen from Figures 2(a–c) and 3(a)–(c) both versions of CFS are able to simulate rainfall maxima over western coast and North Bay of Bengal as seen in the verification climatology in Figure 1; however, there are large dry biases of rainfall over most parts of main land of India (Figures 2(d)–(f) and 3(d)–(f)). CFSv2 is able to simulate the west coast maximum. However, in CFSv1, it is extended much to the west into the Arabian Sea, leading to large positive bias of rainfall over the Arabian Sea. Overall, the distribution of seasonal rainfall over the Indian monsoon region is better simulated in CFSv2 compared to CFSv1. An another study by Saha et al. (2013a), using the 30 years of free runs from both versions of model, has also shown that CFSv2 simulates the right location of equatorial rainfall maxima compared to that of CFSv1. However, the fundamental problem of dry bias over Indian land mass still persists and it is further enhanced in CFSv2 as indicated by lower mean rainfall over Indian land mass in CFSv2 compared to that of CFSv1 with all lead times (Table 1). The coefficient of variability (CV) of forecast AISMR in CFSv2 is very close to the corresponding observed CVs and is comparatively higher than the corresponding CVs in CFSv1 (Table 1). The study by Saha et al. (2013b) using a free run of CFSv2 model also found dry bias over Indian land mass. Their study attributed this dry bias over Indian land mass to European snow bias. It is also observed that both versions of CFS indicated wet biases of rainfall over Indian Ocean with positive bias over the Arabian Sea in CFSv1 (Figure 3(d)–(f)) and wet bias mainly over equatorial Indian Ocean in CFSv2 (Figure 2(d)–(f)). With respect to the positive bias over the Arabian Sea in CFSv1, Pattanaik and Kumar (2010) have shown that the excessive precipitation over the Arabian Sea region is consistent with the bias in the lower level circulation patterns with the presence of anomalous circulation over the Arabian Sea. Although, with respect to the CFSv2, the large positive bias over the Arabian Sea is not seen, the positive bias is observed over the equatorial Indian ocean (Figure 2(d)–(f)). Pokhrel et al. (2012) have suggested the positive bias in CFSv2 over the equatorial Indian Ocean is due to the local evaporation.

3.2. Simulation of interannual variability of monsoon rainfall

In order to see the year-to-year departure of AISMR rainfall in both versions of the model, the forecast AISMR departures in CFSv2 and CFSv1 along with observed departure of AISMR is shown in Figure 4(a) and (b) respectively, where the anomaly is calculated by subtracting the respective hindcast climatologies. The CFSv2 forecast captured the year-to-year variation of AISMR better compared to that for CFSv1 with all lead times. The correlation coefficient (CC) between the forecast and observed rainfall departure in case of CFSv1 is found to be significant at 95% level with only April initial condition, whereas, in CFSv2, the

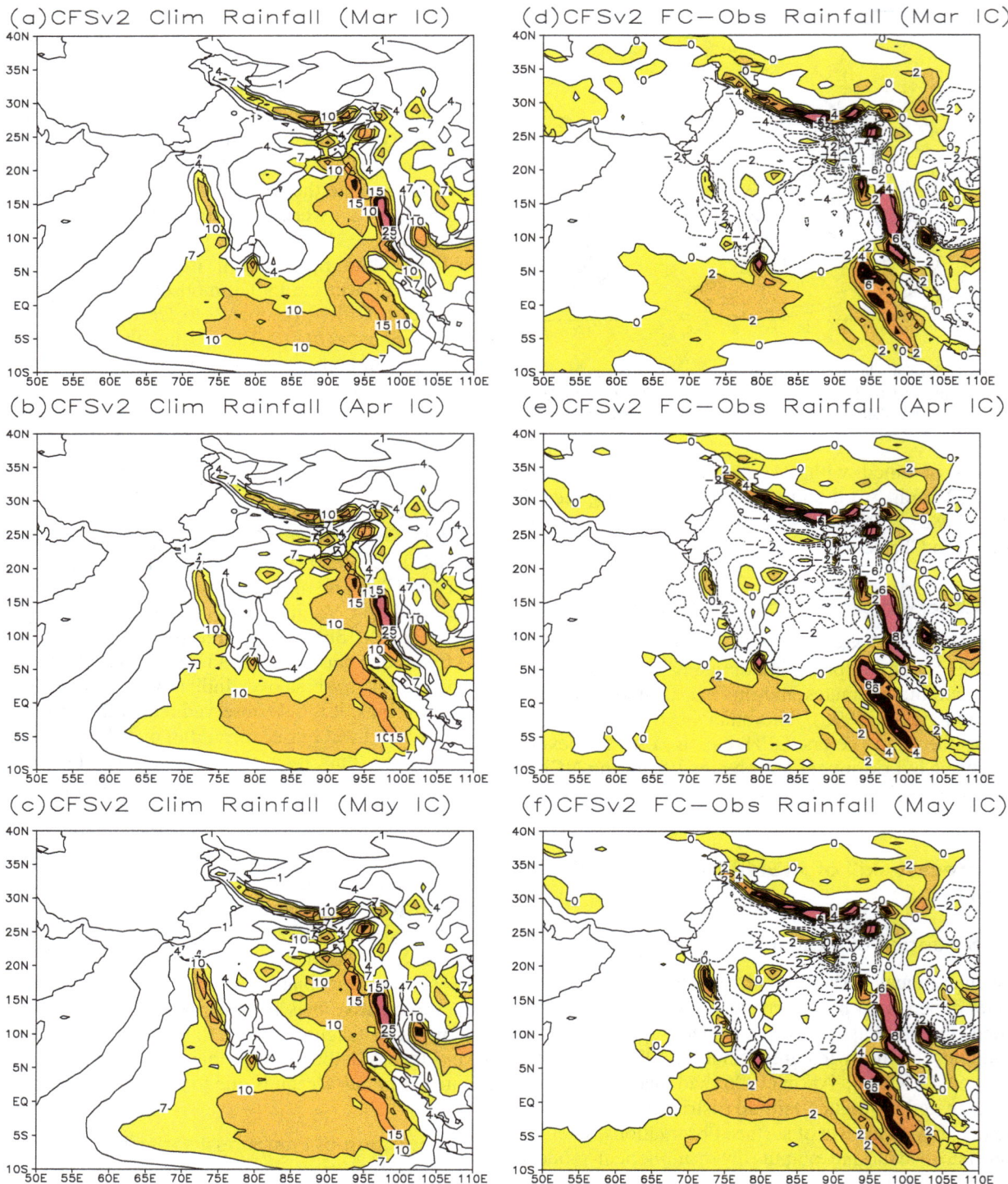

Figure 2. Spatial climatological rainfall (mm/day) obtained from CFSv2 for 28-year period from 1982 to 2009 valid for JJAS with more than 7 mm/day is shaded. (a) March ensembles (b) April ensemble and (c) May ensemble. The corresponding difference between the forecast and observed climatology is shown in (d)–(f) with positive values are shaded.

CC is found to be significant at 95% level with March initial condition and significant at 99% level with April and May initial conditions. It is further seen from Figure 4(a) that in many years the CFSv2 forecast rainfall departure matches with the corresponding observed rainfall departure with recent three drought years (2002, 2004 and 2009) captured well, whereas, in CFSv1 the

deficient rainfall of 2002 and 2004 during the hindcast periods was not captured. The recent drought year of 2009 was also not captured in CFSv1 (Janakiraman et al., 2011; Pattanaik et al., 2012). Thus, it is found that, spatial pattern of seasonal mean rainfall and the year-to-year variability of Indian monsoon rainfall is well simulated and is more realistic in CFSv2 compared

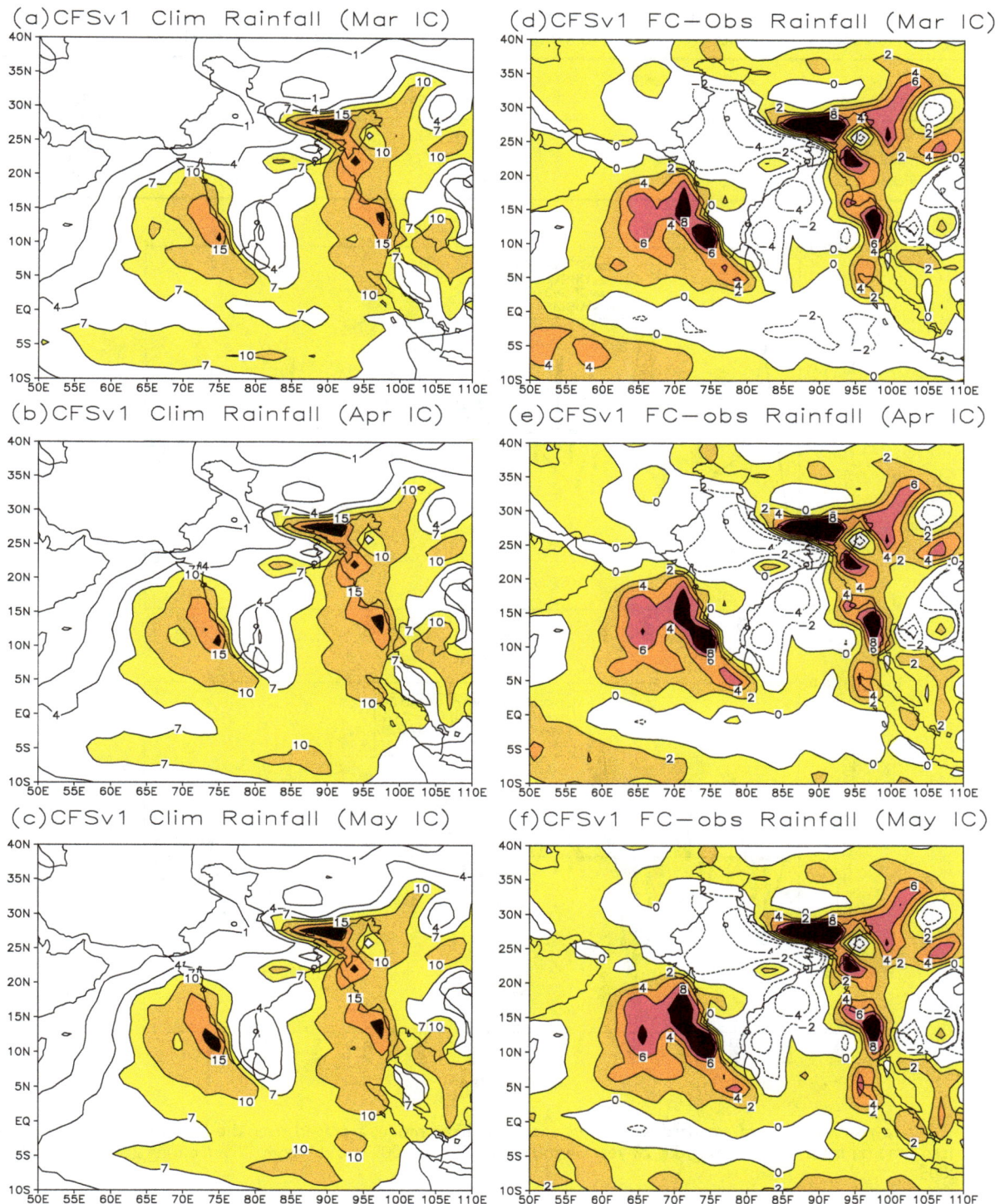

Figure 3. Spatial climatological rainfall (mm/day) obtained from CFSv1 for 25-year period from 1981 to 2005 valid for JJAS with more than 7 mm/day is shaded. (a) March ensembles (b) April ensemble and (c) May ensemble. The corresponding difference between the forecast and observed climatology is shown in (d)–(f) with positive values are shaded.

to CFSv1. Thus, there is an overall increase in the skill of seasonal monsoon forecast in CFSv2. The improvement in prediction of CFSv2 is a result of increase in resolution of atmospheric and ocean models, improvement in assimilation and also improvement in the modeling components (Saha *et al.*, 2014). Above all, CFSv2 was designed to improve consistency between

the model states and the initial states produced by the data assimilation system. Further, the atmospheric model used in the CFSv2 is the NCEP operational system that is 7 years later compared to the atmospheric model in CFSv1. During that time numerous model improvements have been incorporated; CFSv2 has a higher horizontal resolution; initial atmospheric

Table I. The mean, coefficient of variability (CV) between IMD rainfall climatology and CFSv1 and CFSv2 model hindcast climatology over the Indian land points valid for June to September (JJAS) with initial conditions of March, April and May.

JAS Indian land Rainfall	Obs. AISMR (IMD) Rainfall	March ICs CFSv1 (CFSv2)	April ICs CFSv1 (CFSv2)	May ICs CFSv1 (CFSv2)
JJAS (mean) rainfall (mm)	7.21	5.21 (3.98)	5.33 (4.49)	5.38 (4.97)
JJAS (CV) rainfall (%)	≈10.0 %	5.13% (11.83%)	5.68% (9.53%)	5.20% (7.97%)
Correlation coefficient (CC)	–	0.24 (0.40)	0.44 (0.48)	0.30 (0.47)

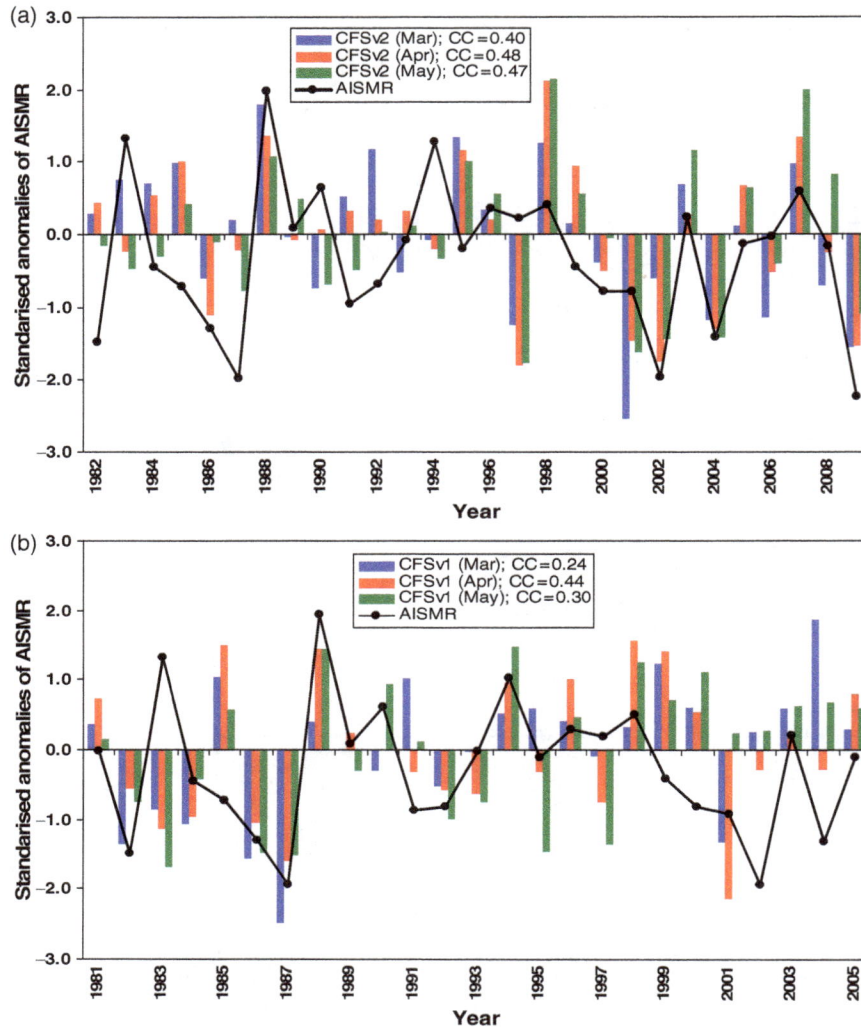

Figure 4. Year-to-year variation of standardized AISMR anomalies from model averaged over the Indian land region only with different lead time (a) from CFSv2 during the 28-year period from 1982 to 2009 and (b) from CFSv1 during 25-year period from 1981 to 2005.

conditions for CFSv2 are from the CFSR while the initial conditions for CFSv1 were from NCEP/DOE Reanalysis 2. The CFSR has a much higher resolution and the data assimilation system incorporates many advances.

4. A new hybrid (dynamical–empirical) model for AISMR prediction

Although, there are improvements in CFSv2 as compared to CFSv1, the highest CC between the observed

and forecast AISMR in CFSv2 during the hindcast period is still 0.48 for the initial condition of April (Figure 4(a)). In order to use this model for real-time operational forecast there is a need to further improve the CC. Pattanaik and Kumar (2010) and Pattanaik *et al.* (2012) found that the skill of El Niño prediction in CFSv1 is much higher than the skill of AISMR prediction. However, as demonstrated by Pattanaik and Kumar (2010) using the Niño3.4 forecast SST from CFSv1 as predictor of AISMR in a regression model, CC was improved to 0.47 between the observed and forecast AISMR during the hindcast period of

25 years from 1981 to 2005, which does not indicate significant increase in CC from its raw value shown in Figure 4(b).

Further, a recent study has shown that the teleconnections of El Niño Southern Oscillation (ENSO) and Indian summer monsoon rainfall in terms of Niño3 SST and monsoon rainfall correlation are more realistic in the latest version of the model, i.e. CFSv2 (Saha et al., 2013a). Therefore, the forecast skill of AISMR related to ENSO is expected to be better in CFSv2. The forecast SSTs from CFSv2 valid for JJAS with March, April and May ensembles are correlated with the observed AISMR during the training period of first 24 years (1982–2005) of whole hindcast period of 28 years from 1982 to 2009. The CC maps are shown in Figure 5(a)–(c). As seen from Figure 5(a)–(c) the anti-correlation between AISMR and SSTs over the eastern and central Pacific is seen in the CFSv2 model with CC of about 0.4 with March and April initial conditions (Figure 5(a) and (b)). The relationship is weaken for the May initial conditions (Figure 5(c)). Although the recent study by Saha et al. (2013a) has found more realistic teleconnection patterns between El Niño and Indian summer monsoon rainfall in CFSv2, when the Niño3.4 forecast SST from CFSv2 is used as the predictor in regression model, the CCs between the observed and predicted AISMR with March, April and May ensembles are found to be 0.47, 0.48 and 0.40 during the whole hindcast period of 28 years (1982–2009). Hence, like with the CFSv1 the raw skill of CFSv2 shown in Figure 4(a) was not improved by using forecast Niño3.4 SST from CFSv2 model as the predictor and in fact it reduces in the month of May. Thus, other teleconnection parameters with Indian monsoon rainfall need to be examined to develop a new hybrid model like it was done for CFSv1 by Pattanaik and Kumar (2010). They had developed a hybrid model for the real-time forecast of AISMR by taking many other forecast variables from CFSv1 like SSTs, zonal wind, rainfall, etc., which had much higher forecast skill compared to the raw forecast (Pattanaik and Kumar, 2010; Pattanaik et al., 2012).

Many recent studies (Kripalani and Kulkarni, 1997; Krishna Kumar et al., 1999) have identified weakening in ENSO–monsoon relationship. Thus, in spite of having a better simulation of ENSO in CFSv2, the better forecast skill of ENSO in CFSv2 has not translated into a better forecast skill of AISMR. Some studies (Clark et al., 2000; Sahai et al., 2003) have found the other oceanic regions (other than the region of ENSO) to have a more stable relationship with AISMR. Thus, to find out the variability of AISMR in CFSv2, the other forecast variables from CFSv2 based on March ensembles and valid for JJAS are correlated with simultaneous AISMR over the entire globe.

Recent studies by Mujumdar et al. (2007) and Pattanaik and Rajeevan (2007) have indicated robust features of northwest (NW) Pacific circulation and rainfall over the Indian monsoon region. The NW Pacific anticyclone during JJAS induces positive precipitation anomalies over most of Indian monsoon region,

whereas, NW Pacific cyclone during JJAS induces negative precipitation anomalies. This indicates a strong feedback between NW Pacific circulation and convection over the Indian monsoon region. Chowdary et al. (2009) have demonstrated better skill of coupled models in capturing major modes of atmospheric variability over the NW Pacific during the monsoon season. In order to use the higher predictive skill of NW Pacific circulation features, the composite GPH anomalies at 850 hPa during deficient (1982, 1987, 2002, 2004 and 2009) and excess (1983, 1988 and 1994) AISMR years are prepared from the March ensembles of CFSv2 hindcasts of 28 years (Figure 6(a) and (b)). The excess (deficient) years are identified based on the departure of AISMR ≥ 1 standard deviation (≤ 1 standard deviation), where 1 standard deviation of AISMR is $\approx 10\%$ of the mean. It is seen from Figure 6(a) and (b) that there are very distinct patterns of geopotential anomalies over NW Pacific and Indian regions with CFSv2 model able to capture the weakening (strengthening) of anticyclone over NW Pacific during the deficient (excess) AISMR years.

For using the GPH variable over NW Pacific in the hybrid model for skillful prediction of AISMR, the CCs map between forecast 850-hPa GPH from CFSv2 of March ensembles valid for JJAS with the simultaneous observed AISMR during the training period of 24 years (1982–2005) is shown in Figure 6(c). Based on the region of significant CC (Figure 6(c)), a linear regression model is developed between GPH averaged over 157.5–172.5°E; 30–40°N, known as Z850-index with observed AISMR. The CC between the observed and predicted AISMR based on the regression model using Z-850 index is found to be 0.60 during the training period (Table 2 column 'b'). Thus, when the forecast GPH over NW Pacific is used the skill of hybrid model for the forecast of AISMR increased. Zhu et al. (2008), using 500-hPa model predicted GPH, found that the downscaled precipitation has higher forecast skill in the South China Sea, western North Pacific and the East Asia Pacific regions, where the ACC has been improved, corresponding to the conventional MME forecast.

In addition to the GPH over the NW Pacific, relationships for many other forecast variables, such as zonal wind at 850 and 200 hPa, GPHs at 200 and 500 hPa, rainfall, vertical wind shear (U200–U850) etc. were investigated to identify suitable CFSv2 model forecast variables for preparing downscaled AISMR forecast. Two additional variables, i.e.(1) the forecast rainfall averaged over 70–50°W and 50–40°S, known as R-index and (2) 850 hPa zonal wind averaged over 65–50°W and 67–71°N, known as U850-index are also found to have significant CC with the observed AISMR. These two additional indices are used separately to generate the hybrid forecast AISMR as given in Table 2. The physical relationship to explain these high correlation centers in CFSv2 is not very clear at present, but when these two indices are used in the linear regression model for the prediction of AISMR

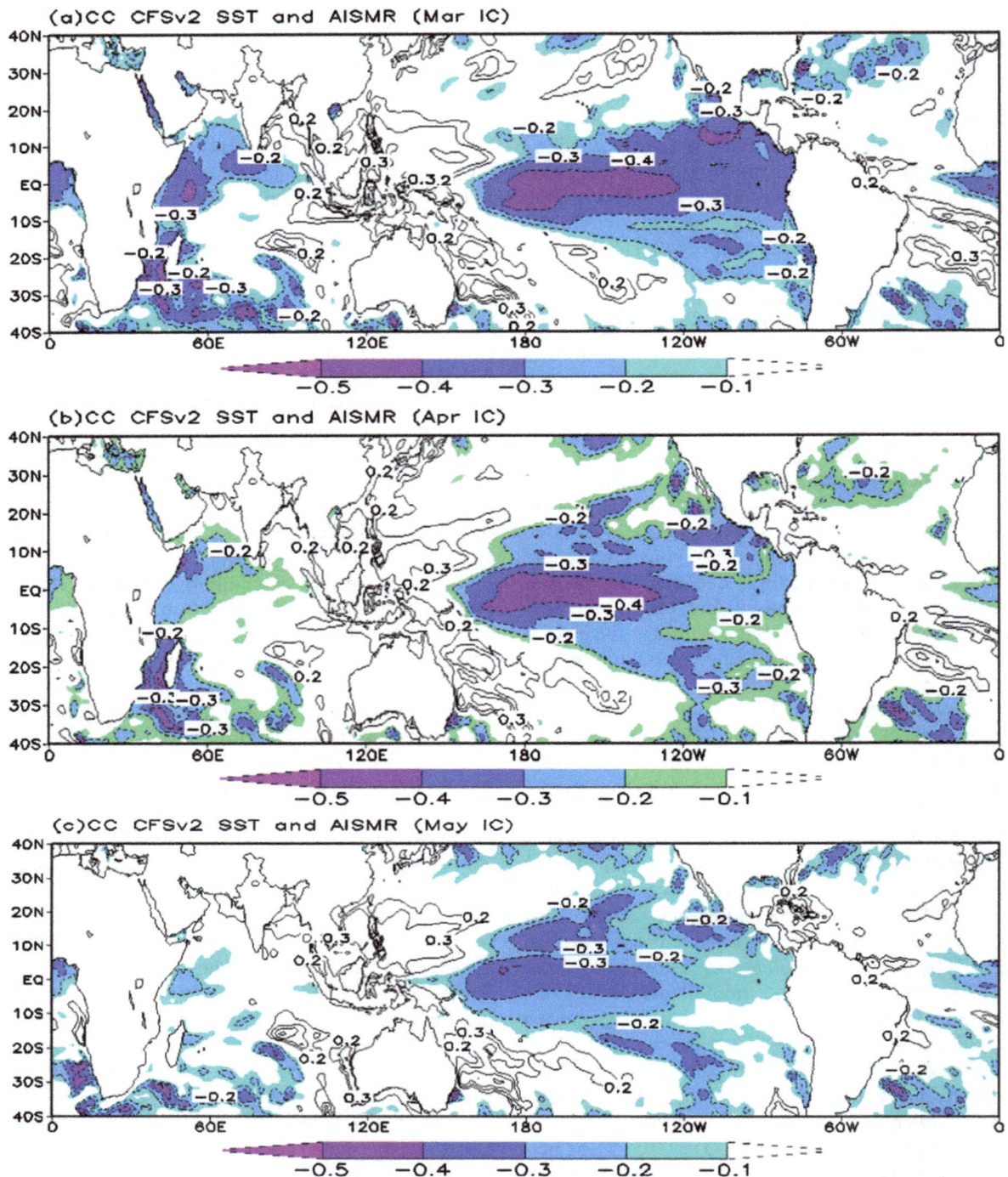

Figure 5. Correlation maps between the observed all India summer monsoon rainfall (AISMR) over India and the corresponding CFSv2 forecast SST during JJAS at each grid point based on (a) March, (b) April and (c) May initial conditions during the training period from 1982 to 2005.

the predicted AISMR is found to be having CCs of 0.60 and 0.64 respectively with the observed AISMR during the whole period of 28 years (1982–2009) as given in Table 2. Such indices were also used by Pattanaik and Kumar (2010) and Pattanaik *et al.* (2012) for downscaling of AISMR forecast using CFSv1 model forecast variables. The mean AISMR hybrid forecast is calculated (Equation 1) by taking the average of AISMR forecasts based on the three linear regression

models using the three indices as given in Table 2.

$$AISMR_{mean} = \left(AISMR_{z850\text{-index}} + AISMR_{R\text{-index}} \right.$$
$$\left. + AISMR_{u850\text{-index}} \right) / 3 \qquad (1)$$

The final forecast (AISMR$_{final}$) as shown in Figure 7 is the normalized variance inflated/deflated average forecast (AISMR$_{mean}$), where the normalization of variance is as per the method used by Sahai and Satyan (2000).

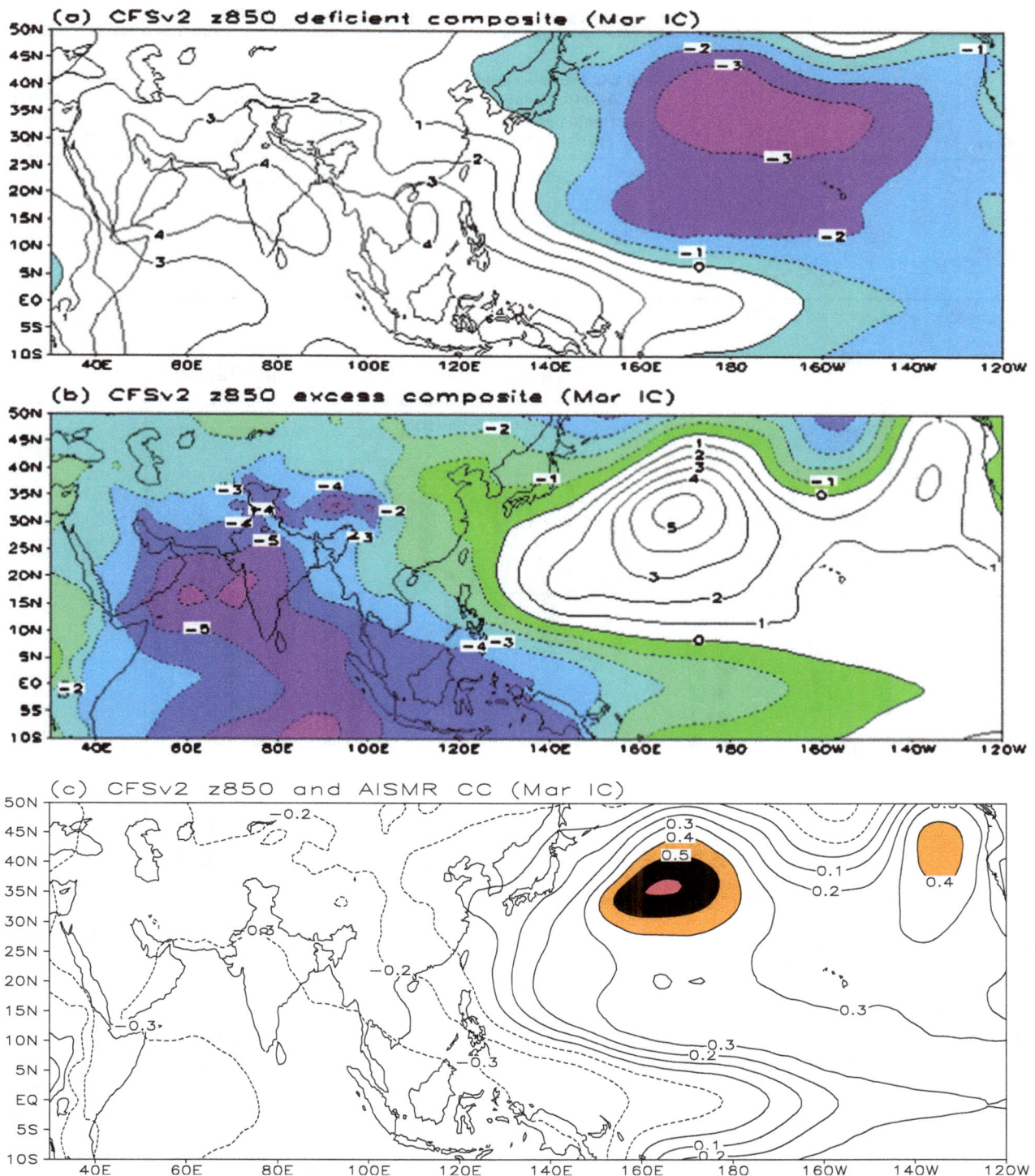

Figure 6. Geopotential height composite anomalies at 850 hPa from CFSv2 hindcasts valid for JJAS. (a) Deficient AISMR years, (b) excess AISMR year, (c) the correlation coefficient map between the observed AISMR and CFSv2 forecast 850 hPa GPH during JJAS at each grid point based on March ensembles during the training period from 1982 to 2005.

Although, the attempt was made to develop similar hybrid model with April and May ensembles of CFSv2, the forecast skill of the hybrid model with the first index (Z850-index) completely vanishes with April and May ensembles and the final forecast with March ensembles was found to be the best. The final forecast shows CC of 0.69 (significant at 99.9%) with the observed AISMR

and is found to be much higher than the raw skill of CFSv2 shown in Figure 4(a). It is also observed that during most of the years the predicted and actual AISMR are in phase even during the test period of 2006–2009. Other reasons for having the only March ensembles for the hybrid forecast was to provide the timely input for the operational forecast of AISMR, which is issued by

Table 2. The three indices identified along with the regression equations for the prediction of AISMR.

Indices	(a) Region	(b) CC with observed AISMR during training period (1982–2004)	(c) Linear regression equation for AISMR departure	(d) CC between observed and predicted AISMR for whole period (1982–2009)
Z850-index	157.5–172.5°E; 30–40°N	0.60	$AISMR_{Z850\text{-}index} = 1.47X - 2231.04$	0.56[99]
R-index	70–50°W; 50–40°S	−0.68	$AISMR_{R\text{-}index} = -72.54X + 120.48$	0.60[99]
U850-index	65–50°W; 67–71°N	0.46	$AISMR_{U850\text{-}index} = 70.13X - 21.71$	0.64[99.9]

The superscript in column 'd' is the significance level.

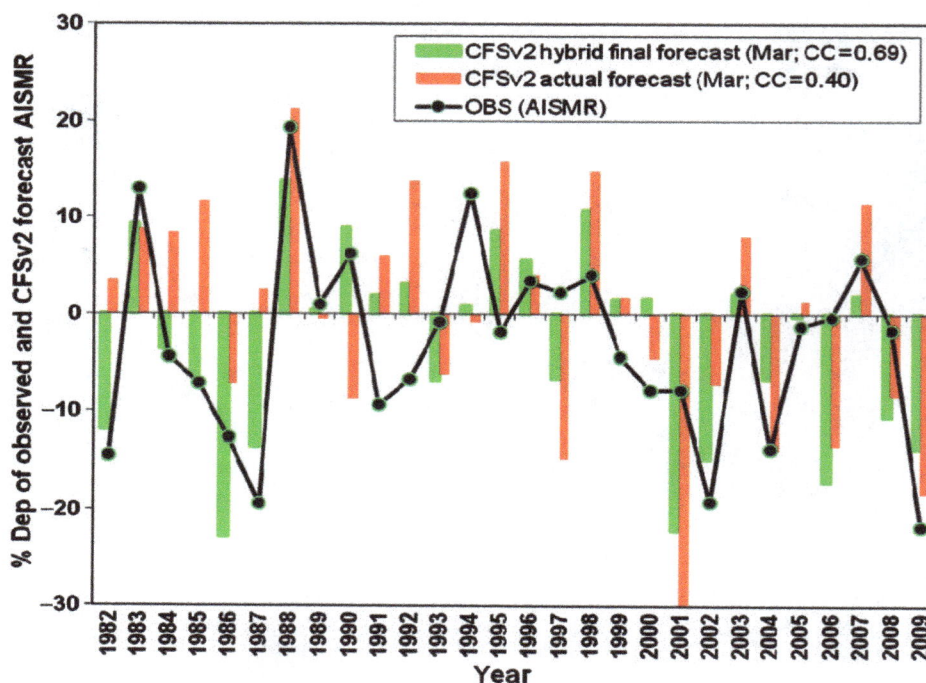

Figure 7. The variance inflated/deflated final forecast of AISMR based on the hybrid model out of CFSv2 forecast variables of March ensembles along with actual CFsv2 forecast and observed AISMR during the whole period of 28 years (1982–2009).

IMD during the middle of April every year and the skill of raw CFSv2 forecast with March ensemble was lowest compared to April and May ensembles (Figure 4(a)). Thus, when the hybrid model is used there could be a correction not only to the sign of the AISMR departure from the raw CFSv2 forecast but also to its magnitude.

The final forecast for AISMR 2013 based on the new hybrid model of CFSv2 of March ensemble was found to be 105.4% of LPA, which was very close to the observed AISMR of 106% of LPA (Pai and Bhan 2014). Thus, this study demonstrated that the new hybrid model forecast based on latest generation coupled model (CFSv2) could become a useful tool for a skillful prediction of AISMR in the real time.

5. Conclusions

It was found that the spatial pattern of JJAS mean rainfall was more realistic in CFSv2 as compared with

CFSv1. There was also improvement in the forecast skill of interannual variability of AISMR in CFSv2 with CC between the observed and forecast AISMR found to be 0.40 with March ensemble (significant at 95% level) and 0.48 and 0.47 with April and May ensembles, respectively (significant at 99% level). To further improve the forecast skill of AISMR in CFSv2, a new hybrid model was developed on the basis of the teleconnections of Indian monsoon rainfall with forecast variables of CFSv2. Based on the regions of significant CCs between the observed AISMR and the corresponding simultaneous forecast variables (forecast GPH and forecast zonal winds at 850 hPa; forecast rainfall) from CFSv2 three linear regression models were developed and the final forecast was the variance inflated/deflated mean forecast. The skill of final forecast of AISMR based on the hybrid model of March ensembles of CFSv2 demonstrated highly significant CCs (99.9% or more) and was found to be much higher than the raw skill of CFSv2. The AISMR forecast for 2013 monsoon was found to be very close to the observed value of AISMR. Thus, this study demonstrated that the new hybrid model forecast based on latest generation coupled model (CFSv2) could become a useful tool for a skillful prediction of AISMR in the real time.

Acknowledgements

We sincerely acknowledge NCEP, USA, for making available the real-time forecast for 2013 monsoon along with the hindcast data. The first author is thankful to the Director General, India Meteorological Department (IMD), Additional Director General of Meteorology, IMD Pune and Deputy Director General of Meteorology, NWP division, IMD, for providing all facility to carry out this work in IMD. The authors are also thankful to the anonymous reviewer for providing valuable comments and suggestions for the improvement of this paper.

References

Behringer DW, Xue Y. 2004. Evaluation of the global ocean data assimilation system at NCEP: The Pacific Ocean. In Eighth Symposium on Integrated Observing and Assimilation Systems for Atmosphere, Oceans, and Land Surface, AMS 84th Annual Meeting, Washington State Convention and Trade Center, Seattle, WA; 11–15.

Blanford HF. 1884. On the connection of the Himalayan snowfall with dry winds and seasons of droughts in India. *Proceedings of the Royal Society of London* 37: 3–22.

Chen TC, Yen MC. 1994. Interannual variation of the Indian monsoon simulated by the NCAR Community Climate Model: effect of the tropical Pacific SST. *Journal of Climate* 7: 1403–1415.

Chowdary JS, Xie SP, Luo JJ, Jan Hafner SB, Masumoto Y, Yamagata T. 2009. Predictability of Northwest Pacific climate during summer and the role of the tropical Indian Ocean. *Climate Dynamics* 36: 607–621.

Clark CO, Cole JE, Webster PJ. 2000. Indian Ocean SST and Indian summer rainfall: predictive relationships and their decadal variability. *Journal of Climate* 13: 2503–2519.

DelSole T, Shukla J. 2012. Climate models produce skillful predictions of Indian summer monsoon rainfall. *Geophysical Research Letters* 39: L09703, DOI: 10.1029/2012GL051279.

Ek M, Mitchell KE, Lin Y, Rogers E, Grunmann P, Koren V, Gayno G, Tarpley JD. 2003. Implementation of Noah land-surface model advances in the NCEP operational mesoscale Eta model. *Journal of Geophysical Research* 108(D22): 8851, DOI: 10.1029/2002 JD003296.

Fennessy MJ, Kinter JL III, Kirtman B, Marx L, Nigam S, Schneider E, Shukla J, Straus D, Vernekar A, Xue X, Zhou J. 1994. The simulated Indian monsoon: a GCM sensitivity study. *Journal of Climate* 7: 33–43.

Gadgil S, Rajeevan M, Nanjundiah R. 2005. Monsoon prediction - why yet another failure? *Current Science* 88: 1389–1400.

Gowariker V, Thapliyal V, Kulshrestha SM, Mandal GS, Sen Roy N, Sikka DR. 1991. A power regression model for long range forecast of southwest monsoon rainfall over India. *Mausam* 42: 125–130.

Janakiraman S, Mohit Ved RN, Laveti PY, Gadgil S. 2011. Prediction of the Indian summer monsoon rainfall using a state-of-the art coupled ocean–atmosphere model. *Current Science* 100: 354–362.

Kalnay E, Kanamitsu M, Kistler R, Collins W, Deaven D, Gandin L, Iredell M, Saha S, White G, Woollen J, Zhu Y, Chelliah M, Ebisuzaki W, Higgins W, Janowiak J, Mo KC, Ropelewski C, Wang J, Leetmaa A, Reynolds R, Jenne Roy, Joseph Dennis. 1996. The NCEP/NCAR 40 year re-analysis. *Bulletin of the American Meteorological Society* 77: 437–471.

Kanamitsu M, Ebisuzaki W, Woollen J, Yang S-K, Hnilo JJ, Fiorino M, Potter GL. 2002. NCEP–DOE AMIP-II Reanalysis (R-2). *Bulletin of the American Meteorological Society* 83: 1631–1643, DOI: 10.1175/BAMS-83-11-1631.

Kripalani RH, Kulkarni A. 1997. Climate impact of El Nino La-Nina on the Indian monsoon: a new prespective. *Weather* 52: 39–46.

Krishna Kumar K, Rajagopalan B, Cane MA. 1999. On the weakening relationship between the Indian monsoon and ENSO. *Science* 284: 2156–2159.

Krishnamurti TN, Kumar V. 2012. Improved seasonal precipitation forecasts for the Asian monsoon using 16 atmosphere–ocean coupled models. Part II: Anomaly. *Journal of Climate* 25: 65–88.

Krishnan R, Sundaram S, Swapna P, Vijay K, Ayantika DC, Mujumdar M. 2011. Crucial role of ocean–atmosphere coupling on the Indian monsoon anomalous response during dipole events. *Climate Dynamics* 37: 1–17, DOI: 10.1007/s00382-010-0830-2.

Kumar V, Krishnamurti TN. 2012. Improved seasonal precipitation forecasts for the Asian monsoon using 16 atmosphere–ocean coupled models. Part I: Climatology. *Journal of Climate* 25: 39–64.

Mamoria BC. 1978. *Geography of India*. Shivalal Agarwala & Company: Agra, India; 123–125.

Mitchell KE, Wei H, Lu S, Gayno G, Meng J. 2005. NCEP implements major upgrade to its medium range global forecast system, including land surface component. GEWEX newsletter, May 2005.

MoAg. 2009. *Agricultural Statistics at a Glance*. Ministry of Agriculture: New Delhi.

Mujumdar M, Kumar V, Krishnan R. 2007. The Indian summer monsoon drought of 2002 and its linkage with tropical convective activity over northwest Pacific, 2006. *Climate Dynamics* 28: 743–758, DOI: 10.1007/s0038Z-006-0208-7.

Pai DS, Bhan SC. 2014. Monsoon 2013 – a report. India Meteorological Department Met. Monograph. Synoptic Meteorology No. ESSO/IMD/SYNOPTIC MET/01-2014/15; 1–222.

Palmer TN, Brankowic C, Viterbo P, Miller MJ. 1992. Modelling interannual variations of summer monsoons. *Journal of Climate* 5: 399–417.

Pattanaik DR, Kumar A. 2010. Prediction of summer monsoon rainfall over India using the NCEP climate forecast system. *Climate Dynamics* 34: 557–572.

Pattanaik DR, Rajeevan M. 2007. North-West Pacific tropical cyclone activity and July rainfall over India. *Meteorology and Atmospheric Physics* 95: 63–72.

Pattanaik DR, Kalsi SR, Hatwar HR. 2005. Evolution of convection anomalies over the Indo-Pacific region in relation to Indian monsoon rainfall. *Mausam* 56: 811–824.

Pattanaik DR, Tyagi A, Kumar A. 2012. Dynamical-empirical forecast for the Indian monsoon rainfall using the NCEP coupled modelling system – application for real time monsoon forecast. *Mausam* 63: 433–448.

Pokhrel S, Rahaman H, Parekh A, Saha SK, Dhakate A, Chaudhari HS, Gairola RM. 2012. Evaporationprecipitation variability over Indian Ocean and its assessment in NCEP Climate Forecast System (CFSv2). *Climate Dynamics* 39(9–10): 2585–2608, DOI: 10.1007/s00382-012-1542-6.

Preethi B, Kripalani RH, Krishna Kumar K. 2010. Indian summer monsoon rainfall variability in global coupled ocean-atmospheric models. *Climate Dynamics* 35: 1521–1539.

Rajeevan M, Pai DS, Dikshit SK, Kelkar RR. 2003. IMD's new operational models for long-range forecast of southwest monsoon rainfall over India and their verification for 2003. *Current Science* 86: 422–430.

Rajeevan M, Unnikrishnan CK, Preethi B. 2012. Evaluation of the ENSEMBLES multi-model seasonal forecasts of Indian summer monsoon variability. *Climate Dynamics* 38: 2257–2274.

Saha S, Moorthi S, Pan H-L, Wu X, Wang J, Nadiga S, Tripp P, Kistler R, Woollen J, Behringer D, Liu H, Stokes D, Grumbine R, Gayno G, Wang J, Hou Y-T, Chuang H-Y, Juang H-MH, Sela J, Iredell M, Treadon R, Kleist D, Van Delst P, Keyser D, Derber J, Ek M, Meng J, Wei H, Yang R, Lord S, Van Den Dool H, Kumar A, Wang W, Long C, Chelliah M, Xue Y, Huang B, Schemm J-K, Ebisuzaki W, Lin R, Xie P, Chen M, Zhou S, Higgins W, Zou C-Z, Liu Q, Chen Y, Han Y, Cucurull L, Reynolds RW, Rutledge G,

Goldberg M. 2010: The NCEP Climate Forecast System Reanalysis. *Bulletin of American Meteorological Society* **91**: 1015–1057, DOI: 10.1175/2010BAMS3001.1.

Saha S, Nadiga S, Thiaw C, Wang J, Wang W, Zhang Q, Van den Dool HM, Pan HL, Moorthi S, Behringer D, Stokes D, Pena M, Lord S, White G, Ebisuzki W, Peng P, Xie P. 2006. The NCEP climate forecast system. *Journal of Climate* **19**: 3483–3517.

Saha SK, Pokhrel S, Chaudhari HKS, Dhakate A, Shewale S, Sabeerali CT, Salunke K, Hazra A, Mahapatra S, Suryachandra Rao A. 2013a. Improved simulation of Indian summer monsoon in latest NCEP climate forecast system free run. *International Journal of Climatology* **34**: 1628–1641, DOI: 10.1002/joc.3791.

Saha SK, Pokhrel S, Chaudhari HS. 2013b. Influence of Eurasian snow on Indian summer monsoon in NCEP CFSv2 free run. *Climate Dynamics* **41**: 1801–1815, DOI: 10.1007/s0038201216174.

Saha S, Moorthi S, Wu X, Wang J, Nadiga S, Tripp P, Pan H-L, Behringer D, Hou Y-T, Chuang H-y, Iredell M, Ek M, Meng J, Yang R, van den Dool H, Zhang Q, Wang W, Chen M. 2014. The NCEP Climate Forecast System Version 2. *Journal of Climate* **27**: 2185–2208, DOI: 10.1175/JCLI-D-12-00823.1.

Sahai A. K. and V. Satyan. 2000. Reduction of AGCM systematic error by artificial neural network: a new approach for dynamical seasonal prediction of Indian summer monsoon rainfall. IITM Research Report No. RR088, ISSN 0252-1075, Pune, India.

Sahai AK, Grimm AM, Satyan V, Pant GB. 2003. Long-lead prediction of Indian summer monsoon rainfall from global SST evolutions. *Climate Dynamics* **20**: 855–863.

Shukla J. 1984. Predictability of a large atmospheric model. In *Predictability and Turbulence*, Holloway G, West BJ (eds). American Institute of Physics: College Park, MD; 449–456.

Shukla J, Mooley DA. 1987. Empirical prediction of the summer monsoon rainfall over India. *Monthly Weather Review* **115**: 695–703.

Shukla J, Anderson J, Baumhefner D, Brankovic C, Chang Y, Kalnay L, Marx L, Palmer TN, Paolino DA, Ploshay J, Schubert S, Straus DM, Suarez M, Tribbia J. 2000. Dynamical seasonal prediction. *Bulletin of the American Meteorological Society* **81**: 2593–2606.

Sperber KR, Palmer T. 1996. Interannual tropical rainfall variability in general circulation model simulations associated with the Atmospheric Model Inter-comparison Project. *Journal of Climate* **9**: 2727–2750.

Xie P, Arkin PA. 1996. Analyses of global monthly precipitation using gauge observations, satellite estimates, and numerical model predictions. *Journal of Climate* **9**: 840–858.

Zhu C, Park C-K, Lee W-S, Yun W-T. 2008. Statistical downscaling for multi-model ensemble prediction of summer monsoon rainfall in the Asia-Pacific region using geopotential height field. *Advances in Atmospheric Sciences* **25**: 867–884.

16

Improving the estimation of the true mean monthly and true mean annual air temperatures in Greece

N. K. Sakellariou and H. D. Kambezidis*

Atmospheric Research Team, Institute for Environmental Research & Sustainable Development, National Observatory of Athens, Greece

Correspondence to:
H. D. Kambezidis,
Atmospheric Research Team,
Institute for Environmental
Research & Sustainable
Development, National
Observatory of Athens,
GR-11810 Athens, Greece.
E-mail: *harry@noa.gr*

Abstract

The true mean monthly/annual air-temperature estimations are usually based on the monthly averages of meteorological observations performed at standard hours. The performance of this combination varies as it depends on local climate of the particular meteorological station; it consequently allows for small/large deviations from its exact value. This study examines the possible deviations in Greece and suggests ways of improving the derived estimates. Two cases are considered. (1) An area has one thermograph and a number of thermometric stations. (2) An area has several thermographs and many thermometric stations. Solutions for minimizing the estimation error in both cases are provided.

Keywords: temperature estimation; true average; climate normals; Greece

1. Introduction

At meteorological stations equipped with thermographs, continuous air-temperature recordings are available. From such observations, 24-h air-temperature values can be deduced. Their arithmetic mean is defined as the true mean temperature or the true daily mean (Conrad and Pollak, 1950; Weiss and Hays, 2005; Dall' Amico and Hornsteiner, 2006), while the monthly average of the true daily means yields the true monthly mean.

When only a few (usually three) daily air-temperature observations are available, a linear combination of the monthly average temperatures at the standard hours of the observations is used in order to obtain the closest possible approximation to the true monthly mean. These combinations depend on the climate of the station; Conrad and Pollak (1950) have listed the formulas that obtain best results for the greater part of Europe as well as for tropical and subtropical climates.

All the formulas can give an accurate answer, provided that there is no daily temperature variation. This implies that the formulas tend to express the average daily air-temperature variation of the stations that corresponds to the climate; they consequently fail when the daily air-temperature variation at a particular station strongly deviates from its assumed pattern. This failure may yield miscalculations concerning the heating demands of buildings (e.g. Kaufmann *et al.*, 2013) as well as predictions of crop production (e.g. Rosenzweig and Parry, 1994; Olesen and Bindi, 2002).

The formula that has been suggested by the aforementioned authors as very advantageous for the greater part

of Europe is:

$$\overline{T} = \frac{\left(\overline{T_7} + \overline{T_{14}} + 2\overline{T_{21}}\right)}{4} \quad (1)$$

where $\overline{T}$ is the approximate true average monthly air temperature and $\overline{T_7}$, $\overline{T_{14}}$, and $\overline{T_{21}}$ are the monthly mean air temperatures at the 0700, 1400, and 2100 of observation, respectively. The hours refer to mean local time, UTC+2.

A modified version of this formula, with $\overline{T_8}$ replacing $\overline{T_7}$, has been used by Aeginitis (1907) and Mariolopoulos (1938) for the estimation of the true mean monthly temperature in Greece. The validity of the modified formula has been tested with the help of the thermographic recordings taken by a Richard-type thermograph at the meteorological station of the National Observatory of Athens (hereafter NOA) from 1894 to 1903 (Aeginitis, 1907). The error of the approximation was found to be small at NOA (Aeginitis, 1907) and of the order of $0.1\,^{\circ}\mathrm{C}$ in most cases (Mariolopoulos, 1938). The formula:

$$\overline{T} = \frac{\left(\overline{T_8} + \overline{T_{14}} + 2\overline{T_{21}}\right)}{4} \quad (2)$$

was used all over Greece for the estimation of the true monthly air temperature under the implied assumption that it performs equally well all over the country.

The latter constitutes a rather optimistic point of view, considering the climatic diversification of Greece. The aim of this study is, therefore, to investigate the possible errors that may arise as a result of the discriminatory application of this formula and to suggest ways of improving the true mean monthly temperature estimates.

It should be noted that Equation (2) is still in use by the Hellenic National Meteorological Service (HNMS) in a slightly different form: the observations taken at 2100 local time have been replaced by those at 2000 from 1930 onwards.

Two different cases are considered here for this problem. First, when only one thermograph operates in a region but thermometric observations are available at a number of locations. This case resembles the situation in Greece from 1894 to 1930. Second, thermographic observations are available at a limited number of sites and thermometric ones exist over a denser network of meteorological stations. This resembles the situation in Greece from 1950 onwards. Solutions to the problem are suggested in this study for both cases and the results are inter-compared. The objective of this study is to provide corrections to the pre-1930 temperature measurements when thermometric stations mainly existed.

2. Case 1. True mean air-temperature estimations with one thermograph

It should be mentioned here that a typical temperature reading difference between a thermometer and a thermograph may be around $0.5-1\,^{\circ}$C (Srivastava, 2008). At NOA's meteorological station a 2-year analysis of the temperature difference between the dry-bulb thermometer and the bimetallic thermograph has shown this to be at $0.5\,^{\circ}$C on average, with the thermograph giving lower temperature values.

In order to improve the air-temperature estimates deduced by Equation (2), it is required that the following are secured:

1 The average hourly and daily values for every month and for a number of years should be available from the operating thermograph; its location is called *principal location* hereafter.
2 The average hourly values for each month should be available at a number of meteorological stations, where air-temperature observations are taken at standard hours every day; these sites are called *secondary locations* hereafter.
3 A procedure that takes into account the aforementioned thermographic and thermometric data and produces improved true mean monthly estimates at the secondary locations should also be developed.
4 Supplementary thermographic data at the secondary locations should be available, in order to validate the new estimates.

NOA has been chosen as the primary location; its registered thermographic observations taken in the period 1916–1930 have been used in this study. The secondary locations considered belong to the HNMS network and there has been given attention to be distributed evenly across Greece; their data are used in this study.

Table 1. Secondary locations of HNMS stations and years of observations.[a]

No.	HNMS station	Years
1	Agrinio	1992, 1993
2	Arta	1992, 1993
3	HNMS headquarters at Elliniko (Athens area)	1993
4	New Philadelphia (Athens area)	1992
5	Ierapetra	1992, 1993
6	Ioannina	1992, 1993
7	Kastoria	1992, 1993
8	Kerkyra (Corfu)	1992, 1993
9	Kozani	1992, 1993
10	Larissa	1992, 1993
11	Trikala	1992, 1993

[a]The location of the stations is shown in Figure 1.

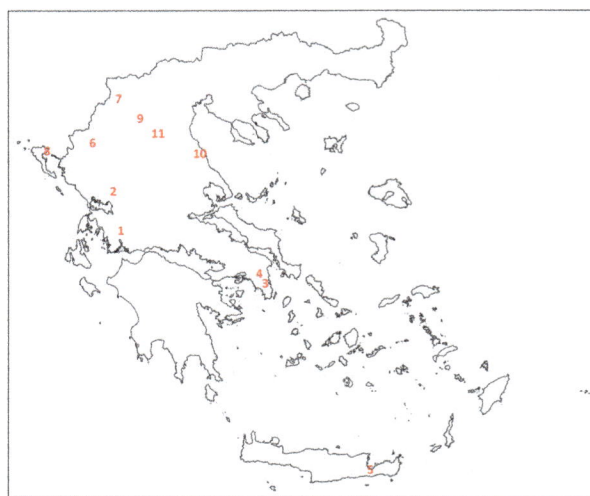

Figure 1. Location of HNMS stations referred to in Table 1.

Another requirement of the study is the availability of thermographic recordings for validation purposes, in parallel to the availability of thermometric observations at the same stations, even for a small number of years. Table 1 and Figure 1 give a full list of the meteorological stations and the corresponding years of measurements used in this study. These stations have thermographic recordings. The low number of years for the HNMS data used in this study depends upon their verified reliability.

The proposed algorithm takes into account the true mean monthly air-temperature value (T) and its monthly averages at the standard times of observations (T_8, T_{14}, T_{21}) at the principal location. These standard times were in use in Greece during the period of 1894–1930.

At a secondary location, X, the corresponding true mean monthly air temperature, T_X, is assumed equal to T, when the corresponding monthly averages at the standard hours of observations are equal to those at the principal location. If one of those observations differs and the remaining two are equal, it is assumed that the difference decreases linearly with time and becomes zero at the other two observational hours. If, for example, T_{X14} is greater than T_{14}, their difference is:

$$D_{14} = T_{X14} - T_{14} \qquad (3)$$

which decreases linearly to zero at 0800 and 2100; then T_X becomes greater than T:

$$T_X - T = D_{14}\frac{6.5}{24} \text{ or}$$

$$T_X = T + D_{14}\frac{6.5}{24} \qquad (3.1)$$

Similarly:

$$T_X = T + D_8\frac{8.5}{24} \qquad (3.2)$$

and:

$$T_X = T + D_{21}\frac{9}{24} \qquad (3.3)$$

If all monthly averages at the standard hours of observations differ, then, by combining Equations (3.1)–(3.3), we deduce:

$$T_X = T + \frac{8.5D_8 + 6.5D_{14} + 9D_{21}}{24} \qquad (4)$$

Equation (4) is the mathematical expression of the proposed algorithm; it implies that maximum differences of the averaged hourly air temperatures between station X and NOA occur at the standard hours of observations. This is rather optimistic, but as the actual daily course of the air temperature at station X in unknown, Equation (4) is expected to yield better approximation estimates of the true monthly means at station X than that of Equation (1). In practice, Equation (2) can be regarded as the first-order approximation to the true monthly mean at station X, and Equation (4) can serve as the second-order approximation.

3. Validation of the algorithm

Equations (2) and (4) were applied to the data from all stations and years listed in Table 1 and Figure 1. For every month, year, and station, the differences between the actual true mean air temperatures deduced from thermographic recordings were compared with the estimations derived from Equations (2) and (4).

The obtained differences, averaged over all stations and years of observations, are listed in Table 2; it is noticed that from late spring until autumn, Equation (2) overestimates the true mean monthly air temperatures far more than previously anticipated. The averaged differences exceed $0.8\,^\circ\text{C}$ in June; occasionally, the difference can become higher than $1.0\,^\circ\text{C}$ at inland stations. Application of Equation (4) alleviates this problem; it corrects more than 70% of the occurring error from March until September. During these months (except for March and September), the results from Equation (4) are found to produce better estimates of the true mean monthly temperatures than those received from Equation (2) at the significance level of 99%.

Table 2. Differences between true mean monthly air temperatures derived from thermographic recordings and corresponding estimations deduced from Equations (2) and (4).

Month	Difference from Equation (2) (°C)	Difference from Equation (4) (°C)
January	0.074	−0.580
February	0.167	−0.115
March	0.274	−0.072
April	0.504	0.060
May	0.602	0.116
June	0.846	0.287
July	0.792	0.162
August	0.657	0.139
September	0.382	−0.042
October	0.130	−0.202
November	0.064	−0.110
December	0.098	0.008
Year	0.385	0.011

During the remaining months, Equation (4) performs worse than Equation (2), with an exception for February and December. Equation (2) gives moderate overestimated true mean monthly temperatures, while Equation (4) has a tendency toward considerable underestimation. This behavior of Equation (4) can partly be attributed to the breakdown of the hypothesis for a linear decrease of the observed differences D_8, D_{14}, and D_{21} with time. The shorter length of the day may modify and complicate the daily course of the temperature during the months of October through February.

True mean annual temperatures can be deduced from true mean monthly values. As Equation (4) provides both overestimated values for some months and underestimated for others, it approaches, on the average, the true annual mean with a very good accuracy (0.011), while Equation (2) gives an average overestimation by almost $0.385\,^\circ\text{C}$. This overestimation varies from year to year and, thus, considerable noise can be introduced into the time-series analysis that tries to estimate temperature-related climate change, when Equation (2) is used in order to deduce the mean annual temperature values.

4. Case 2. True mean air-temperature estimations with a number of thermographs

In this scenario, a limited number of thermograph-equipped stations exist (more than one principal location). The problem to be solved is the estimation of the true mean monthly air temperatures at a number of secondary locations, where only daily temperature observations at standard hours are available.

As more thermographs are now available, it is possible to express their true mean monthly T as a function of the monthly temperature averages T_8, T_{14}, and T_{21}, taken at the corresponding standard observational hours with the help of linear multiple regression analysis. Consequently, 12 equations, one for each month, are

Table 3. Multiple regression analysis equations derived from the observations made at the stations and years listed in Table 1.

Month	Equation
January	$T = -0.58 + 0.32T_8 + 0.35T_{14} + 0.35T_{21}$
February	$T = -0.95 + 0.26T_8 + 0.34T_{14} + 0.46T_{21}$
March	$T = -0.16 + 0.28T_8 + 0.27T_{14} + 0.45T_{21}$
April	$T = -0.01 + 0.32T_8 + 0.25T_{14} + 0.41T_{21}$
May	$T = -0.53 + 0.34T_8 + 0.26T_{14} + 0.40T_{21}$
June	$T = 0.02 + 0.34T_8 + 0.23T_{14} + 0.40T_{21}$
July	$T = 0.67 + 0.42T_8 + 0.17T_{14} + 0.38T_{21}$
August	$T = 0.87 + 0.37T_8 + 0.25T_{14} + 0.25T_{21}$
September	$T = -0.14 + 0.31T_8 + 0.23T_{14} + 0.47T_{21}$
October	$T = 0.26 + 0.35T_8 + 0.32T_{14} + 0.31T_{21}$
November	$T = 0.07 + 0.40T_8 + 0.31T_{14} + 0.29T_{21}$
December	$T = -0.32 + 0.34T_8 + 0.31T_{14} + 0.37T_{21}$

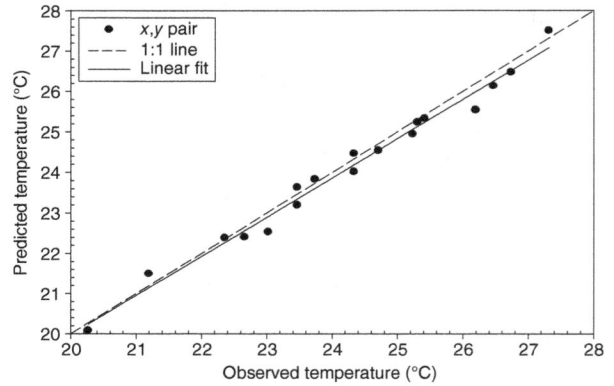

Figure 3. As in Figure 2, but for July. The linear fit has $R^2 = 0.983$ and equation $y = 0.596 + 0.969x$ (x = observed temperature, y = predicted temperature).

Table 4. HNMS stations and years of observations employed for the inter-comparison of the results obtained from Equations (2) and (4) and those from the regression analysis of Table 3.[a]

No.	HNMS station	Years
1	Kalamata	1991, 1992, 1993
2	Argostolion	1991, 1992
3	Aliartos	1991, 1992, 1993
4	Alexandroupolis	1991, 1992, 1993
5	Argos	1991, 1992, 1993
6	Araxos	1991, 1992, 1993
7	Aghialos	1991, 1992, 1993
8	Zakynthos (Zante)	1991, 1993
9	Iraklion	1992, 1993, 1994

[a] The location of the stations is shown in Figure 4.

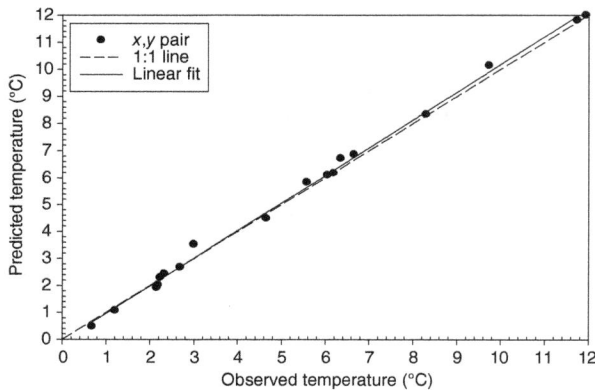

Figure 2. Predicted (corrected) versus observed air temperatures for January for all stations and years listed in Table 1. The linear fit has $R^2 = 0.997$ and equation $y = -0.062 + 1.024x$ (x = observed temperature, y = predicted temperature).

derived. Such equations, deduced from the observations made at the stations and years listed in Table 1, are presented in Table 3. Comparing the coefficients of these equations to the coefficients of Equation (2) (not shown here), it is realized that only the coefficient at noon is similar in all equations. This is another indication of the inadequacy of Equation (2).

Observed versus predicted values based on the aforementioned equations and for the months of January and July are presented in Figures 2 and 3, respectively, in order to depict the accuracy of the fit. From these equations and using the temperature observations at the same observational times at the secondary locations, the true monthly means at these locations can be estimated. It is implied that the daily temperature course at the secondary locations is more or less similar to the average daily temperature variation of the stations at the primary locations. Otherwise, the estimated values are of poor quality.

It should be pointed out although that the procedure of Section 2 can also be employed for the true mean monthly temperature estimations at the secondary locations. This entails retaining only one primary location and omitting the thermographic recordings of

the others. Consequently, the question about the most promising procedure arises.

To resolve this, another data set made available by HNMS stations with simultaneous thermographic recordings and temperature observations was employed. The stations and years of observations used in the subsequent analysis are listed in Table 4 and Figure 4.

True monthly means were estimated for every particular month, for all years and stations employed with the help of Equations (2) and (4), as well as with the help of equations listed in Table 3. Differences between observed and estimated true monthly means for every particular month and estimation procedure, averaged over all stations and years of observations are listed in Table 5, along with their seasonal and annual values.

Comparing Tables 2 and 5, it is observed that Equation (4) yields better estimates than Equation (2) from April to September. During the same period, the regression analysis equations yield more or less results of equal quality to those obtained through Equation (4). The results obtained from the regression analysis equations tend to be rather symmetrically distributed around the true monthly averages, while those deduced from Equations (2) and (4) tend toward overestimated values. It is, therefore, expected that the seasonal and annual averages obtained through the regression

Figure 4. Location of HNMS stations referred to in Table 4.

Table 5. Differences between observed and estimated true mean monthly air temperatures.

Month	Difference from Equation (2) (°C)	Difference from Equation (4) (°C)	Difference from regression analysis equations (°C)
January	−0.00	−0.06	−0.05
February	0.03	0.09	0.11
March	0.12	−0.14	−0.08
April	0.37	0.06	−0.01
May	0.53	0.14	0.06
June	0.87	0.45	0.13
July	0.75	0.27	−0.30
August	0.60	0.25	0.22
September	0.34	0.02	0.07
October	0.07	−0.16	−0.05
November	0.03	−0.08	0.07
December	0.00	0.02	0.00
Spring	0.34	0.02	−0.01
Summer	0.74	0.32	0.02
Autumn	0.15	−0.07	−0.03
Winter	0.01	0.00	0.02
Year	0.31	0.05	0.00

Table 6. Stations and corresponding record lengths used for the evaluation of possible errors in the climatic estimates of the true mean air temperatures in Greece.[a]

Number	Station	Altitude amsl (m)	Period of observations	Summer error (°C)	Annual error (°C)
1	Preveza	4	1915–1929	−0.42	−0.16
2	Patras	12	1910–1929	−0.61	−0.24
3	Larissa	75	1910–1929	−0.80	−0.40
4	Ioannina	465	1915–1929	−0.98	−0.43
5	Chalkis	12	1910–1929	−0.43	−0.21
6	Kythira	160	1910–1929	−0.60	−0.28
7	Arta	58	1910–1929	−0.83	−0.31
8	Anogeia	745	1915–1929	−0.26	−0.18
9	Volos	3	1910–1929	−0.49	−0.19
10	Syros	42	1910–1929	−0.46	−0.18
11	Kozani	700	1915–1929	−0.73	−0.38
12	Tripolis	665	1915–1929	−0.73	−0.35
Average				−0.61	−0.28

[a]The location of the stations is shown in Figure 5.

analysis equations should yield better estimates than those obtained through Equations (2) and (4). The annual averages listed in Table 5 demonstrate this. Equation (2) can be retained for winter and late autumn months, as it performs more or less equally well with the other two methods.

5. Possible effect on climatic temperatures of Greece

The first tabulation of the mean monthly temperatures, based on observations taken over a long-term period from many meteorological stations operating all over Greece, appears for the first time in the 'Climate of Greece' by Mariolopoulos (1938). Mean temperatures were derived with the help of Equation

(2) and it was assumed that they represent the true means with great accuracy. The climatic period was 1900–1929, but as many of the stations started their operation later than 1900 (about 1915) the means of the period 1915–1929 had to be adjusted to yield the 1900–1929 values. The adjustment was based on the already derived mean values from the stations with full records.

To assess the possible error introduced in the true mean climatic estimations, observations taken at a number of stations listed in Table 6 and for periods of observations varying from 1910–1929 and 1915–1929 were used in order to obtain averages at the standard hours of observations for every month. Their data were taken from the NOA archives. It should be mentioned that before 1930, NOA was responsible for all meteorological observations in Greece.

From these averages, the monthly means were estimated with the help of Equation (2) as well as with the help of the regression analysis equations. The latter are considered representing the true means; the differences between their estimates and the corresponding estimates from Equation (2) are considered representing the error in the climatic true mean air-temperature estimates.

Comparing Tables 5 and 6, it is noticed that the general characteristics of the overestimated values obtained from Equation (2) are exhibited in both tables, implying that the overestimations of Equation (2) concern not only individual years but also the large observational periods as well. Maximum errors are found at stations located in northern Greece and away from the sea, while minimum ones occur at stations located in the southern part of the country and close to the sea, as can be seen in Figure 5, where the locations of the stations are shown. The annual error equals approximately 50% of the error of the summer season in both tables. As mentioned in Section 1, the meteorological observations at 2100 local time were replaced by those at 2000 local

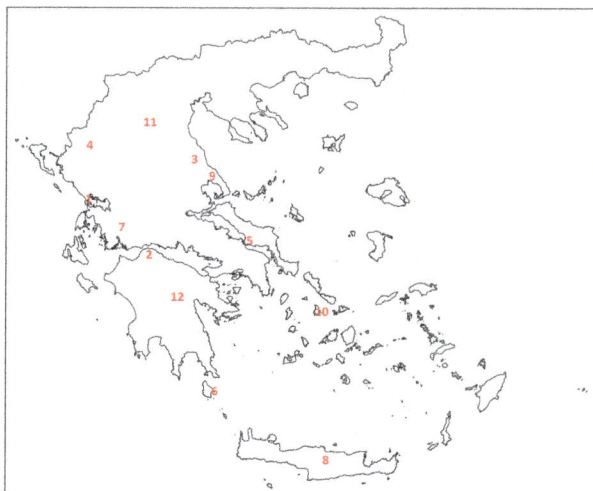

Figure 5. Location of HNMS stations referred to in Table 6.

time from 1930 onwards. It is, therefore, expected that the overestimation error has been augmented from the estimated values so far.

6. Conclusions

From the present study, it is evident that Equation (2) largely overestimates the true mean air temperatures, with the exception of winter and late autumn months. Overestimations occur for individual months, years, or climatic periods. They reach maximum values in the summer months and for stations located in northern Greece and away from the sea. The opposite holds for stations located in the southern part of the country and close to the sea. Two solutions for the problem have been suggested.

The first is applied when only one thermograph is operating in an area and its recordings should be used along with their air-temperature observations taken at other locations, in order to obtain the true mean air temperature at these locations. Equation (4) is employed for the solution to the problem and corrects more than 50% of the overestimation error in spring, summer, and the annual true mean air temperatures.

The second solution is applied when more than one thermographs is operating in an area. Multiple regression analysis is employed and the true mean air temperatures for these stations are expressed in the form of equations having the true mean value as dependent variable and the mean monthly averages at the standard observation times as independent ones (see equations in Table 3). These equations yield the true mean values at locations, where only air-temperature observations at the standard observation hours are available.

The two solutions yield comparable results for individual spring and summer months, while the second solution yields better seasonal and annual estimates. The annual averages obtained through the second solution are more accurate than those obtained from Equation (2). Therefore, the time series of the mean annual air-temperature values derived from Equation (2) may contain noise, thus obscuring the detection of any climate change signal.

Acknowledgements

Thanks to the HNMS are expressed for providing the air-temperature data (thermometric and thermographic) for the mentioned stations and years in the study. The authors also declare that they have no conflict of interest.

References

Aeginitis D. 1907. *The Climate of Greece. Part A: The Climate of Athens.*: P. Sakkelariou Editions, Athens; 229 pp.

Conrad V, Pollak WL. 1950. *Methods in Climatology*, 2nd ed. Harvard University Press: Cambridge, Massachusetts; 154–160.

Dall' Amico M, Hornsteiner M. 2006. A simple method for estimating daily and monthly mean temperatures from daily minima and maxima. *International Journal of Climatology* **26**(13): 1929–1936.

Kaufmann RK, Gopal S, Tang X, Raciti SM, Lyons PE, Geron N, Craig F. 2013. Revisiting the weather effect on energy consumption: implications for the impact of climate change. *Energy Policy* **62**: 1377–1384.

Mariolopoulos H. 1938. *The Climate of Greece*. Papaspyrou Press, Athens; 19–21.

Olesen JE, Bindi M. 2002. Consequences of climate change for European agriculture productivity, land use and policy. *European Journal of Agronomy* **16**(4): 239–262.

Rosenzweig C, Parry ML. 1994. Potential impact of climate change on world food supply. *Nature* **367**: 133–138.

Srivastava GP. 2008. *Surface Meteorological Instruments and Measurement Practices*. Atlantic Publishers and Distributors Ltd, New Delhi, 978-81-269-0968-1.

Weiss A, Hays CJ. 2005. Calculating daily mean air temperatures by different methods: implications from a non-linear algorithm. *Agricultural and Forest Meteorology* **128**: 57–65.

17

Seasonal winter forecasts and the stratosphere

A. A. Scaife,[1]* A. Yu. Karpechko,[2] M. P. Baldwin,[3] A. Brookshaw,[1] A. H. Butler,[4,5] R. Eade,[1] M. Gordon,[1] C. MacLachlan,[1] N. Martin,[1] N. Dunstone[1] and D. Smith[1]

[1] Met Office Hadley Centre, Exeter, UK
[2] Arctic research, Finnish Meteorological Institute, Helsinki, Finland
[3] Department of Mathematics and Computer Science, University of Exeter, UK
[4] Cooperative Institute for Research in Environmental Sciences (CIRES), Boulder, CO, USA
[5] Earth System Research Laboratory, NOAA, Boulder, CO, USA

*Correspondence to:
A. A. Scaife, Met Office Hadley
Centre, FitzRoy Road, Exeter
EX1 3PB, UK.
E-mail:
adam.scaife@metoffice.gov.uk

Abstract

We investigate seasonal forecasts of the winter North Atlantic Oscillation (NAO) and their relationship with the stratosphere. Climatological frequencies of sudden stratospheric warming (SSW) and strong polar vortex (SPV) events are well represented and the predicted risk of events varies between 25 and 90% from winter to winter, indicating predictability beyond the deterministic range. The risk of SSW and SPV events relates to predicted NAO as expected, with NAO shifts of −6.5 and +4.8 hPa in forecast members containing SSW and SPV events. Most striking of all is that forecast skill of the surface winter NAO vanishes from these hindcasts if members containing SSW events are excluded.

Keywords: seasonal forecast; NAO; stratosphere

1. Introduction

Previous studies have quantified the deterministic limit on the predictability of the *timing* of sudden stratospheric warming (SSW) events. These studies typically indicate significant forecast skill out to around 2 weeks (Mukougawa *et al.*, 2005; Stan and Straus, 2009; Marshall and Scaife, 2010; Sigmond *et al.*, 2013) and very occasionally 1 month (Kuroda, 2008) ahead. However, there could be additional forecast skill beyond this timescale if we consider the probabilistic *risk* of an event occurring. In addition, there could also be forecast skill for strong polar vortex (SPV) events which have received relatively little attention compared with SSW events despite the apparent symmetry between the tropospheric impacts of both types of event (Baldwin and Dunkerton, 1999). Given the stratospheric influence on winter surface climate (e.g. Boville, 1984; Scaife and Knight, 2008; Kolstad *et al.*, 2010; Mitchell *et al.*, 2013; Sigmond *et al.*, 2013), we also investigate the relationship with winter seasonal forecasts of the surface North Atlantic Oscillation (NAO).

Here, we investigate retrospective forecasts of the risk of winter SSW and SPV events and the resulting impact on seasonal forecasts of surface winter climate from the Met Office Global Seasonal forecast system GloSea (Arribas *et al.*, 2011). We use ensembles of 24 forecasts starting in early November for each of the 20 winters from 1992/1993 to 2011/2012 from the fifth generation of GloSea (MacLachlan *et al.*, 2014). These winter forecasts have statistically significant forecast skill for the surface NAO (Scaife *et al.*, 2014). We define stratospheric events using the daily zonal Arctic winds at 10 hPa and averaged around 60 N in each forecast ensemble member. SSW events are defined to occur if this wind decreases below zero at some time in the winter, while SPV events are defined to occur if this wind increases above 48 m s^{-1} on some day in the winter. This SPV (upper) threshold is chosen as it is broken with the same frequency as the lower SSW threshold in our forecasts.

2. Predictability of stratospheric events beyond the deterministic range

To eliminate predictability on the timescale of weeks described above, we discard forecast data from the first month (November) and include only data for the December to February period. The 20 winters from 1992/1993 to 2011/2012 are included, with 24 member ensemble forecasts for each winter, making a total of 480 winter forecasts. This total number of realizations is an order of magnitude greater than the number of stratospheric winters we have in the observational record since the advent of comprehensive satellite data for the stratosphere (e.g. Pawson and Fiorino, 1999; Scaife *et al.*, 2000). This allows more stable statistics to be calculated than are possible from the observational record alone.

Figure 1 shows the ensemble of 24 member forecasts for the winters 1997/1998 and 1999/2000.

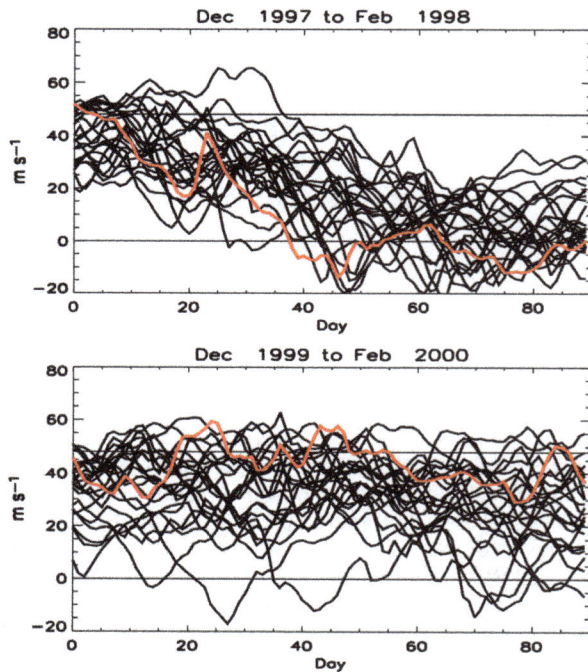

Figure 1. Ensemble forecast evolution of zonal wind in the stratosphere. Zonal mean U winds are shown (10 hPa, 60 N) in two different winters, with 24 ensemble members per winter and an example member colored red. The horizontal lines show the threshold for SSW (0 ms^{-1}) events and SPV (48 ms^{-1}) events. The two winters shown exhibit quite different predicted probability of a SSW or SPV event, indicating the potential for forecasting the risk of these events well beyond the deterministic range of a few weeks.

Forecast members highlighted in red illustrate example evolutions of the stratospheric wind. In the case of 1997/1998, the sample member undergoes a SSW event around day 37 (6 January) when it crosses the zero wind line. Similarly, while the sample member for 1999/2000 does not show any sudden warmings, it does go through an SPV event around day 18 (18 December) when it crosses the upper threshold. Counting whether at least one event occurs in each forecast member yields a forecast probability of either a SSW or SPV event for each winter. The climatological frequency of SSW and SPV events averaged over all 20 winters and all forecast members is 0.53 and 0.55, respectively. This matches the observed frequency of 0.45 and 0.60 for the same period to within statistical sampling uncertainty, using the same criteria. GloSea5 evidently produces a good simulation of stratospheric climate variability in this sense.

The winter ensembles in Figure 1 illustrate the very different probability of SSW and SPV events between winter 1997/1998 and winter 1999/2000 in these long-lead forecasts. This variation in the predicted risk of such events changes from 1 year to the next for both SSW and SPV events and varies between 25 and 90% across the 20 winters considered. Given that we discard the first month of the forecasts, these events therefore show potential predictability, at least in a probabilistic sense, well beyond the previously

identified deterministic limit of a few weeks. To quantify the predictability further, we would ideally take many observed events from a very long sequence of winters and quantify the frequency of an observed event as a function of forecast probability. However with the limited number of observed events over the period used here, this is not possible due to the small sample size. Instead, we create a more statistically stable estimate of predictability using the forecast ensemble data alone (sometimes called perfect predictability) and resampling individual members for each year to create 1000 series of proxy observations. We then calculate how much the risk of a SSW event or SPV event changes on average (with replacement of the observed proxy member) in the years when an event occurred in the proxy observations. The result of this calculation is almost identical for both SSW and SPV events and the forecast probability of an event rises by 12% on average from 47% in winters in which there is no event to 59% in winters in which an event occurs. Although this is a modest shift in probability, the large number of samples involved means that this potential probabilistic forecast skill is significant beyond the 95% level.

3. Relationship with the surface NAO

The latest generation of seasonal forecast systems has begun to show consistent and statistically significant skill for seasonal predictions of the winter NAO (Scaife et al., 2014) and its hemispheric equivalent of the Arctic Oscillation (AO; Riddle et al., 2013; Athanasiadis et al., 2014; Kang et al., 2014; Stockdale et al., 2015). The stratosphere is expected to contribute to this skill (Orsolini et al., 2011; Folland et al., 2012; Smith et al., 2012) due to its influence on the tropospheric NAO and AO, which is present in our model (e.g. Fereday et al., 2012). The seasonal forecasts examined here are consistent with this idea because the predicted surface NAO shows a significant interannual correlation of −0.43 with the predicted frequency of SSW events (Figure 2). As expected, the correlation with the predicted frequency of SPV events shows the opposite sign, with an increase in predicted SPV events coinciding with an increase in predicted NAO. However, the correlation in this case is only 0.23 and is not statistically significant for the 20 years available here.

The impact of the predicted occurrence of a SSW event or SPV event on predicted winter NAO values is shown in Figure 3. Distributions of predicted NAO, conditioned on the occurrence or absence of a SSW, are shown in the upper panels of Figure 3 and indicate a mean shift of −6.5 hPa when a SSW occurs. This is a large value given that the interannual standard deviation of the NAO is around 8 hPa. A similar result holds for the SPV events, with a mean shift of +4.8 hPa when a SPV event occurs in forecast members. Both shifts are statistically significant beyond the 99% level due to the large sample size. While very low or very high NAO values can still occur irrespective of

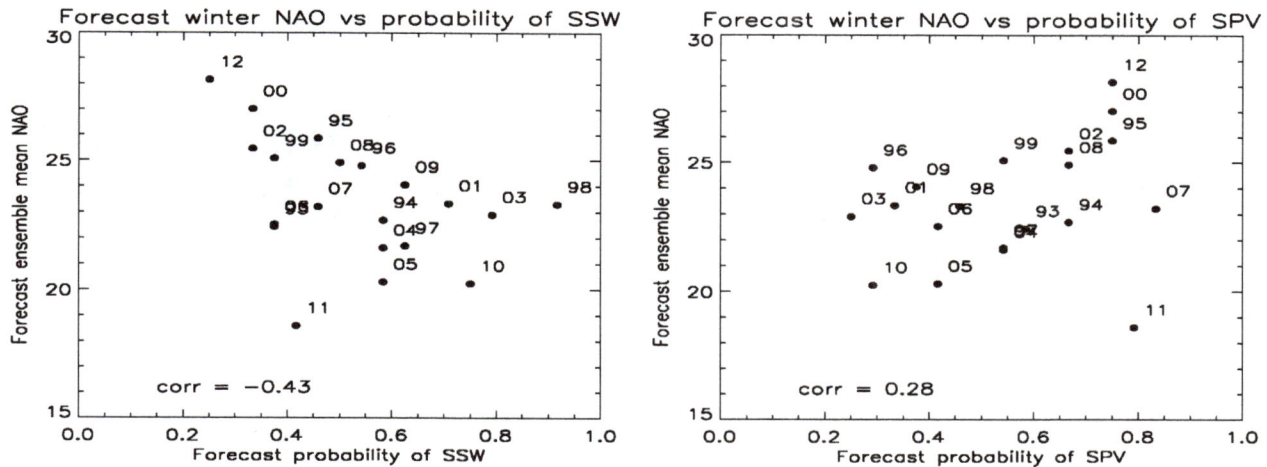

Figure 2. Correlation between ensemble mean NAO forecast and risk of SSW (left), and between ensemble mean NAO forecast and risk of SPV (right) events. Results are over the 20 winters from 1992/1993 to 2011/2012 and are significant at the 95% level in the SSW case. The correlation with SPV risk is not statistically significant but does show the expected sign. Winters are labeled by the year of the corresponding January.

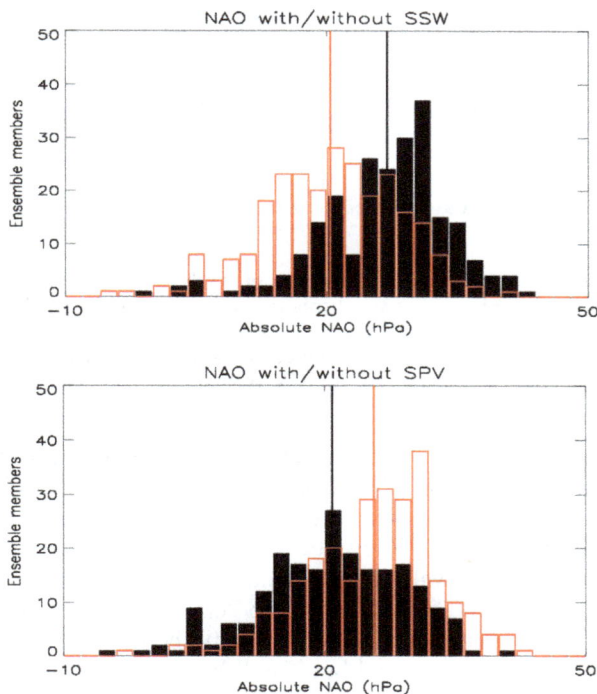

Figure 3. Distribution of ensemble forecasts of the surface NAO. Upper panels show NAO forecasts with (red) and without (black) SSW events. Lower panels show NAO forecasts with (red) and without (black) SPV events. There are 480 individual forecasts in the distributions (20 winters × 24 members). Vertical lines show the distribution means. Differences between the means with and without events are significant beyond the 99% level according to a one-sided t-test.

the occurrence of a stratospheric event, and this does not mean that forecast NAO signals originate in or are driven by the stratosphere, these large shifts in surface climate confirm that the ensemble forecast surface winter NAO is strongly conditional on events in the stratosphere.

4. Impact on surface forecast skill

We have seen that on average, the occurrence of a strong (SPV) or weak (SSW) stratospheric event gives large conditional changes in the predicted surface NAO toward positive or negative values. Here, we quantify how this affects forecast skill. To do this, we calculate the forecast skill with and without members that include a SSW (Figure 4). In the case of SSW events, this produces a striking result: the full ensemble mean correlation skill of the NAO (0.62, Scaife et al., 2014) vanishes if SSW events are excluded. The remaining ensemble mean shows a correlation skill of just 0.09 which is statistically insignificant. A smaller (but statistically insignificant) reduction occurs if SPV events are instead omitted from the ensemble. Skill in these NAO forecasts is therefore conditional on the inclusion of SSWs.

While this result is striking, we must interpret it carefully. Given the downward propagating and lagged influence of the stratosphere on the troposphere, the lack of NAO skill without SSW events is consistent with the stratosphere playing a key role in NAO predictability. However, it does not necessarily mean that the source of NAO forecast skill originates in the stratosphere (see Sun et al., 2012). Second, after members containing SSW events are removed, the ensemble size is reduced by about half (the frequency of SSW events in the hindcasts) and so the skill will inevitably drop due to the smaller ensemble size (Scaife et al., 2014, Figure 3). We therefore tested whether the correlation of 0.09 represents a significant reduction in skill given the smaller ensemble size. To do this, we resampled ensembles of the same average reduced size and calculated their ensemble mean correlation with the observed NAO. Using 1000 resampled ensembles of the same average reduced size, 99% resulted in a correlation that exceeds the case with no SSW events, confirming that

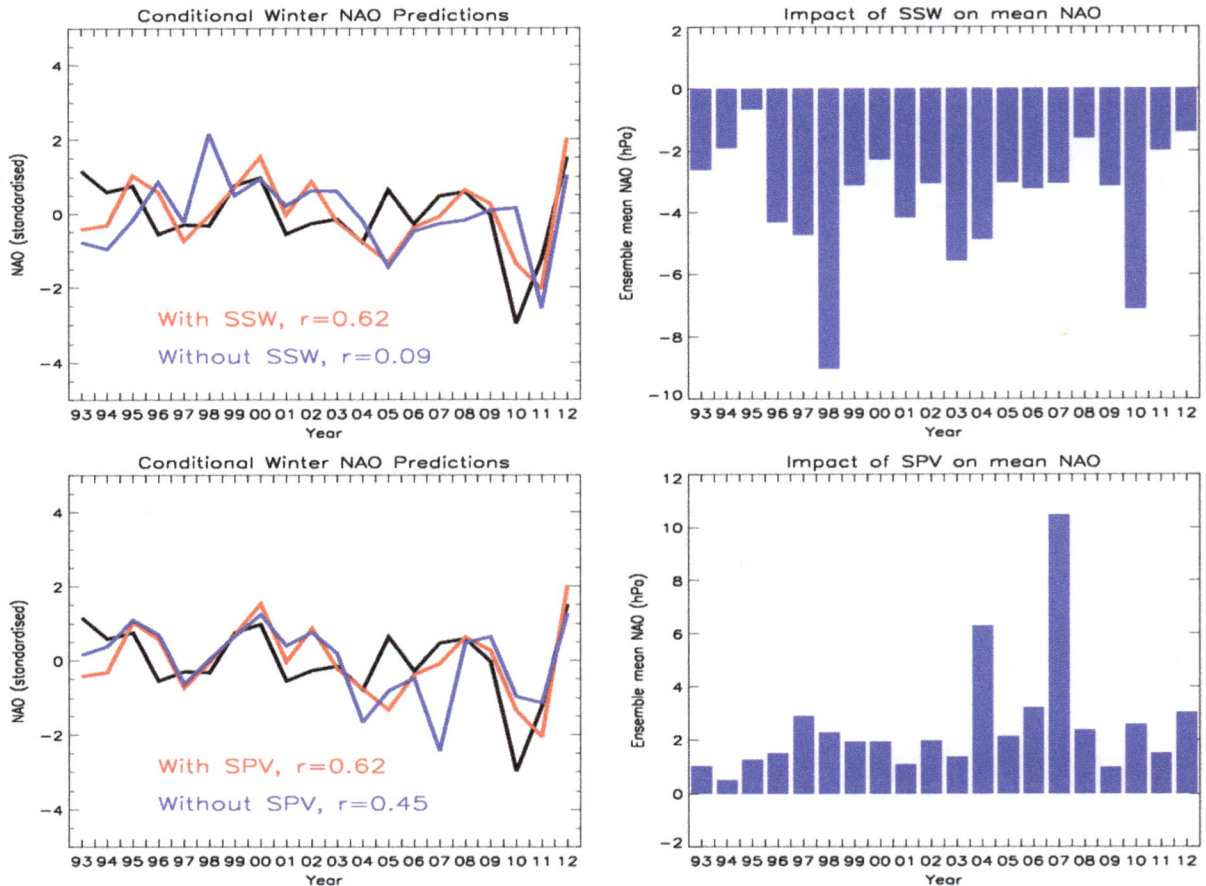

Figure 4. Relationship between SSW (upper) and SPV (lower) events and seasonal winter forecasts of the NAO. Panels on the left show standardized ensemble mean forecasts for each winter including members with stratospheric events (red) and excluding members with stratospheric events (blue). Observed NAO values are in black. Panels on the right show the difference in ensemble mean NAO predictions when SSW events are included (upper right) and when SPV events are included (lower right). Large forecast impacts are seen in 1997/1998 and 2009/2010 for SSW events and 2003/2004 and 2006/2007 for SPV events.

removing SSW events very likely leads to a genuine reduction in forecast skill.

Some winter forecasts are much more strongly influenced than others by the occurrence of stratospheric events. The influence of stratospheric SSW or SPV events on each winter NAO forecast is shown in the histograms in Figure 4. Members containing SSW events decrease the ensemble mean NAO forecast in all winters, as expected. However, forecasts for the winters of 1997/1998 and 2009/2010 are particularly strongly affected. There was a very high forecast probability of SSW events in these two winters (92 and 75% respectively) and both are El Niño winters. We can therefore explain this result by the well-known increase in stratospheric sudden warmings during El Niño (Brönnimann et al., 2004; Taguchi and Hartmann, 2006; Bell et al., 2009; Butler and Polvani, 2011; Domeisen et al., 2015) and the need for this stratospheric response in order to generate a strong surface NAO signal in the Atlantic (Toniazzo and Scaife, 2006; Cagnazzo and Manzini, 2009; Ineson and Scaife, 2009; Butler et al., 2014). It is interesting to note that the very negative NAO forecast for winter 2010/2011 was unaffected and this winter is also an outlier in the correlation plot in Figure 2.

This is consistent with evidence for an Atlantic Ocean rather than stratospheric origin of the negative NAO in this particular winter (Maidens et al., 2013). The occurrence of SPV events is associated with higher forecast NAO values in all winters but the effect was largest in 2003/2004 and 2006/2007 (Figure 4). We have no simple explanation for why these winters are particularly influenced but it will be interesting to see if the same winters are highlighted in future seasonal forecast systems.

5. Conclusions

Deterministic forecast skill for stratospheric sudden warmings has been demonstrated out to around 2 weeks lead time in several previous studies. Here, we use a seasonal forecast system with a good simulation of climate variability in the stratosphere to provide evidence of potential predictability of SSW and SPV occurrence well beyond this deterministic limit, with the risk of an event varying between 25 and 90% between different winters in forecasts out to 4 months ahead. Perfect predictability is however still modest and demonstration of actual skill by comparing against observed occurrence of events requires further work and more cases.

The impact of stratospheric variability on surface winter forecasts is as expected, with SSW/SPV events being associated with lower/higher ensemble mean surface NAO forecasts for winter. The occurrence or absence of a stratospheric event is associated with a shift of several hPa in the surface NAO, a substantial fraction of the interannual variability of the NAO itself. A striking result is that the previously reported skill for seasonal NAO predictions (Scaife *et al.*, 2014) is conditional on the inclusion of forecast members containing SSW events. This is consistent with the idea (though does not conclusively demonstrate) that a well-resolved stratosphere is important for winter forecast skill in the Atlantic basin. This result may also be robust across forecast systems and timescales (e.g. Sigmond *et al.*, 2013; Stockdale *et al.*, 2015).

The El Niño winters of 1997/1998 and 2009/2010 were periods of high risk of a SSW event, whereas the winters of 2003/2004 and 2006/2007 were winters with a high risk of a SPV event. Winter 2009/2010 is already known to have been highly disturbed in the stratosphere and this was important for seasonal winter forecasts of this event (Fereday *et al.*, 2012). However, the observed 1997/1998 winter did not quite reach the threshold for SSW categorization. Assuming a good simulation of the teleconnection and the background wave field (Fletcher and Kushner, 2011) so that the modeled ENSO teleconnection is realistic (e.g. Fereday *et al.*, 2012), the multiple seasonal realizations used here suggest that we were unlucky not to see a SSW in that winter. More generally, the results shown here suggest that the year to year risk of SSW and SPV events in future winters is likely to be predictable on seasonal timescales, months ahead of the actual event.

Acknowledgements

This work was supported by the Joint DECC/Defra Met Office Hadley Centre Climate Programme (GA01101), the UK Public Weather Service research program and the European Union Framework 7 SPECS project. The contribution of AYK is funded by FMI's tenure track program and the Academy of Finland under grant 286298.

References

Arribas A, Glover M, Maidens A, Peterson KA, Gordon M, MacLachlan C, Graham R, Fereday D, Camp J, Scaife AA, Xavier P, McLean P, Colman A, Cusack S. 2011. The GloSea4 ensemble prediction system for seasonal forecasting. *Monthly Weather Review* **139**: 1891–1910.

Athanasiadis PJ, Bellucci A, Hermanson L, Scaife AA, MacLachlan C, Arribas A, Materia S, Borrelli A, Gualdi S. 2014. The representation of atmospheric blocking and the associated low-frequency variability in two seasonal prediction systems (CMCC, Met-Office). *Journal of Climate* **27**: 9082–9100, doi: 10.1175/JCLI-D-14-00291.1.

Baldwin MP, Dunkerton TJ. 1999. Propagation of the Arctic Oscillation from the stratosphere to the troposphere. *Journal of Geophysical Research* **104**(D24): 30937–30946, doi: 10.1029/1999JD900445.

Bell CJ, Gray LJ, Charlton-Perez AJ, Joshi M, Scaife AA. 2009. Stratospheric communication of El Niño teleconnections to European Winter. *Journal of Climate* **22**: 4083–4096.

Boville BA. 1984. The influence of the polar night jet on the tropospheric circulation in a GCM. *Journal of the Atmospheric Sciences* **41**: 1132–1142.

Brönnimann S, Luterbacher J, Staehelin J, Svendby TM, Hansen G, Svenøe T. 2004. Extreme climate of the global troposphere and stratosphere in 1940–42 related to El Niño. *Nature* **431**: 971–974, doi: 10.1038/nature02982.

Butler AH, Polvani LM. 2011. El Niño, La Niña, and stratospheric sudden warmings: a re-evaluation in light of the observational record. *Geophysical Research Letters* **38**: L13807, doi: 10.1029/2011 GL048084.

Butler AH, Polvani LM, Deser C. 2014. Separating the stratospheric and tropospheric pathways of El Nino-Southern Oscillation teleconnections. *Environmental Research Letters* **9**: 024014, doi: 10.1088/1748-9326/9/2/024014.

Cagnazzo C, Manzini E. 2009. Impact of the stratosphere on the winter tropospheric teleconnections between ENSO and the North Atlantic and European region. *Journal of Climate* **22**: 1223–1238.

Domeisen DIV, Butler AH, Fröhlich K, Bittner M, Müller WA, Baehr J. 2015. Seasonal predictability over Europe arising from El Niño and stratospheric variability in the MPI-ESM seasonal prediction system. *Journal of Climate* **28**: 256–271, doi: 10.1175/JCLI-D-14-00207.1.

Fereday D, Maidens A, Arribas A, Scaife AA, Knight JR. 2012. Seasonal forecasts of Northern Hemisphere Winter 2009/10. *Environmental Research Letters* **7**: 034031, doi: 10.1088/1748-9326/7/3/034031.

Fletcher CG, Kushner PJ. 2011. The role of linear interference in the annular mode response to tropical SST forcing. *Journal of Climate* **24**: 778–794.

Folland CK, Scaife AA, Lindesay J, Stephenson DB. 2012. How potentially predictable is northern European winter climate a season ahead? *International Journal of Climatology* **32**: 801–808, doi: 10.1002/joc.2314.

Ineson S, Scaife AA. 2009. The role of the stratosphere in the European climate response to El Niño. *Nature Geoscience* **2**: 32–36.

Kang D, Lee MI, Im J, Kim D, Kim H-M, Kang H-S, Schubert SD, Arribas AA, MacLachlan C. 2014. Prediction of the Arctic Oscillation in boreal winter by dynamical seasonal forecasting systems. *Geophysical Research Letters* **10**: 3577–3585, doi: 10.1002/2014GL060011.

Kolstad EW, Breiteig T, Scaife AA. 2010. The association between stratospheric weak polar vortex events and cold air outbreaks. *The Quarterly Journal of the Royal Meteorological Society* **136**: 886–893.

Kuroda Y. 2008. Role of the stratosphere on the predictability of medium-range weather forecast: a case study of winter 2003–2004. *Geophysical Research Letters* **35**: L19701, doi: 10.1029/2008 GL034902.

MacLachlan C, Arribas A, Peterson KA, Maidens A, Fereday D, Scaife AA, Gordon M, Vellinga M, Williams A, Comer RE, Camp J, Xavier P. 2014. Description of GloSea5: the Met Office high resolution seasonal forecast system. *The Quarterly Journal of the Royal Meteorological Society* **141**: 1072–1084, doi: 10.1002/qj.2396.

Maidens A, Arribas A, Scaife AA, MacLachlan C, Peterson KA, Knight JR. 2013. The influence of surface forcings on the North Atlantic Oscillation regime of winter 2010/11. *Monthly Weather Review* **141**: 3801–3813.

Marshall A, Scaife AA. 2010. Improved predictability of stratospheric sudden warming events in an AGCM with enhanced stratospheric resolution. *Journal of Geophysical Research* **115**: D16114, doi: 10.1029/ 2009JD012643.

Mitchell DM, Gray LJ, Anstey J, Baldwin MP, Charlton-Perez AJ. 2013. The influence of stratospheric vortex displacements and splits on surface climate. *Journal of Climate* **26**: 2668–2682, doi: 10.1175/ JCLI-D-12-00030.1.

Mukougawa H, Sakai H, Hirooka T. 2005. High sensitivity to the initial condition for the prediction of stratospheric sudden warming. *Geophysical Research Letters* **32**: L17806, doi: 10.1029/2005GL022909.

Orsolini YJ, Kindem IT, Kvamstø NG. 2011. On the potential impact of the stratosphere upon seasonal dynamical hindcasts of the North Atlantic Oscillation: a pilot study. *Climate Dynamics* **36**: 579–588, doi: 10.1007/s00382-009-0705-6.

Pawson S, Fiorino M. 1999. A comparison of reanalyses in the tropical stratosphere. Part 3: inclusion of the pre-satellite data era. *Climate Dynamics* **15**: 241–250.

Riddle EE, Butler AH, Furtado JC, Cohen JL, Kumar A. 2013. CFSv2 ensemble prediction of the wintertime Arctic Oscillation. *Climate Dynamics* **41**: 1099–1116.

Scaife AA, Knight JR. 2008. Ensemble simulations of the cold European winter of 2005/6. *The Quarterly Journal of the Royal Meteorological Society* **134**: 1647–1659, doi: 10.1002/qj.312.

Scaife AA, Austin J, Butchart N, Pawson S, Keil M, Nash J, James IN. 2000. Seasonal and interannual variability of the stratosphere diagnosed from UKMO TOVS analyses. *The Quarterly Journal of the Royal Meteorological Society* **126**: 2585–2604.

Scaife AA, Arribas A, Blockley E, Brookshaw A, Clark RT, Dunstone N, Eade R, Fereday D, Folland CK, Gordon M, Hermanson L, Knight JR, Lea DJ, MacLachlan C, Maidens A, Martin M, Peterson AK, Smith D, Vellinga M, Wallace E, Waters J, Williams A. 2014. Skilful long range prediction of European and North American Winters. *Geophysical Research Letters* **41**: 2514–2519, doi: 10.1002/2014 GL059637.

Sigmond M, Scinocca JF, Kharin VV, Shepherd TG. 2013. Enhanced seasonal forecast skill following stratospheric sudden warmings. *Nature Geoscience* **6**: 98–102, doi: 10.1038/ngeo1698.

Smith D, Scaife AA, Kirtman B. 2012. What is the current state of scientific knowledge with regard to seasonal and decadal forecasting? *Environmental Research Letters* **7**: 015602, doi: 10.1088/1748-9326/7/1/015602.

Stan C, Straus DM. 2009. Stratospheric predictability and sudden stratospheric warming events. *Journal of Geophysical Research* **114**: D12103, doi: 10.1029/2008JD011277.

Stockdale TN, Molteni F, Ferranti L. 2015. Atmospheric initial conditions and the predictability of the Arctic Oscillation. *Geophysical Research Letters* **42**: 1173–1179, doi: 10.1002/2014GL062681.

Sun L, Robinson WA, Chen G. 2012. The predictability of stratospheric warming events: more from the troposphere or the stratosphere? *Journal of the Atmospheric Sciences* **69**: 768–783, doi: 10.1175/JAS-D-11-0144.1.

Taguchi M, Hartmann DL. 2006. Increased occurrence of stratospheric sudden warmings during El Niño simulated by WACCM. *Journal of Climate* **19**: 324–332, doi: 10.1175/JCLI3655.1.

Toniazzo T, Scaife AA. 2006. The influence of ENSO on winter North Atlantic climate. *Geophysical Research Letters* **33**: L24704, doi: 10.1029/2006GL027881.

18

An initial assessment of observations from the Suomi-NPP satellite: data from the Cross-track Infrared Sounder (CrIS)

Andrew Smith,* Nigel Atkinson, William Bell and Amy Doherty

Met Office, Exeter, UK

*Correspondence to:
A. Smith, Met Office, FitzRoy Road, Exeter, EX1 3 PB, UK.
E-mail:
andrew.smith@metoffice.gov.uk

This article is published with the permission of the Controller of HMSO and the Queen's Printer for Scotland.

Abstract

This paper gives an appraisal of the hyperspectral Cross-track Infrared Sounder (CrIS) on board the Suomi-NPP satellite, for use in numerical weather prediction (NWP) at the Met Office. The quality of CrIS data is assessed by comparison with short-range NWP forecasts, and the impact on forecasts is evaluated using results from assimilation experiments. The impact of assimilating CrIS and the complementary microwave sounder ATMS (Advanced Technology Microwave Sounding Unit) together is also presented.

Keywords: CrIS; Suomi-NPP; hyperspectral sounder; numerical weather prediction; data assimilation

1. Introduction

A comprehensive description of Cross-track Infrared Sounder (CrIS) and the calibration and validation of its data can be found in the study by Han *et al.* 2013. CrIS data received at the Met Office are calibrated, geolocated and apodised radiances. The ATOVS and AVHRR Pre-processing Package – AAPP (NWP SAF, 2011) is used to thin the data to one field-of-view (the warmest) per field-of-regard, and reduce the number of channels for potential use from 1305 to 399, corresponding to the channel selection of Gambacorta and Barnet (2013).

Data from two hyperspectral sounders are already assimilated operationally at the Met Office: AIRS (Atmospheric Infrared Sounder) has been used since 2004 (Cameron *et al.*, 2005) and IASI (Infrared Atmospheric Sounding Interferometer) since 2007 (Hilton *et al.*, 2009). CrIS has broadly similar spectral coverage to AIRS, but less than IASI. The spectral resolution is lower than for both of those instruments but a compensating factor is that CrIS has a very low-noise profile (Zavyalov *et al.*, 2013). The aim with CrIS is to make use of the data in the same way as for AIRS and IASI.

2. Data quality assessment

Before a satellite instrument can be included in the assimilation system it is essential to assess the quality of the data, firstly to determine that the instrument is behaving as expected in a consistent manner, and secondly to make decisions on a suitable configuration for the assimilation. The most appropriate way of doing this for a numerical weather prediction (NWP) system is to compare against the model fields from a short-range forecast (known as the *background*) by looking at statistics of departures between observed radiances (or brightness temperatures, O) and those from a radiative-transfer model that takes those background fields as input (B). This has been used successfully for validation of several recent missions, e.g. ATMS (Advanced Technology Microwave Sounding Unit) (Bormann *et al.*, 2012), FY3A (Lu *et al.*, 2011). The data used to produce these O-B statistics must be free from major sources of uncertainty such as cloud and inaccurate surface emission values. Observation data may also be biased with respect to the background. The static bias-correction scheme (Harris and Kelly, 2001), used for all sounding data at the Met Office, is also employed for CrIS.

To understand the data in the context of the full observing system, it is instructive to compare O-B statistics against those for the other assimilated hyperspectral sounders, AIRS and IASI. For the key temperature sounding channels in the long-wave ($655-750\,\mathrm{cm}^{-1}$) CO_2 absorption band, the background errors are usually very small and therefore the instrument noise is a major component of the O-B values. These channels are of particular interest in assessing CrIS data quality because of the expected improvement in noise performance and the fact that they provide important information for NWP.

Figure 1 shows O-B statistics for CrIS, IASI and AIRS in the long-wave CO_2 absorption region, using global-coverage data accumulated over a 3-month period. Data are bias corrected and screened for cloud

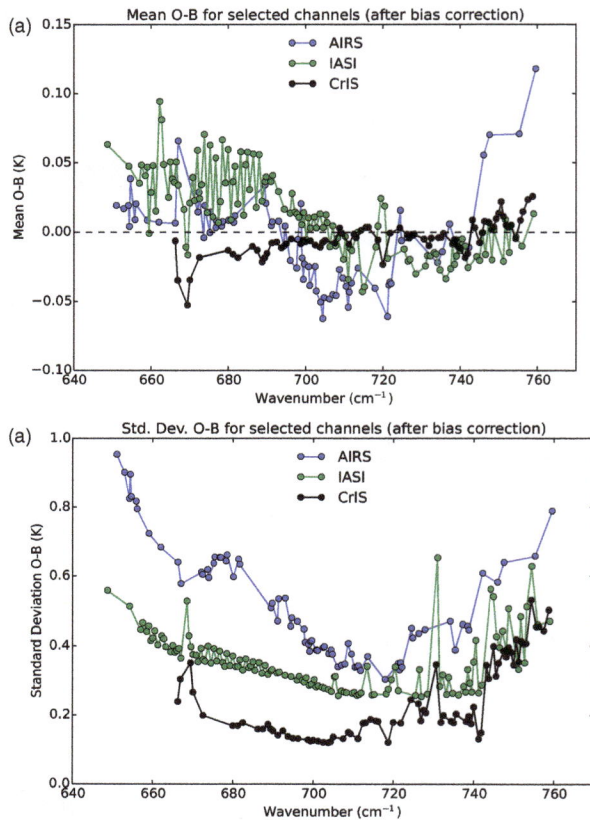

Figure 1. Mean and standard deviation of O-B in the long-wave CO_2 band for AIRS, IASI and CrIS.

(see Section 3.2), surface and other detrimental effects. Only channels selected for assimilation are included.

The residual global mean bias is consistently very small, with a magnitude less than 0.02 K for most of these channels. A few channels around 670 cm^{-1} which have weighting-function peaks in the stratosphere have a negative residual bias, in contrast to the high peaking channels for many other sounders which usually exhibit positive biases with the same bias-correction method. The bias is still relatively small and overall the bias-correction scheme is performing well for CrIS.

The plot of O-B standard deviation shows very clearly the effect of the low-noise profile implicit in the CrIS instrument design. Values in the important temperature sounding channels in the 690–710 cm^{-1} band that sense the troposphere and lower stratosphere are around half those of equivalent IASI channels and just one third that of AIRS.

Figure 2 shows global maps of bias corrected O-B for a single channel from each instrument close to 700 cm^{-1}. These channels have similar weighting functions, peaking in the lower stratosphere at around 14 km. They have low biases and amongst the lowest standard deviations in O-B, allowing a 'clean' comparison of the data. The CrIS map exhibits a much smoother field than the other two, which look comparatively noisy. It is easy to pick out areas, particularly in the tropics, that are likely meteorological features that are not represented accurately in the background. This

is somewhat true with IASI but much less so with AIRS. There are no strong indications of any regional bias except around the Antarctic sea-ice edge, but this can also be seen in other instruments and does not appear to be as great with CrIS.

Overall, the data look to be of excellent quality, indicating that CrIS should be a beneficial addition to the observing system.

3. The assimilation system

A comprehensive description of the Met Office 4D-Var assimilation system is described by Rawlins *et al.* (2007). Here we outline the specific processing applied to CrIS data.

3.1. 1D-Var pre-processor

A 1D-Var pre-processing step is employed before CrIS data are passed to the assimilation system. This includes a number of important functions such as quality control checks, bias correction and channel selection. The 1D-Var is an optimal estimation technique used for the retrieval of atmospheric profiles that are consistent with the observation data and a short-range forecast as the background. The retrievals are not assimilated, but retrieved surface skin temperature, cloud-top-height and cloud fraction are passed to 4D-Var with the observed radiances and used as an additional constraint in the assimilation as they are unavailable in the background fields. The 1D-Var is also useful as a final quality control check – any observation for which the retrieval fails is excluded.

3.2. Channel selection

It is not feasible to assimilate all of the available channels in an NWP system as the computational resources required are too great. Moreover, many channels do not provide useful information for the analysis of temperature or humidity and there is redundancy where channels have overlapping weighting functions. A subset of channels is therefore required that captures most of the apposite information in the observations. This has been partially addressed with the channel selection of Gambacorta and Barnet (2013) but further screening is still necessary for use in NWP.

Additional screening draws on the experience gained with assimilating AIRS and IASI. The equivalent pre-storage channel selection for IASI consists of just 314 channels (from the original 8461), selected according to an information content measure tailored for an NWP system (Collard, 2007). For operational assimilation this is further reduced to a total of 138 channels following data quality assessments and forecast impact trials.

The CrIS channel selection is made on a similar basis: channels that exhibit anomalously large O-B statistics, or are difficult to forward model, are removed from the

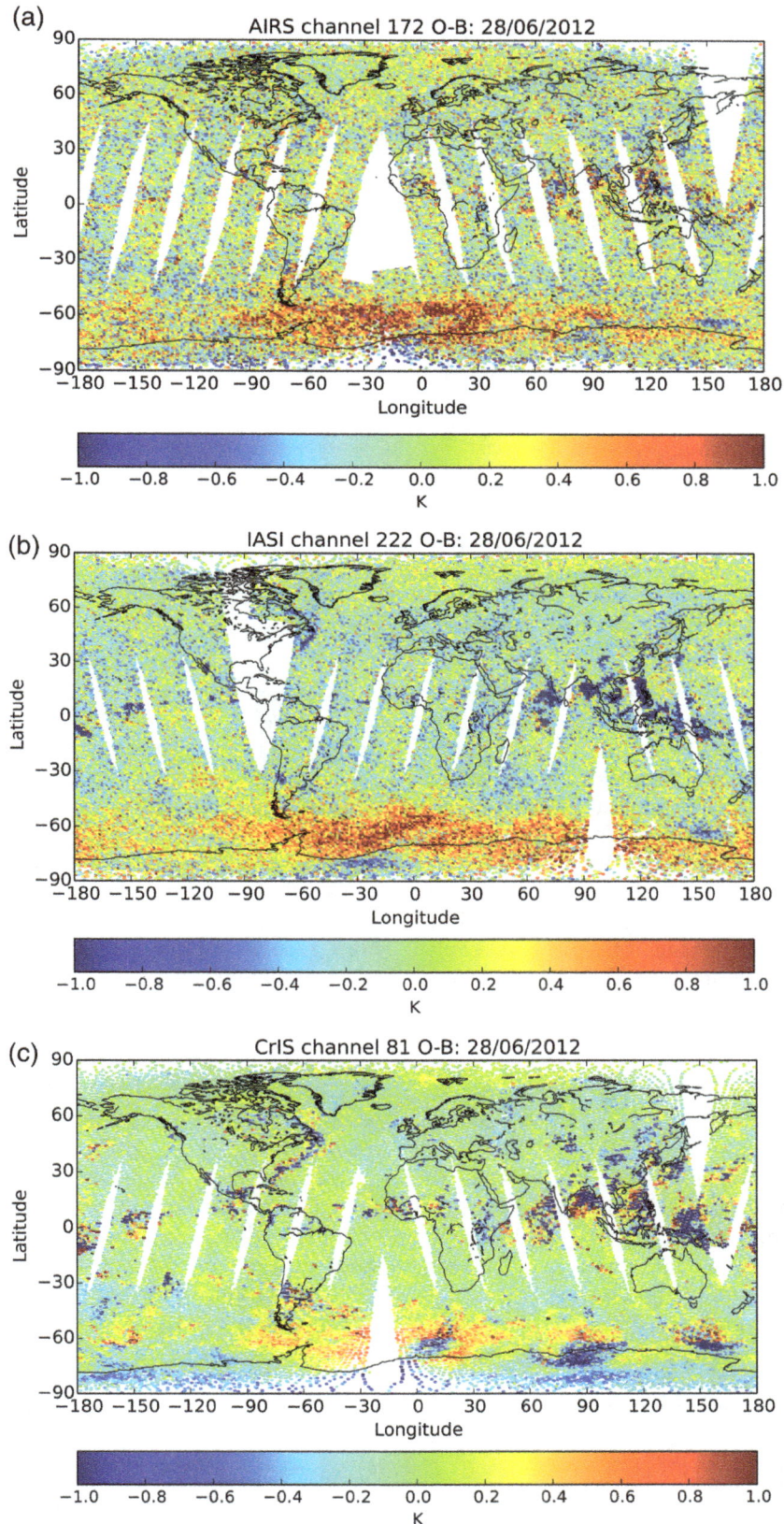

Figure 2. O-B maps for a lower stratospheric channel at approximately 700 cm^{-1} from each of AIRS, IASI and CrIS. Data are for two consecutive 6-h assimilation windows on 28 June 2012.

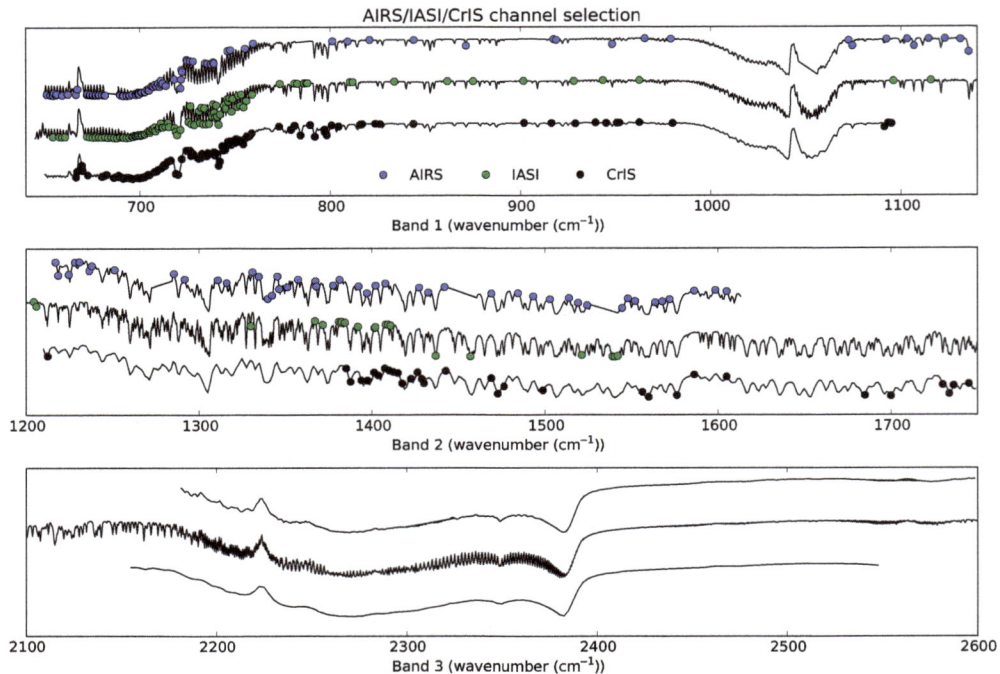

Figure 3. CrIS, IASI and AIRS channel selections overlaid on simulated spectra.

processing. These include any with trace-gas sensitivity, all short-wave (band-3) channels, some water-vapour channels and some high peaking temperature channels. A further reduction in the number of water-vapour channels, to obtain a similar proportion of the total as that of the IASI selection, was achieved by removing one from each pair of spectrally adjacent channels.

The remaining set of 134 CrIS channels then consists of 76 temperature, 45 water-vapour and 13 surface channels. The spectral coverage of this basis set can be seen in Figure 3 along with the AIRS and IASI channel selections. The selections are broadly similar in distribution, the main differences being in the water-vapour selection in band-2.

During processing, further screening is usually necessary to account for the effects of cloud and the underlying surface. This is simplified for our initial implementation by only using observations over the sea (where errors in surface emission are small). Most cloud-affected channels have to be excluded from the assimilation but a limited set can be used in cloudy conditions by retrieving cloud-top pressure and cloud fraction and keeping those channels which have weighting functions predominantly (90% or more) above the cloud (Pavelin *et al.*, 2008). This approach is used for AIRS and IASI and has also been applied here for CrIS.

3.3. Observation errors

For the 1D-Var retrieval, the combined forward model and observation error profile is comprised of an estimated radiative-transfer error of 0.2 K for all channels and the instrument noise which is derived from the pre-launch instrument specification.

The errors used in 4D-Var are inflated to account for representativeness error, misspecification of the background errors and the assumption that channels have uncorrelated errors (a diagonal error covariance matrix is used). Initially, values were chosen following a similar pattern to that used with IASI; 0.5 K for lower atmosphere sounding channels, 1 or 2 K for other temperature or surface-sensing channels and 4 K for water-vapour channels which are given a low weighting to reduce the likelihood of convergence problems in the assimilation.

The adverse effects of correlated error can be lessened by avoiding pairs of adjacent channels which are known in advance to be highly correlated (by approximately 63%) as a result of apodisation. However, adjacent temperature channels present in the original selection are retained in this case. Although there are a significant number of these (17 sequences of 2 or more consecutive channels in total), the effects can be mitigated by the use of conservative error estimates in the initial implementation. A better treatment of correlated error, such as that recently introduced for IASI (Weston *et al.*, 2014) can then be introduced in a later upgrade.

4. Assimilation experiments

An experimental forecasting suite with an operational configuration was used to assess the impact of CrIS, covering the period 28 June to 28 August 2012.

CrIS data, as noted earlier, is thinned before storage to one field-of-view per field-of-regard. For data assimilation this is thinned further to one field-of-view within a 125 km radius in the extra-tropics and 154 km in the

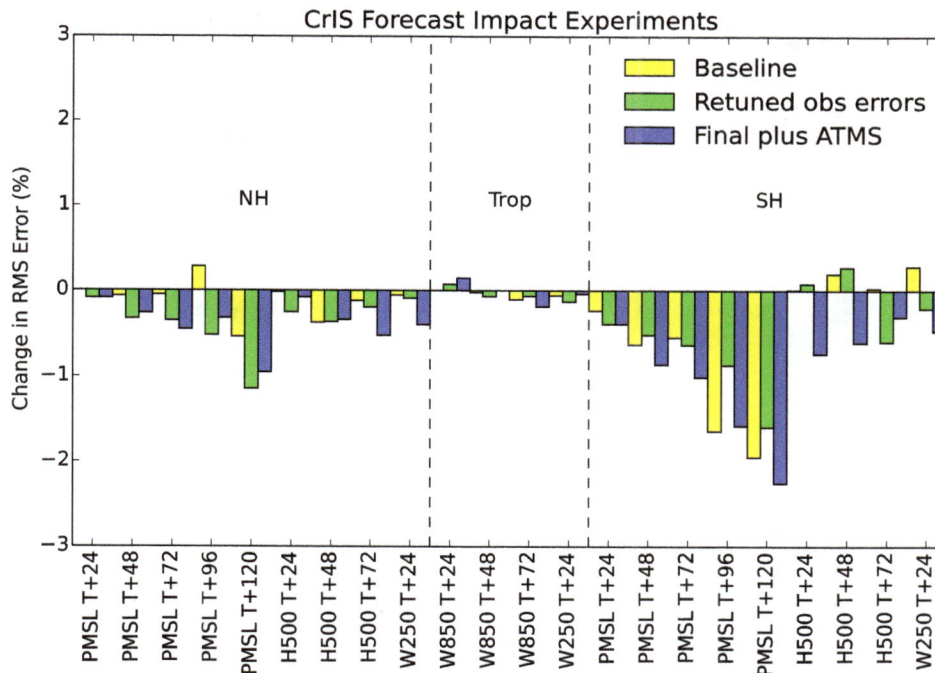

Figure 4. Changes in RMS error for selected forecast fields verified against observations (PMSL, pressure at mean sea level; H500, 500 hPa geopotential height; W250, wind velocity at 250 hPa) at selected forecast times (in hours, T = analysis time) for three experiments against a standard control: the baseline configuration, the final configuration (retuned observation errors) and the final configuration with ATMS. The mean RMS error across all fields shown is 0.25, 0.36 and 0.53% respectively.

tropics, the same spatial separation used for AIRS and IASI. CrIS observations were used only over the sea in these experiments.

The first experiment used the set of 134 channels (76 temperature, 45 water-vapour and 13 surface) and a conservative set of observation errors, as described above. This closely replicates the setup used for IASI but with more water-vapour channels and is referred to here as the baseline configuration. A number of variants were then tested to assess the impact of using different channel selections, and retuning the observation errors. Finally, an experiment was also run with both CrIS and ATMS included, to confirm that the two instruments are complementary when introduced simultaneously.

The impact on forecasts is presented here in terms of the change in RMS error for a selection of fields which are relevant to the large-scale circulation of the atmosphere. Forecast error is measured by verification against selected observations such as surface synoptic observation (SYNOP) stations for PMSL (pressure at mean sea level), and sondes for geopotential height and upper-level winds.

Figure 4 shows the impact of adding CrIS to a standard control experiment using the baseline configuration. The main improvements are to PMSL in the southern hemisphere, at all forecast ranges, but particularly at $T+96$ and $T+120$ where the RMS error is reduced by close to 2.0%. Changes to other fields are small and the mean change for all these fields combined is a reduction in RMS error of 0.25%. In terms of overall impact the baseline configuration gives a slight improvement in forecasting capability.

4.1. Effect of changing the channel selection

The baseline configuration contained 45 water-vapour channels. Previous experience with AIRS and IASI has shown that the assimilation of many such channels can be a problem. In early experiments with IASI (Hilton *et al.*, 2009), the assimilation system failed to converge properly when a large number of water-vapour channels were included. Only 30 water-vapour channels are assimilated for IASI, so a smaller set of 26 water-vapour channels was tested for CrIS. These were chosen by removing those with large departures from the mean O-B value, and those that have Jacobians with awkward features such as double peaks or unusually long tails into the stratosphere. An experiment was also performed with no water-vapour channels at all.

The results showed that removing water-vapour channels has an adverse effect. Using a reduced (26-channel) set gave a combined root-mean-square error (RMSE) reduction of only 0.1%, significantly lower than the original baseline experiment. Excluding all water-vapour channels lowers the impact further. Much of the observed improvement in the original experiment thus appears to come from the inclusion of water-vapour channels. The expected impact from the low-noise temperature channels is not very evident, likely because the observation errors are too conservative in this instance for them to have much of an effect on the system.

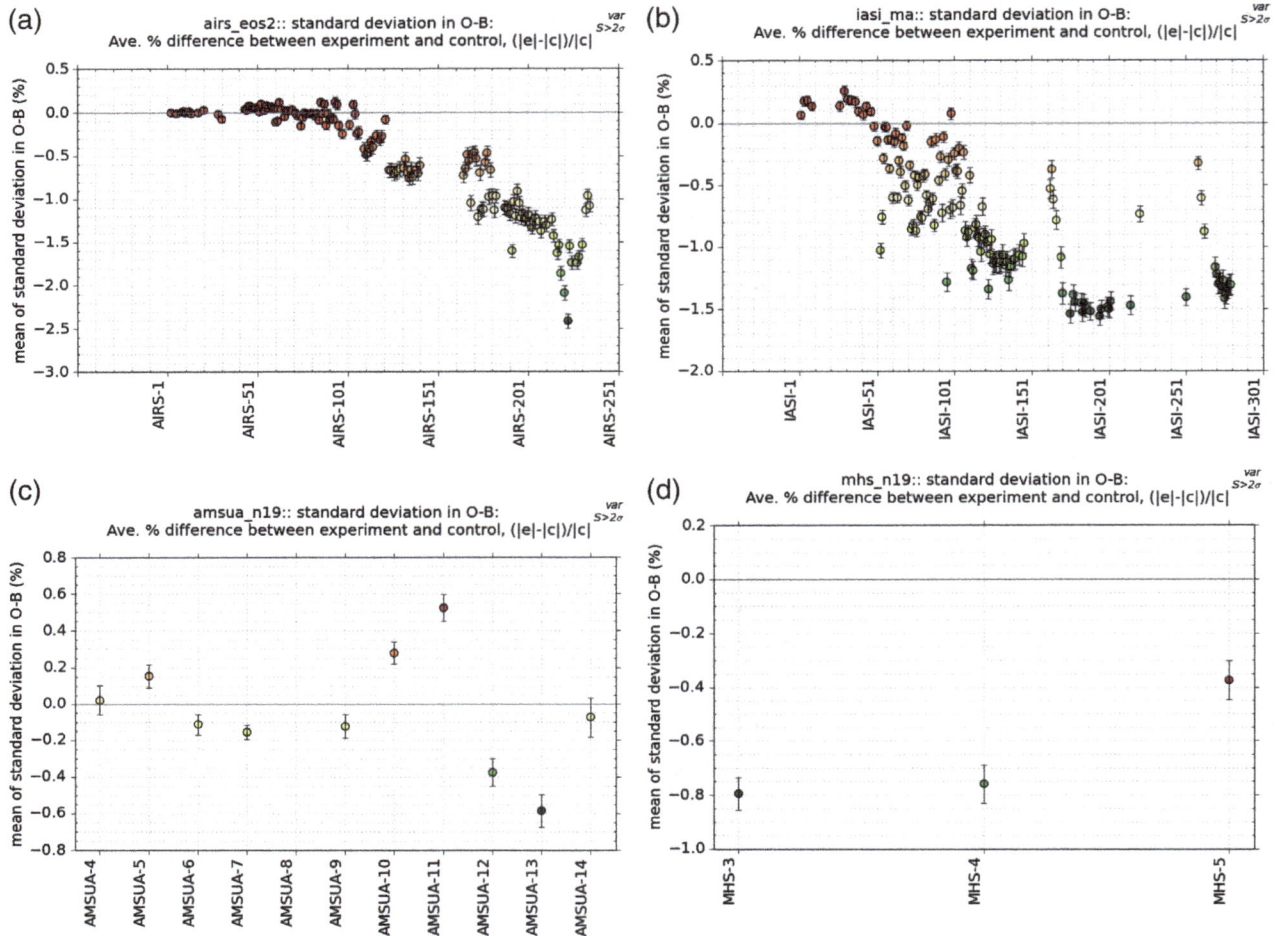

Figure 5. Changes in the standard deviation of O-B for AIRS, IASI, NOAA-19 AMSU-A and MHS for the final configuration experiment compared to the standard control. Values are obtained for each forecast cycle and averaged over the trial period. Note: channel numbers for AIRS/IASI refer to indices of stored subsets.

It was decided on this evidence that the full set of 134 baseline channels would be best for operational implementation.

4.2. Effect of tuning observation errors

The assumed errors in the baseline configuration were set to imitate those used for IASI, but the departures of O-B values from the mean for the temperature sounding channels are much smaller for CrIS than for IASI. On the basis of this we can justify reducing the observation errors for these channels, thereby increasing their weight in the assimilation.

Error estimates are always suboptimal for the reasons listed in Section 3.3, so in practice, a useful approximation that has been employed for other satellite data in the Met Office system, is that the standard deviation in the observation error should be no lower than twice the standard deviation in (O-B). The result of tuning by this method is that 44 of the temperature channels then have an error value of either 0.3 or 0.4 K. The error in a small number of window channels was inflated from 1 to 2 K based on the same reasoning.

The results of the experiment with retuned observation errors are also displayed in Figure 4. Overall, they show an improvement over the original experiment, with a mean RMS error reduction of 0.36% across the verification metrics shown. There is a slight loss of impact at mid-range in the southern hemisphere (where the statistical significance limits are large) but there are notable improvements in the Northern Hemisphere where the background errors are typically smaller and hence the analysis fields are more difficult to improve.

A study of the statistical significance of the individual verification metrics in the Met Office system (Weston, personal communication) shows that for a 47-day experiment the average confidence intervals range from ~0.5% for T+24 to ~2% for T+120, at a 95% confidence level. This suggests that some observed impacts may not be statistically significant for a 2-month experiment. Nevertheless, there is a general indication of improvement.

Another way of assessing the impact shown in this experiment is to look at the effect on O-B statistics for other instruments, for which significance limits are very small. Plots of the change in standard deviation of O-B for AIRS, IASI and NOAA-19 Advanced Microwave

Sounding Unit (AMSU-A) and Microwave Humidity Sounder (MHS) are shown in Figure 5, with significance limits shown. A clear improvement (up to 1.5%) can be seen for most AIRS and IASI channels, particularly those sensitive to water-vapour (on the right of each plot) and also for MHS. For D-A the change is more mixed but relatively small (less than 0.2% in the troposphere, up to 0.6% in the stratosphere) and starting from a small absolute value. Similar patterns can be seen for the microwave instruments on other satellites.

This configuration shows a small but clear improvement in forecast skill and so it was chosen for operational implementation.

4.3. In combination with ATMS

A further trial with both CrIS and ATMS included was also run, for which the results can be seen in Figure 4. Adding ATMS to CrIS (with 134 channels and retuned observation errors) gives a further improvement in forecast skill, mostly confined to the southern hemisphere. This is in line with results from ATMS-only trials (Doherty *et al.*, 2012). The mean RMSE reduction in this case is 0.53%. The improvement in forecast skill obtained by adding both instruments simultaneously is higher than that of either instrument on its own, but lower than their cumulative individual effects. This is to be expected as there is an overlap in atmospheric sensitivity.

5. Summary

The quality of CrIS data has been assessed by comparison with simulated observations derived from NWP model fields. Some of the key temperature sounding channels in the long-wave CO_2 band have a standard deviation in background departures as low as 0.15 K, around half that of IASI, and one third that of AIRS.

Forecast impact experiments using only observations over sea show a modest improvement in forecast error verified against observations when CrIS is added to the system. They also show significantly improved fit to background for AIRS, IASI and MHS channels. Further improvements are seen when ATMS is also included. However, more work is required to include data over land and to characterize the observation errors and correlations to fully exploit the low-noise characteristics of CrIS.

The final configuration chosen for operational implementation comprises 134 channels, and uses observation errors for the temperature sounding channels that

are reduced significantly compared with those for AIRS and IASI. Both CrIS and ATMS were made operational in the Met Office assimilation system on 30 April 2013.

References

Bormann N, Fouilloux A, Bell W. 2012. Evaluation and assimilation of ATMS data in the ECMWF system. ECMWF Technical Memorandum No. 689. http://old.ecmwf.int/publications/library/do/references/list/14 (accessed 11 December 2014).

Cameron JRN, Collard AD, English SJ. 2005. Operational use of AIRS observations at the Met Office. In Proceedings of ITSC-XIV, Beijing, China, 25–31 May 2005. http://cimss.ssec.wisc.edu/itwg/itsc/itsc14 (accessed 11 December 2014).

Collard AD. 2007. Selection of IASI channels for use in numerical weather prediction. *Quarterly Journal of the Royal Meteorological Society* **133**: 1977–1991.

Doherty A, Atkinson N, Bell W, Candy B, Keogh S, Cooper C. 2012. An initial assessment of data from the Advanced Technology Microwave Sounder. Met Office Forecast Research Technical Report No. 569. http://www.metoffice.gov.uk/learning/library/publications/science/weather-science-technical-reports (accessed 11 December 2014).

Gambacorta A, Barnet CD. 2013. Methodology and information content of the NOAA NESDIS operational channel selection for the Cross-Track Infrared Sounder (CrIS). *IEEE Transactions on Geoscience and Remote Sensing* **51**(6): 3207–3216.

Han Y, Revercomb H, Cromp M, Gu D, Johnson D, Mooney D, Scott D, Strow L, Bingham G, Borg L, Chen Y, DeSlover D, Esplin M, Hagan D, Jin X, Knuteson R, Motteler H, Predina J, Suwinski L, Taylor J, Tobin D, Tremblay D, Wang C, Wang L, Wang L, Zavyalov V. 2013. Suomi NPP CrIS measurements, sensor data record algorithm, calibration and validation activities, and record data quality. *Journal of Geophysical Research: Atmospheres* **118**(22): 12734–12748.

Harris BA, Kelly G. 2001. A satellite radiance-bias correction scheme for data assimilation. *Quarterly Journal of the Royal Meteorological Society* **127**: 1453–1468.

Hilton F, Atkinson NC, English SJ, Eyre JR. 2009. Assimilation of IASI at the Met Office and assessment of its impact through observing system experiments. *Quarterly Journal of the Royal Meteorological Society* **135**: 495–505.

Lu Q, Bell W, Bauer P, Bormann N, Peubey C. 2011. Characterising the FY-3A microwave temperature sounder using the ECMWF model. *Journal of Oceanic and Atmospheric Technology* **28**: 1373–1389.

NWP SAF. 2011. Annex to AAPP scientific documentation: pre-processing of ATMS and CrIS, document NWPSAF-MO-UD-027. http://nwpsaf.eu/deliverables/aapp/index.html (accessed 11 December 2014).

Pavelin EG, English SJ, Eyre JR. 2008. The assimilation of cloud-affected infrared satellite radiances for numerical weather prediction. *Quarterly Journal of the Royal Meteorological Society* **134**: 737–749.

Rawlins F, Ballard SP, Bovis KJ, Clayton AM, Li D, Inverarity GW, Lorenc AC, Payne TJ. 2007. The Met Office global four-dimensional variational data assimilation scheme. *Quarterly Journal of the Royal Meteorological Society* **133**: 347–362.

Weston PP, Bell W, Eyre JR. 2014. Accounting for correlated error in the assimilation of high resolution sounder data. *Quarterly Journal of the Royal Meteorological Society*, doi: 10.1002/qj.2306.

Zavyalov V, Esplin M, Scott D, Esplin B, Bingham G, Hoffman E, Lietzke C, Predina J, Frain R, Suwinski L, Han Y, Major C, Graham B, Phillips L. 2013. Noise performance of the CrIS instrument. *Journal of Geophysical Research: Atmospheres* **118**(23): 13108–13120.

Comparative assessment of evapotranspiration derived from NCEP and ECMWF global datasets through Weather Research and Forecasting model

Prashant K. Srivastava,*Dawei Han, Miguel A. Rico Ramirez and Tanvir Islam

Water and Environment Management Research Centre, Department of Civil Engineering, University of Bristol, UK

*Correspondence to:
P. K. Srivastava, Water and Environment Management Research Centre, Department of Civil Engineering, University of Bristol, Bristol, BS8 1TR, UK.
E-mail: cepks@bristol.ac.uk*

Abstract

In many hydro-meteorological applications, it is not always possible to get access to *in situ* weather measurements, especially for the ungauged catchments. This study explores the performances of downscaled weather data for Reference Evapotranspiration (ET_o) retrieval using the global European Centre for Medium Range Weather Forecasts (ECMWF) ERA interim and National Centers for Environmental Prediction (NCEP) reanalysis data, simulated through Weather Research and Forecasting (WRF) mesoscale model. The range of the Nash-Sutcliffe efficiency calculated for the ECMWF pooled datasets derived ET_o varies from 0.31 to 0.87, while for NCEP it is found to be 0.11 to 0.38. Bias and Root Mean Square Error (RMSE) are also indicating a very high discrepancy in the NCEP ET_o (Bias = −0.05; RMSE = 0.11) as compared to ECMWF (Bias = 0.00; RMSE = 0.06). The overall findings reveal that ECMWF downscaled products have a much better performance than the NCEP's counterparts.

Keywords: evapotranspiration; Weather Research and Forecasting model; NCEP; ECMWF; meteorological variables

1. Introduction

Reference Evapotranspiration (ET_o) is an important variable for hydro-meteorological applications (Liguori and Rico-Ramirez, 2013). ET_o has significant effect on catchment water balance and hence water yield and groundwater recharge (Al-Shrafany et al., 2012) and its reliable estimates from cropped surfaces are required for efficient irrigation management and scheduling (Lorite et al., 2012). Hence, monitoring ET_o at local, regional or global scales is important for assessing climate and human-induced effects on natural and agricultural ecosystems (Kustas and Norman, 1996; Thakur et al., 2012). However, to reduce the uncertainty in the ET_o estimates, appropriate meteorological data selection is important if derived from mesoscale models (Evans et al., 2011). There are now choices of global data available for ET_o estimation with reliable mesoscale model like the Weather Research and Forecasting (WRF) model for meteorological applications (Niyogi et al., 2009; Bromwich et al., 2009).

Many methods for estimating ET_o have been developed, and accurate estimates of ET_o are becoming available through the use of ground-based observations using the Penman Monteith equation (Novák, 2012). But these ground-based observations can cover only a smaller area. However, larger areas require a large number of observation sites because of the heterogeneity of landscapes and very high variations in the energy transfer processes (Detto et al., 2006). Nevertheless, the above-mentioned approaches are very expensive and labour intensive, so other approaches are required such as mesoscale models to estimate ET_o at large scales. Recently, efforts have been made to determine the spatial and temporal variability of ET_o through mesoscale model like the MM5 (Ishak et al., 2010; Niyogi et al., 2009) but there is a lack of appropriate studies available with the WRF model, especially for temperate maritime climate.

There are very rare studies available which demonstrated the accuracy of data chosen for the ET_o estimation from mesoscale model such as WRF using the global European Centre for Medium Range Weather Forecasts (ECMWF) ERA interim and NCEP (National Centers for Environmental Prediction) reanalysis data to determine the spatial and temporal distribution of ET_o (Buizza et al., 2005). Therefore hydro-meteorologists would like to know how well the downscaled global data products are as compared to ground-based measurements and whether it is possible to use the downscaled data for ungauged catchments. Even with gauged catchments, most of the stations have only rain and flow gauges installed. Measurements of other weather hydro-meteorological variables such as solar radiation, wind speed, air temperature and dew point are usually missing and thus complicate the problems. Hence, the foremost objective of this article is to compare the ET_o products estimated from the downscaled ECMWF and NCEP global datasets using the WRF model, centred over the Brue catchment in south west of the England. The validations of the products are made by using the ground-based measurements retrieved from meteorological weather station located in the Brue catchment.

2. Materials and methodology

2.1. Study area and datasets

The Brue catchment (135.5 km^2) is chosen as the study area which is located in the south-west of England, 51.11°N and 2.47°W. The major land use/land cover is pasture land on clay soil with some patches of woodland in the higher eastern catchment. The land use/land cover of Brue is illustrated in Figure 1 with the Digital Elevation Model and land use. The meteorological datasets are provided by the British Atmospheric Data Centre (BADC), UK that includes wind speed, net radiation, surface temperature and dew point. Despite the increasing computing power from desktop PCs, downscaling by the WRF at a high spatial and temporal resolution is still quite time consuming. As a result, only 4 months (January 2011, April 2011, July 2011 and October 2011) of data have been analysed in this study corresponding to four seasons in UK that is winter, spring, summer and autumn. The data provided by BADC are used for evaluating the hourly downscaled meteorological data from the WRF model Version 3.1. The global ECMWF ERA interim and NCEP reanalysis data can be downloaded from their respective websites (ECMWF-http://www.ecmwf.int/ and NCEP-http://rda.ucar.edu/). The ERA interim dataset is updated in monthly batches with 3 months' delay and has a resolution of T255 (triangular truncation at 255), N128 (128 latitude circles, pole to equator), L60 (model levels), 37 pressure levels and 15/16 isentropic levels (www.ecmwf.int/products/data/). the NCEP FNL (Final) Operational Global Analysis data are on $1.0 \times 1.0°$ grids prepared

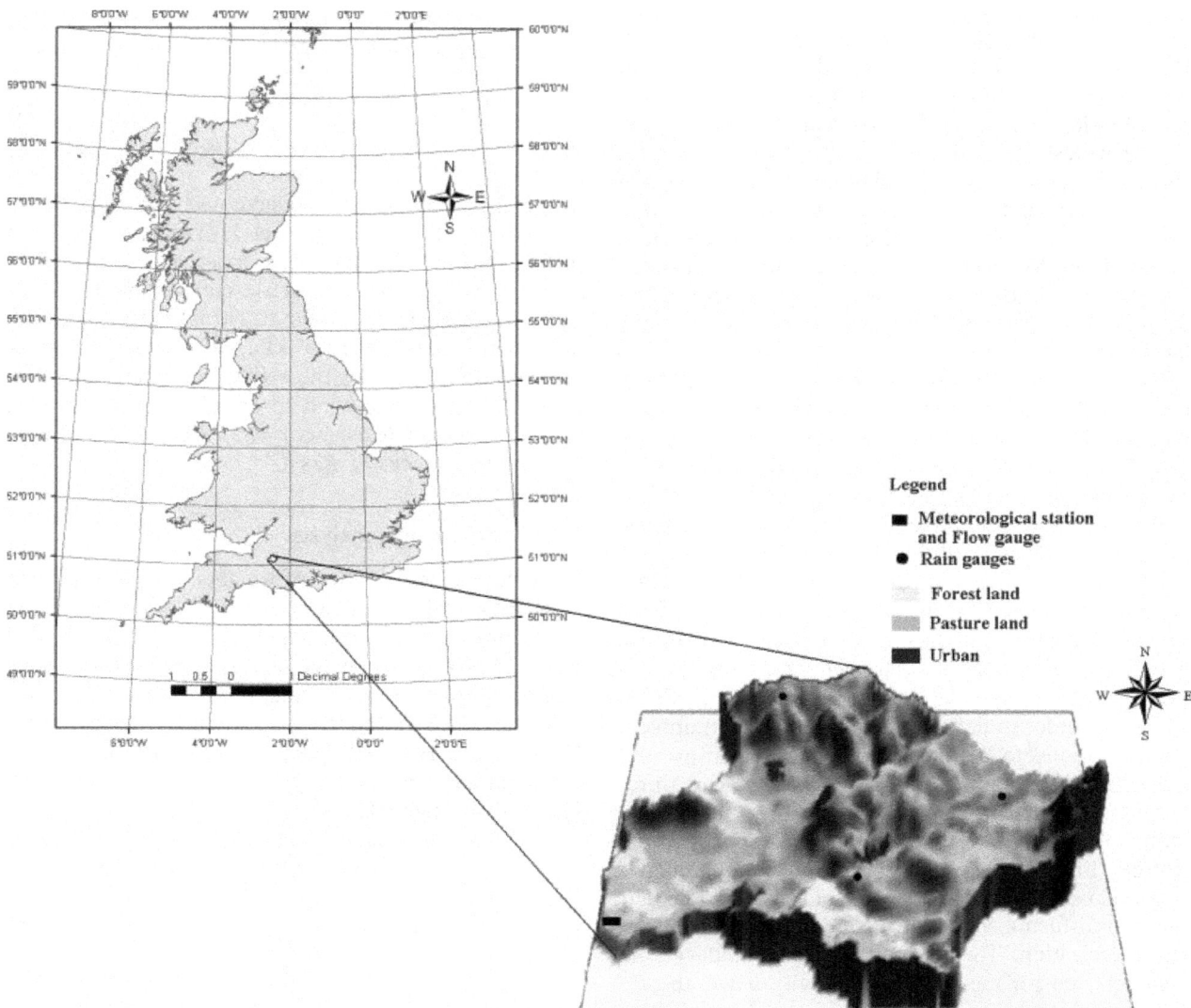

Figure 1. Geographical location of the study area with digital elevation model, land use and observation stations.

operationally every 6 h. This product is from the Global Data Assimilation System, which continuously collects observational data. The archive time series is continuously extended to a near-current date but not maintained in real-time (http://rda.ucar.edu/datasets/ds083.2/). The main purposes of these reanalysis data are to deliver compatible, high-resolution and high quality historical global atmospheric datasets for their use in weather research communities.

2.2. Weather Research Forecasting model

The mesoscale model used in this study is the WRF Model with Advanced Research WRF dynamic core version 3.1 (Powers, 2007; Schwartz et al., 2009). WRF is a next-generation, non-hydrostatic, with terrain following eta-coordinate mesoscale modelling system designed to serve both operational forecasting and atmospheric research needs (Skamarock and Klemp, 2008). We choose WRF in this study because it is being developed and studied by a broad community of government and university researchers and results are quite efficient (Skamarock et al., 2005). The WRF model is centred over the Brue catchment with three nested domains (D1, D2 and D3) of horizontal grid spacing of 81, 27 and 9 km, in which the innermost domain (D3) is the area of interest. These three domains consisted of 18×18, 19×19 and 22×22 horizontal grids points. A two-way nesting scheme is used allowing information from the child domain to be fed back to the parent domain. Imposed boundary conditions are updated every 6 h when using the ECMWF or NCEP Final Analysis ($1° \times 1°$ FNL) dataset.

The WRF model is used to downscale the ECMWF and NCEP data to predict wind speed, solar radiation, surface temperature and dew point temperature. The main physical options used in the WRF setup were the Dudhia shortwave radiation (Dudhia, 1989) and Rapid Radiative Transfer Model long wave radiation (Mlawer et al., 1997) with Lin microphysical parameterization; the Betts–Miller–Janjic (BMJ) Cumulus parameterization schemes; the Yonsei University planetary boundary layer scheme (Hu et al., 2010). The BMJ cumulus parameterization scheme is used because it considers sophisticated cloud mixing scheme in order to determine entrainment/detrainment which is found to be more applicable to non-tropical convection (Gilliland and Rowe, 2007). The third-order Runge–Kutta is used for the time integration while for spatial differencing scheme the sixth-order centred differencing scheme is used. The Arakawa C-grid is used for the horizontal grid distribution. The Thermal diffusion scheme is used for the surface layer parameterization. The top and bottom boundary condition chosen for the study are Gravity wave absorbing (diffusion or Rayleigh damping) and physical or free-slip respectively. The Lambert conformal conic projection is used as the model horizontal coordinates.

The vertical coordinate η is defined as:

$$\eta = \frac{(p_r - p_t)}{(p_{rs} - p_t)} \quad (1)$$

where, P_r is pressure at the model surface being calculated; p_{rs} is the pressure at the surface and P_t is the pressure at the top of the model. In the vertical 28 terrain following eta levels (eta levels = 1.000, 0.990, 0.978, 0.964, 0.946, 0.922, 0.894, 0.860, 0.817, 0.766, 0.707, 0.644, 0.576, 0.507, 0.444, 0.380, 0.324, 0.273, 0.228, 0.188, 0.152, 0.121, 0.093, 0.0.069, 0.048, 0.029, 0.014, 0.000) from surface were used. These eta levels are used in this study because of their better representation of the topography (Routray et al., 2010).

2.3. Evapotranspiration estimation

The ET_o from WRF downscaled meteorological and observation stations variables is calculated using the Penman and Monteith (PM) method proposed and developed by (Penman, 1956; Monteith, 1965) as given in FAO56 report (Allen et al., 1998). The ET_o (in mm) according to the PM equation is as follows:

$$ET_o = \frac{0.408\Delta (R_n - G) + \gamma \frac{37}{T+273} U_2 (e_s - e_a)}{\Delta + \gamma \left(1 + \frac{r_c}{r_a}\right)} \quad (2)$$

where Δ is the slope of the saturated vapour pressure curve (kPa $°C^{-1}$); R_n the net radiation at the crop surface (MJ m^{-2} h^{-1}); G the soil heat flux density (MJ m^{-2} h^{-1}); γ the psychrometric constant (kPa $°C^{-1}$); T the mean air temperature at 2 m height ($°C$); e_s the saturation vapour pressure (kPa); e_a the actual vapour pressure (kPa); $e_s - e_a$ the saturation vapour pressure deficit (kPa); U_2 the wind speed at 2 m height m s^{-1}; r_a (aerodynamic resistance) = $208/U_2$ s m^{-1}; and r_c (canopy resistance) = 70 s m^{-1}.

2.4. Performance analysis

The detailed investigation of weather variables derived from mesoscale model is compared to ground-based measurements. The estimated ET_o from the NCEP and ECMWF is compared with in situ observations. The three performance statistics: the Nash-Sutcliffe efficiency (NSE) (Nash and Sutcliffe, 1970), Root Mean Square Error (RMSE) and Absolute Bias (Bias) are taken into account for performance measurements. The NSE is calculated using:

$$\text{NSE} = 1 - \frac{\sum_{i=1}^{n} [y_i - x_i]^2}{\sum_{i=1}^{n} [x_i - \overline{x}]^2} \quad (3)$$

where x_i is the ground-based measurements and y_i is the estimated measurements.

The RMSE is calculated using the equation:

$$\text{RMSE} = \sqrt{\left(\frac{1}{n}\sum_{i=1}^{n}[y_i - x_i]^2\right)} \qquad (4)$$

The absolute bias (Bias) measures the positive or negative deviation of the measured value from the true value. The optimal value of Bias is 0.0, with low-magnitude values indicating accurate model simulation. It can be calculated using the following relation:

$$\text{Bias} = [(\overline{y} - \overline{x})] \qquad (5)$$

where $\overline{x}$ is the mean of ground-based measurements and $\overline{y}$ is the mean of estimated measurements.

3. Results and discussion

3.1. Performance of hydro-meteorological variables

The WRF is simulated over the Brue catchment in order to calculate the hydro-meteorological variables. As discussed earlier, these calculations are made on an hourly basis representing dominant season in UK during the year 2011. The comparisons of the methods are first made on a seasonal basis and then represented on a combined form (pooled datasets). The observed seasonal and pooled weather variables performance statistics are shown in Table I. The three statistical indices, i.e. NSE, RMSE and Bias are calculated between the WRF downscaled weather variables (wind speed, dew point, surface temperature and solar radiation) and *in situ* measurements. It is seen that on a seasonal basis the ECMWF datasets are giving the smallest discrepancies as compared to the NCEP datasets derived variables. All the four weather variables from ECMWF have a RMSE much lower than the variable from NCEP. Weather variables downscaled from the ECMWF dataset have the minimum bias discrepancies. The modelled wind speed is generally greater than the measured during all the seasons under consideration. Previous studies by (Ishak *et al.*, 2010; Ishak *et al.*, 2013) also indicate that wind speed is the most difficult variable to downscale using the MM5 model with the results showing significant over estimation. On the basis of the NSE statistics, dew point and temperature have the highest values followed by least RMSE. Bias suggested a significant under estimation of radiation in nearly all seasons. Studies by Remesan *et al.* (2008) and Ahmadi *et al.* (2009) also suggested that solar radiation is a difficult parameter to obtain and has a high sensitivity towards ET_o estimation (Ishak *et al.*, 2010). In order to see the performances of the weather variables, the relative scatter plots are shown in Figure 2(a)–(h). All the four seasons are giving good results as compared with the observed datasets. The worst case is the NCEP dataset, which has the largest deviation for equiline and showing a very poor NSE.

Table I. Performance statistics for the seasonal and pooled hourly weather variables.

Variables	ECMWF NSE	ECMWF RMSE	ECMWF Bias	NCEP NSE	NCEP RMSE	NCEP Bias
January						
Dewpoint (°C)	0.78	2.14	−0.80	0.75	2.30	−0.65
Temperature (°C)	0.69	2.52	−0.41	0.53	3.10	−0.45
Wind speed (m s⁻¹)	−2.25	4.96	4.26	−16.24	11.43	9.52
Solar radiation (W m⁻²)	0.19	54.52	−1.91	0.32	49.83	−1.23
April						
Dewpoint (°C)	0.19	2.46	−1.63	−0.72	3.59	−2.53
Temperature (°C)	0.74	2.33	−0.28	−1.30	7.00	−4.97
Wind speed (m s⁻¹)	−1.73	3.82	2.96	−12.59	8.52	7.12
Solar radiation (W m⁻²)	0.84	99.99	−1.79	0.65	150.10	−54.18
July						
Dewpoint (°C)	−0.15	2.43	−1.52	−1.29	3.43	−2.37
Temperature (°C)	0.64	2.03	−1.06	−3.72	7.39	−5.95
Wind speed (m s⁻¹)	−0.61	3.00	2.05	−6.18	6.34	4.80
Solar radiation (W m⁻²)	0.56	171.63	2.36	0.60	163.48	−32.55
October						
Dewpoint (°C)	0.68	2.06	−1.18	0.51	2.53	−1.18
Temperature (°C)	0.65	2.55	−0.47	0.21	3.84	−1.68
Wind speed (m s⁻¹)	−1.86	4.42	3.86	−16.51	10.94	9.38
Solar radiation (W m⁻²)	0.62	84.29	−28.22	0.50	96.76	−25.93
Pooled						
Dewpoint (°C)	0.77	2.28	−1.28	0.59	3.01	−1.67
Temperature (°C)	0.85	2.37	−0.56	0.12	5.65	−3.25
Wind speed (m s⁻¹)	−1.62	4.12	3.28	−13.05	9.54	7.71
Solar radiation (W m⁻²)	0.72	111.37	−7.43	0.66	123.36	−28.27

3.2. Comparative assessment of evapotranspiration products

The relative plots between seasonal ECMWF and NCEP with observed ET_o are shown in Figure 3(a) and (b), while the seasonal and pooled variation in ET_o are depicted in Figure 3(c)–(h). The performance statistics obtained between seasonal and pooled datasets are shown in Table II. A significant difference existed between the NCEP and ECMWF pooled ET_o for all the four seasons under consideration. Two distinct features can be seen from these figures that ECMWF follows the seasonal variation better than the NCEP and secondly, a very high under estimation by NCEP from the ground observed datasets. The simulated and ground-based observation indicates that during the studied year, the ET_o increases during the summer season following spring, and then decreased during the autumn and winter season, exhibiting a bell-shape response in the pooled datasets plots. Though the ET_o increases from winter to summer, this increase is slightly more rapid in spring than other seasons. Bias statistics show that the simulated products, in general, under estimated the ground-based ET_o. However, this under estimation is found to be very high for the NCEP data. The scatter plots obtained for NCEP pooled datasets with the observed ET_o indicates a very high deviation from equiline, revealing a very high under estimation than ECMWF datasets. Again, in case of ECMWF all the four seasons are also giving good results as compared with the observed datasets. The worst case is the

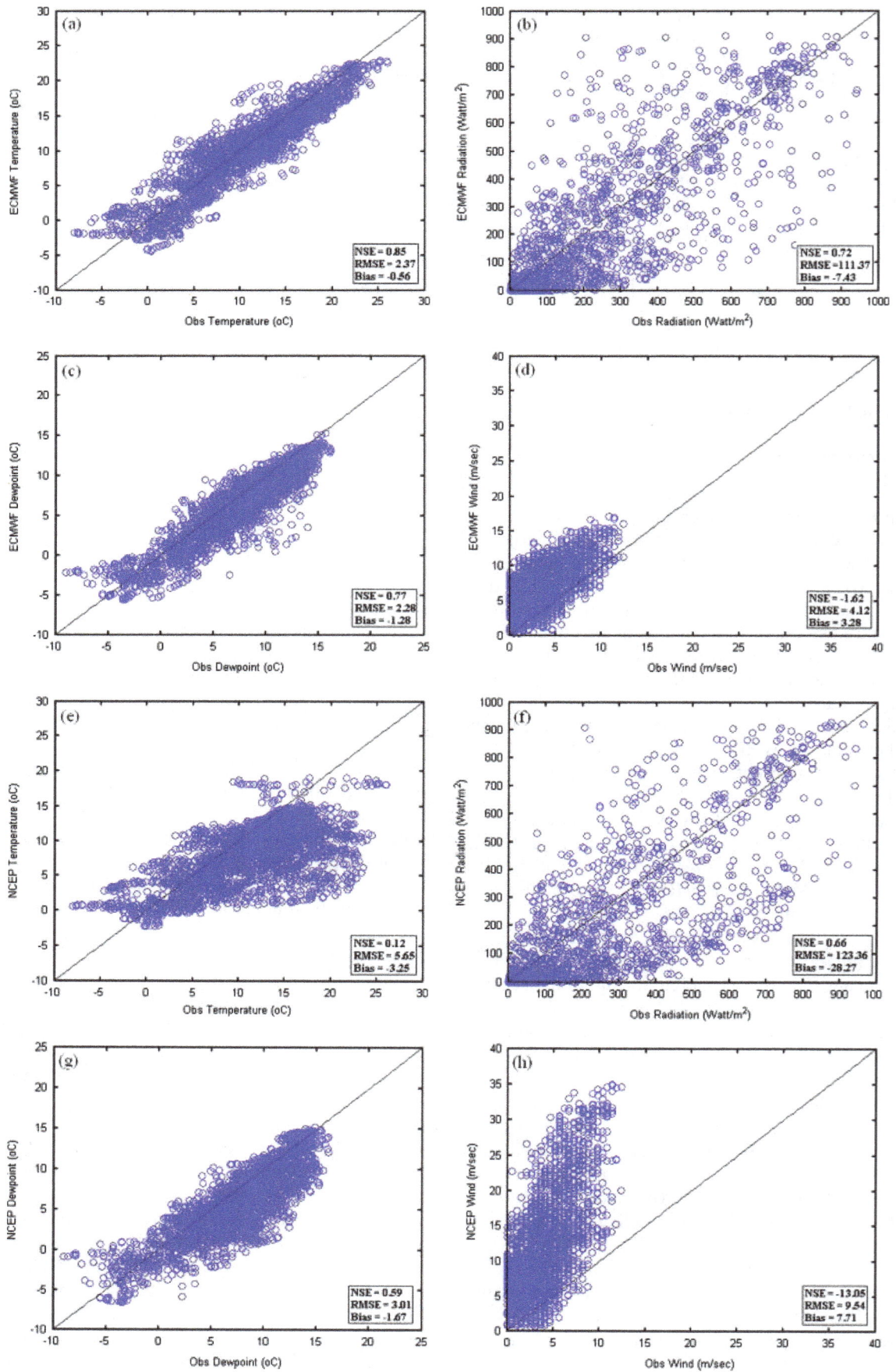

Figure 2. (a–h) Scatter plots representing the ECMWF (a–d) and NCEP (e–h) hourly weather variables.

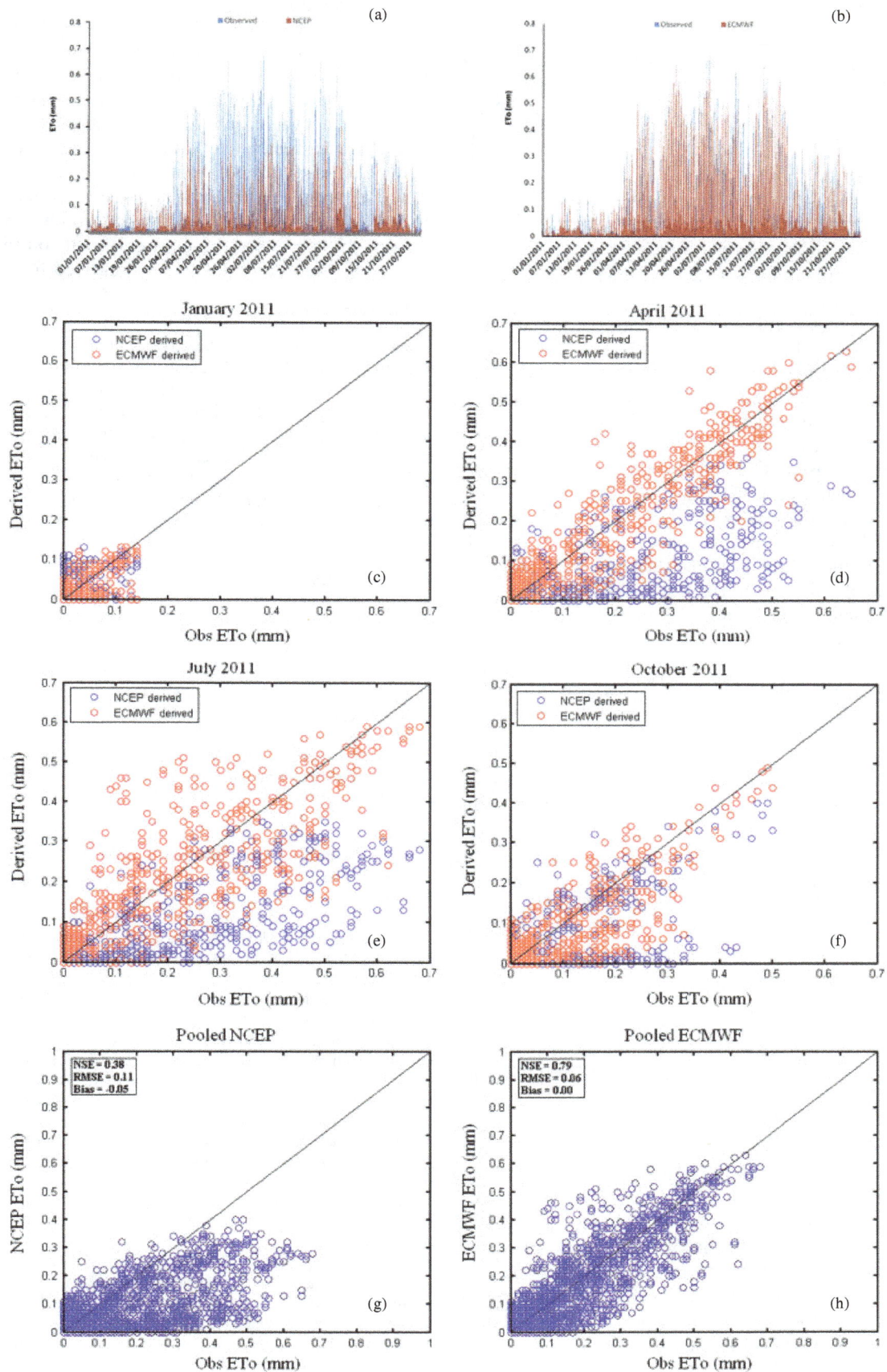

Figure 3. (a–h) Seasonal variations in the estimated ET_o (a–b); scatter plots representing the seasonal (c–f) and pooled (g–h) NCEP and ECMWF hourly ET_o with observed datasets.

Table II. Performance statistics for the seasonal and pooled hourly ET_o.

Data	Indicator	January	April	July	October	Pooled
ECMWF	NSE	0.31	0.87	0.71	0.65	0.79
	RMSE	0.02	0.06	0.09	0.06	0.06
	Bias	0.00	0.01	0.00	−0.01	0.00
NCEP	NSE	0.11	0.23	0.30	0.34	0.38
	RMSE	0.03	0.14	0.15	0.08	0.11
	Bias	0.00	−0.07	−0.08	−0.02	−0.05

NCEP datasets, which has largest deviation for equiline and pitiable NSE observed for pooled one. During all seasons, in general, NCEP ET_o under estimated the ground-based values. In contrast, a small over estimation is observed with the ECMWF datasets during spring and on the other hand autumn is showing an under estimation which seems to be associated with the rapid decrease in air temperature and radiation during this season. Conversely, the over estimations seems to correspond with the increases in air temperature and radiation, particularly during the spring. The RMSE in January is generally lesser than the corresponding values in other seasons, suggesting an influence of climatic variables on model downscaling. The magnitudes of RMSE indicate that ECMWF performed best during all the seasons and the worst performance is observed by NCEP datasets. The NSE values ranged from 0.314 to 0.873. NSE values are positive confirming that all the seasons are giving good agreement with the observed datasets for ET_o except for October. However, for NCEP datasets the range of NSE is found to be from 0.107 to 0.340, which is found to be very small as compared to ECMWF. The most probable reason for the lower efficiency of NCEP ET_o can be attributed to the poor performances of wind and temperature. The high value of NSE obtained for ECMWF shows that ECMWF unambiguously performing better than NCEP and could be used for hydro-meteorological applications.

4. Conclusion

A numerical weather model such as the WRF is able to downscale the global data into much finer resolutions in space and time for hydro-meteorological investigations. However, despite the importance of this valuable data source, there is lack of study in technical literature domain about the quality of such data. The downscaling process generally improves the data quality and provides higher data resolution. The study indicates that downscale products from the global reanalysis data to finer resolutions could be suitable for hydrological and meteorological applications. The ET_o values estimated from the NCEP data are significantly under estimated across all the seasons. The suitability of data for ET_o estimations suggest that ECMWF is giving far better performance than NCEP products. This study provides hydrologists with valuable information on downscaled weather variables from global datasets, and further exploration of this potentially valuable data source by the hydrological community should be encouraged so that useful experience and knowledge could be accumulated for different geographical and climatic conditions. A clear pattern is obtained among some weather variables with *in situ* measurements and seems to be promising for bias corrections in the downscaled data from *in situ* measurements or through regionalization from surrounding weather stations. Hence future research will focus on bias correction of theses global data for improved forecasting.

Acknowledgements

The authors would like to thank the Commonwealth Scholarship Commission, British Council, UK and Ministry of Human Resource Development, Government of India for providing the necessary support and funding for this research. The authors would like to acknowledge the British Atmospheric Data Centre, UK for providing the ground datasets. The author also acknowledges the Advanced Computing Research Centre at University of Bristol for providing the access to supercomputer facility (The Blue Crystal) for some of the analysis.

References

Ahmadi A, Han D, Karamouz M, Remesan R. 2009. Input data selection for solar radiation estimation. *Hydrological Processes* 23(19): 2754–2764.

Allen, R.G, Pereira L.S, Raes D, and Smith M, 1998 Crop evapotranspiration-Guidelines for computing crop water requirements-FAO Irrigation and drainage paper 56. *FAO:* Rome 300:6541.

Al-Shrafany D, Rico-Ramirez MA, Han D. 2012. Calibration of roughness parameters using rainfall–runoff water balance for satellite soil Moisture retrieval. *Journal of Hydrologic Engineering* 17(6): 704–714.

Bromwich DH, Hines KM, Bai LS. 2009. Development and testing of polar weather research and forecasting model: 2. Arctic Ocean. *Journal of Geophysical Research* 114(D8): D08122.

Buizza R, Houtekamer PL, Pellerin G, Toth Z, Zhu Y, Wei M. 2005. A comparison of the ECMWF, MSC, and NCEP global ensemble prediction systems. *Monthly Weather Review* 133(5): 1076–1097.

Detto M, Montaldo N, Albertson JD, Mancini M, Katul G. 2006. Soil moisture and vegetation controls on evapotranspiration in a heterogeneous Mediterranean ecosystem on Sardinia, Italy. *Water Resources Research* 42(8): W08419.

Dudhia J. 1989. Numerical study of convection observed during the winter monsoon experiment using a mesoscale two-dimensional model. *Journal of the Atmospheric Sciences* 46(20): 3077–3107.

Evans JP, McCabe MF, Mueller B, Meng X, Ershadi A. 2011. A comparison of satellite evapotranspiration estimation efforts. Paper read at Proceedings, Water Information Research and Development Alliance Science Symposium. Melbourne, Australia.

Gilliland EK, Rowe CM. 2007. A comparison of cumulus parameterization schemes in the WRF model. Paper read at Proceedings of the 87th AMS Annual Meeting & 21st Conference on Hydrology. San Antonio, TX.

Hu XM, Nielsen-Gammon JW, Zhang F. 2010. Evaluation of three planetary boundary layer schemes in the WRF model. *Journal of Applied Meteorology and Climatology* 49(9): 1831–1844.

Ishak AM, Bray M, Remesan R, Han D. 2010. Estimating reference evapotranspiration using numerical weather modelling. *Hydrological Processes* 24(24): 3490–3509.

Ishak AM, Remesan R, Srivastava PK, Islam T, Han D. 2013. Error correction modelling of wind speed through hydro-meteorological parameters and mesoscale model: a hybrid approach. *Water resources management* 27: 1–23.

Kustas WP, Norman JM. 1996. Use of remote sensing for evapotranspiration monitoring over land surfaces. *Hydrological Sciences Journal* **41**(4): 495–516.

Liguori S, Rico-Ramirez MA. 2013. A practical approach to the assessment of probabilistic flow predictions. *Hydrological Processes* **27**: 18–32.

Lorite IJ, García-Vila M, Carmona MA, Santos C, Soriano MA. 2012. Assessment of the irrigation advisory services' recommendations and farmers' irrigation management: a case study in southern Spain. *Water resources management* **26**: 2397–2419.

Mlawer EJ, Taubman SJ, Brown PD, Iacono MJ, Clough SA. 1997. Radiative transfer for inhomogeneous atmospheres: RRTM, a validated correlated-k model for the longwave. *Journal of Geophysical Research* **102**(D14): 16663–16682.

Monteith JL. 1965. Evaporation and environment *Symposia of the Society for Experimental Biology* **19**: 205–223.

Nash JE, Sutcliffe JV. 1970. River flow forecasting through conceptual models part I—a discussion of principles. *Journal of Hydrology* **10**(3): 282–290.

Niyogi D, Alapaty K, Raman S, Chen F. 2009. Development and evaluation of a coupled photosynthesis-based gas exchange evapotranspiration model (GEM) for mesoscale weather forecasting applications. *Journal of Applied Meteorology and Climatology* **48**(2): 349–368.

Novák V. 2012. Evapotranspiration in the Soil-Plant-Atmosphere System. In *Progress in Soil Science*. Springer Verlag: Berlin.

Penman HL. 1956. Estimating evaporation. *Transactions American Geophysical Union* **37**(1): 43–50.

Powers JG. 2007. Numerical prediction of an Antarctic severe wind event with the Weather Research and Forecasting (WRF) model. *Monthly Weather Review* **135**(9): 3134–3157.

Remesan R, Shamim MA, Han D. 2008. Model data selection using gamma test for daily solar radiation estimation. *Hydrological Processes* **22**(21): 4301–4309.

Routray A, Mohanty UC, Niyogi D, Rizvi SRH, Osuri KK. 2010. Simulation of heavy rainfall events over Indian monsoon region using WRF-3DVAR data assimilation system. *Meteorology and Atmospheric Physics* **106**(1): 107–125.

Schwartz CS, Kain JS, Weiss SJ, Xue M, Bright DR, Kong F, Thomas KW, Levit JJ, Coniglio MC. 2009. Next-day convection-allowing WRF model guidance: a second look at 2-km versus 4-km grid spacing. *Monthly Weather Review* **137**(10): 3351–3372.

Skamarock WC, Klemp JB. 2008. A time-split nonhydrostatic atmospheric model for weather research and forecasting applications. *Journal of Computational Physics* **227**(7): 3465–3485.

Skamarock WC, Klemp JB, Dudhia J, Gill DO, Barker DM, Wang W, Powers JG. 2005. A description of the Advanced Research WRF Version 2. DTIC Document.

Thakur JK, Srivastava PK, Singh SK, Vekerdy Z. 2012. Ecological monitoring of wetlands in semi-arid region of Konya closed Basin, Turkey. *Regional Environmental Change* **12**(1): 133–144, DOI: 10.1007/s10113-011-0241-x.

20

The importance of surface layer parameterization in modeling of stable atmospheric boundary layers

Esa-Matti Tastula,[1]* Boris Galperin,[1] Semion Sukoriansky,[2] Ashok Luhar[3] and Phil Anderson[4]

[1] College of Marine Science, University of South Florida, St. Petersburg, FL, USA
[2] Department of Mechanical Engineering, Ben-Gurion University of the Negev, Beer-Sheva, Israel
[3] CSIRO Marine and Atmospheric Research, Aspendale, Australia
[4] Scottish Association for Marine Science, Oban, UK

*Correspondence to:
E.-M. Tastula, College of Marine Science, University of South Florida, St. Petersburg, FL, USA.
E-mail: tastulae@mail.usf.edu

Abstract

The accuracy of prediction of stable atmospheric boundary layers depends on the parameterization of the surface layer which is usually derived from the Monin–Obukhov similarity theory. In this article, several surface-layer models in the format of velocity and potential temperature Deacon numbers are compared with observations from CASES99, Cardington, and Halley datasets. The comparisons were hindered by a large amount of scatter within and among datasets. Tests utilizing R^2 demonstrated that the quasi-normal scale elimination (QNSE) theory exhibits the best overall performance. Further proof of this was provided by 1D simulations with the Weather Research and Forecasting (WRF) model.

Keywords: constant flux layer; Deacon numbers; stable stratification; turbulence parameterization

1. Introduction

Models of the stable atmospheric boundary layer (SABL), for example, those used in numerical weather prediction (NWP), require a surface layer (SL) parameterization that links them to the near-surface fluxes of momentum, heat, and moisture. Most existing parameterizations rely upon the Monin–Obukhov similarity theory (MOST) which relates the mean profiles of meteorological quantities to their respective surface fluxes (Monin and Obukhov, 1954). MOST pertains to a relatively thin layer, about the lowest 10% of the boundary-layer depth, in which these fluxes are nearly constant, i.e. *the constant flux layer*. Effects of stratification are represented by non-dimensional gradient functions, also known as *similarity functions*. These functions have been scrutinized in a voluminous body of literature; see Högström (1996), Foken (2006), and Mahrt (2014) for appropriate references. The shape of the similarity functions has traditionally been determined from experiments. The extent to which empirical curves differ from each other depends on the prevailing stability: in unstable conditions, all functions are reasonably similar (e.g. Businger *et al.*, 1971) whereas in stable conditions, considerable scatter between different expressions is rather typical (Grachev *et al.*, 2007). These difficulties result from the complicated structure of turbulence in SABL. The unsteadiness of SABL and the weakness of turbulence have been posing multiple problems (Mahrt, 2010, 2014). Slope flows, low-level jets, and gravity waves further complicate the picture. Yet, the ability to properly account for and quantify

SABL's physics is of great practical importance (e.g. Mahrt, 1999).

Among all studies dealing with the flux-gradient relationships in SL, a clear minority concentrates on stable conditions (Grachev *et al.*, 2007). The most commonly used relationships in the stable regime are approximations calculated as a linear interpolation between the neutral and very stable limits (Zilitinkevich and Esau, 2007). However, the applicability of these functions is likely to be limited only to the weakly stable regime as they are potentially incapable of representing surface fluxes under strong stratification (Louis, 1979).

One possible strategy to overcome these difficulties was suggested, among others, by Holtslag and DeBruin (1988), Cheng and Brutsaert (2005), and Zilitinkevich *et al.* (2013) who proposed to use nonlinear expressions for the stability functions. These models will be abbreviated as HDB88, CB05 and ZI13, respectively.

Another approach emerged from the recently developed quasi-normal scale elimination (QNSE) theory of stably stratified turbulence (Sukoriansky *et al.*, 2005) where the MOST similarity functions are derived from first principles in the limit of constant fluxes.

MOST functions serve an important role in NWP systems where they connect the surface boundary conditions with the first computational level. Clearly, the accuracy of NWP critically depends on the accuracy of the MOST functions and so their comprehensive assessment is crucial. The purpose of this article is to undertake such an assessment for HDB88, CB05, ZI13, and QNSE predictions in the constant flux layer and compare them with the data from the Cooperative Atmosphere-Surface Exchange Study in 1999

(CASES99; Poulos *et al.*, 2002), the Cardington dataset (http://badc.nerc.ac.uk/data/cardington/), and the Halley station in Antarctica.

The assessment utilizes a stiff test based upon the Deacon numbers (Viswanadham, 1979; Andreas, 2002; Guo and Zhang, 2007) that involve the second derivatives of the velocity and temperature profiles and are thus quite sensitive to the profiles' curvatures. The ensuing difficulties as well as the biases related to the well-known problem of self-correlation are discussed. Finally, the importance of the SL parameterization in NWP systems is underscored using 1D simulations utilizing the Weather Research and Forecasting (WRF) model.

2. Background information

2.1. The Monin–Obukhov similarity theory

Assuming that the fluxes of momentum and heat in the near-surface region of atmospheric boundary layers are nearly constant with height and using dimensional arguments, Monin and Obukhov (1954) derived the following relationships for the vertical profiles of the mean velocity, U, and mean potential temperature, Θ, within this region:

$$\left(\frac{\kappa z}{u_*}\right)\frac{\partial U}{\partial z} = \frac{K_0}{K_m} = \phi_m(\zeta) \qquad (1)$$

$$\left(\frac{\kappa z}{\theta_*}\right)\frac{\partial \Theta}{\partial z} = \frac{K_0}{K_h} = \phi_h(\zeta) \qquad (2)$$

Here, $\phi_m(\zeta)$ and $\phi_h(\zeta)$ are the MOST functions; u_* is the friction velocity defined as $u_* = (\tau_0/\rho)^{1/2}$; ρ is the air reference density; τ_0 is the surface stress; $\theta_* = H_s/(\rho C_p u_*)$ is the temperature equivalent of the friction velocity; H_s is the surface virtual sensible heat flux (negative downwards); C_p is the specific heat of air at constant pressure; z is the vertical coordinate; κ (=0.4) is the von Kármán constant; $K_0 = \kappa z u_*$ is the vertical eddy viscosity at neutral stratification; $K_m = u_*^2 (\partial U/\partial z)^{-1}$ and $K_h = u_* \theta_* (\partial \Theta/\partial z)^{-1}$ are the vertical eddy viscosity and eddy diffusivity, respectively, in stratified flows; $\zeta = z/L$; L is the Monin–Obukhov length scale,

$$L = -u_*^3 / \left[\kappa \left(g/\Theta_0\right) \overline{w\theta_0}\right] \qquad (3)$$

$-\overline{w\theta_0} = u_* \theta_* = -H_s / \left(\rho C_p\right)$ is the kinematic surface heat flux; w and θ denote the fluctuations of vertical velocity and potential temperature, respectively; g is the acceleration due to gravity, and Θ_0 is the reference potential temperature. The flux–profile relationship for humidity is taken to be identical to that for temperature. This representation is valid in the constant flux layer approximation and is known as the Monin–Obukhov similarity theory, or MOST (Monin and Obukhov, 1954; Monin and Yaglom, 1965).

The MOST functions considered in this study are

HDB88 : $\phi_{m,h} = 1 + 0.7\zeta + 0.75\zeta(6 - 0.35\zeta)e^{-0.35\zeta}$ (4)

$$\text{CB05} : \quad \phi_{m,h} = 1 + a\left\{\frac{\zeta + \zeta^b\left[1 + \zeta^b\right]^{(1-b)/b}}{\zeta + \left(1 + \zeta^b\right)^{1/b}}\right\} \qquad (5)$$

where $a = 6.1$, $b = 2.5$ for ϕ_m, $a = 5.3$, $b = 1.1$ for ϕ_h, and

$$\text{ZI13} : \quad \phi_m = 1 + 1.6\zeta \qquad (6)$$

$$\phi_h = \left(1 + \frac{0.18\zeta + 0.16\zeta^2}{1 + 1.42\zeta}\right)(1 + 1.6\zeta) \qquad (7)$$

The QNSE ϕ_m and ϕ_h used in this study are based on tabulated values from the spectral theory (more details in Section 2.3). The functional forms via fraction-polynomial fits valid for $\zeta \lesssim 3$ are

$$\phi_m = 1 + 2.25\zeta - 0.4\zeta^2 \qquad (8)$$

$$\phi_h = 0.704\left[1 + 2\zeta + 0.7\zeta(\zeta - 0.5)^4\right] \qquad (9)$$

The MOST is known for a spurious self-correlation problem, which arises from the fact that there is an insufficient number of independent scaling parameters to build a completely independent dimensionless groups (Andreas and Hicks, 2002). An example is plotting ϕ_m as a function of ζ in which case both variables contain u_*. Consequently, even random data can yield non-zero correlation (Kim, 1999). A randomization method described by, e.g. Klipp and Mahrt (2004) can be used to quantify these correlations.

2.2. The Deacon numbers

The Deacon numbers for wind and potential temperature are non-dimensional characteristics of the curvature of their respective profiles that provide a stiff test bed for SABL models. They are related to the von Kármán length scale (Tennekes and Lumley, 1972). In the constant flux layer approximation, the Deacon numbers can be expressed in terms of the MOST similarity functions:

$$D_m \equiv -z\frac{d^2U}{dz^2}\bigg/\frac{dU}{dz} = 1 - \frac{\zeta}{\phi_m(\zeta)}\frac{d\phi_m(\zeta)}{d\zeta}$$
$$= 1 - \frac{d\ln\phi_m(\zeta)}{d\ln\zeta} \qquad (10)$$

$$D_h \equiv -z\frac{d^2\Theta}{dz^2}\bigg/\frac{d\Theta}{dz} = 1 - \frac{\zeta}{\phi_h(\zeta)}\frac{d\phi_h(\zeta)}{d\zeta}$$
$$= 1 - \frac{d\ln\phi_h(\zeta)}{d\ln\zeta} \qquad (11)$$

Alternatively, the Deacon numbers can be cast as functions of the gradient Richardson number (Ri_g)

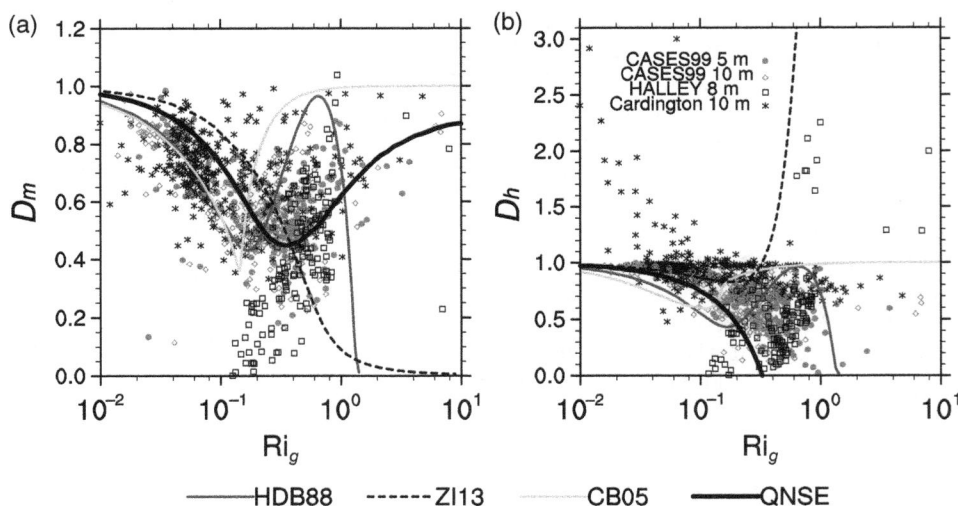

Figure 1. Velocity and potential temperature Deacon numbers ((a) and (b), respectively) as a function of the gradient Richardson number. Observations are from CASES99, Cardington and Halley.

which can also be expressed in terms of the MOST functions,

$$Ri_g \equiv \frac{g}{\Theta_0} \frac{d\Theta}{dz} \left(\frac{dU}{dz} \right)^{-2} = \frac{\zeta \phi_h(\zeta)}{\phi_m^2(\zeta)} \qquad (12)$$

The MOST functions and Deacon numbers are unity for neutral conditions. Owing to the presence of second derivatives, measuring Deacon numbers is very difficult. One cannot, however, refute the importance of these numbers: the shapes of the potential temperature and wind speed profiles are of crucial importance for the SL representation in NWP systems.

2.3. The QNSE theory

In traditional Reynolds-averaged Navier–Stokes (RANS) models, all fluctuating turbulent scales are eliminated by ensemble-averaging and so the models are insensitive to scale-dependent processes taking place on subgrid scales. As a consequence, the subgrid-scale anisotropization and turbulence–wave interactions cannot be captured at the basic level.

QNSE is a spectral theory that employs gradual coarsening of the resolved domain by successively eliminating small shells of unresolved scales. Each shell elimination produces contributions to the eddy viscosity and eddy diffusivity which may differ in the vertical and horizontal directions, hence accounting for flow anisotropization. This process also elucidates the contribution of internal gravity waves which produce complex poles in computations (Sukoriansky et al., 2005). In fact, the internal waves are recognized in QNSE via the dispersion relation that reflects the effect of turbulence and identifies spectral domains where waves or turbulence play predominant roles (Galperin and Sukoriansky, 2010; Sukoriansky and Galperin, 2013).

The QNSE theory produces vertical and horizontal eddy viscosities and eddy diffusivities that can be used in NWP applications. In the constant flux layer

approximation, the QNSE expressions become MOST similarity functions which lend themselves to validation against data and comparison with other models.

2.4. WRF 1D tests

The WRF model is a community-oriented state-of-the-art NWP system developed among many US institutions. It contains two different dynamics solvers, ARW (Advanced Research WRF) and NMM (nonhydrostatic Mesoscale Model), and several physics packages (Skamarock et al., 2008). A complete description of the ARW solver applied in the experiments presented in this paper is given by Skamarock and Klemp (2007). To assess the impact of the QNSE model on the performance of WRF, single column model simulations employing two different vertical resolutions (with a total of 31 and 101 levels) were carried out. Apart from the boundary layer and SL schemes, the only applied physics option was Purdue Lin microphysics (Lin et al., 1983). The test case corresponds to the Beaufort Arctic Stratus Experiment (BASE) described in Kosovic and Curry (2000). The data from BASE was successfully simulated via large eddy simulation (LES) (e.g. Stoll and Porte-Agel, 2008) which are used in this study for comparison with the WRF results.

3. Results

The Deacon numbers were computed using data processing and a profile fitting function Luhar et al. (2009). Figure 1 shows D_m and D_h as functions of Ri_g. For the MOST functions considered in this study, no analytical expressions exist for D_m and D_h in terms of Ri_g. Rather, the curves presented in Figure 1 are obtained by calculating $D_{m,h}(\zeta)$ and $Ri_g(\zeta)$ for a wide range of ζ s and then plotting the results.

There is considerable scatter in the data in Figure 1, but some qualitative trends are discernible. The observations have D_m decreasing up to $Ri_g \simeq 0.2 - 0.3$ beyond which it increases. While all similarity functions agree on the decreasing trend in the weakly stable regime, QNSE is the only one yielding a general increase in the very stable regime. The analytical approximation of the QNSE functions yields $D_m \simeq 0.9$ at large Ri_g which indeed is reflected in the figure and supported by the data. HDB88 and CB05 both increase for Ri_g between 0.15 and 0.2. HDB88 reaches its peak value at $Ri_g \simeq 0.6$ and then falls off sharply, whereas CB05 attains the limit $D_m = 1$ at $Ri_g \simeq 0.3$. In ZI13, $D_m \rightarrow 0$ monotonously at large Ri_g.

The observations show a general decrease in D_h from approximately 1 at $Ri_g = 0.01$ to 0 for $Ri_g = 0.3 - 0.5$. The QNSE curve follows this decrease though there is a pronounced bias toward weaker stabilities. The rest of the similarity functions show relatively poor fit to the observations. Although decreasing when $0.01 < Ri_g < 0.1$, the curves for HDB88, CB05 are both situated below the QNSE curve and the cluster of observations. For stronger stratification, $Ri_g \gtrsim 0.1$, the CB05 quickly returns to the limit of $D_h = 1$, whereas HDB88 approaches $D_h = 1$ at $Ri_g = 0.6$ and then falls to 0. ZI13 is closest to the observations when $Ri_g < 0.25$ and increases rapidly thereafter.

To quantify the accuracy of the MOST similarity functions against observations, we calculate R^2 (coefficient of determination) for each function with demeaned observations and residuals. The R^2 values are then multiplied by 100% to get the percentage of variance explained (pve). Table 1 lists the results of the test for (1) all Ri_g, (2) $Ri_g \leq 0.2$, and (3) $Ri_g > 0.2$. Negative pve (denoted by minuses in Table 1) means that the model has no predictive skills. The results of the pve test for the Deacon numbers are listed in Table 1. When all stabilities are considered, QNSE is the only model featuring positive (albeit small) values for D_m. We split Figure 1 into two parts and first consider pve for the regime $Ri_g \leq 0.2$. For this weakly stable regime, the pve values for QNSE, HDB88 and ZI13 all display moderate skill (34, 25, and 33%) whereas CB05 yields the smallest explained variance (2%). For D_h, all the similarity functions give pve values between 23 and 27%. For $Ri_g > 0.2$, the situation is more problematic. QNSE demonstrates weak skill for D_m with pve about 9%. Other similarity functions stay negative. For D_h, the agreement with the observations is even worse: none of the studied functions yielded a positive pve in this regime.

Would such a seemingly small advantage of the QNSE-based MOST functions over other functions play any role in the accuracy of performance of NWP models? To address this question, we designed a test case comparing two different SL parameterization in the WRF model that employed the QNSE option for the bulk of the boundary layer. One of the SL parameterizations utilized QNSE MOST functions and the other one used HDB88 functions. The WRF model

Table 1. Percentage variance explained for velocity and potential temperature Deacon numbers.

		QNSE	HDB88	CB05	ZI13
All Ri_g	pve D_m	17	–	–	–
	pve D_h	–	–	14	–
$Ri_g \leq 0.2$	pve D_m	34	25	2	33
	pve D_h	25	27	24	23
$Ri_g > 0.2$	pve D_m	9	–	–	–
	pve D_h	–	–	–	–

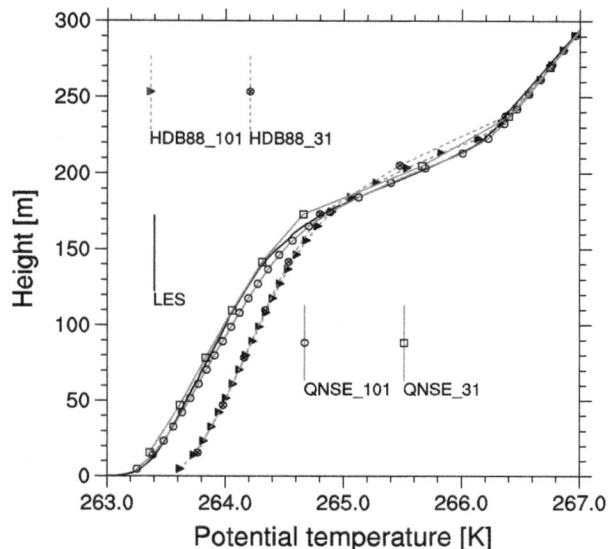

Figure 2. Potential temperature profile from LES and the WRF model using QNSE and HDB88 surface layer parameterizations. The numbers at the end of the labels indicate the number of vertical levels.

results are compared with LES by Stoll and Porte-Agel (2008) in Figure 2. After 9 h of simulation, there was a significant warm bias of the 2-m temperature when the HDB88 MOST functions were used. Applying the QNSE MOST functions completely eliminated this bias. It is important to note that the choice of the MOST functions had significant effect on the temperature up to the height of 150 m. The results were insensitive to the number of vertical levels employed in the model.

4. Conclusions

The results demonstrate the difficulties one faces when trying to validate MOST similarity functions. First, observations from different datasets feature significant scatter due to, e.g. data quality and self-correlation. This situation severely hinders the usage of traditional statistical tools such as pve as shown by Table 1. Second, observations from different datasets lack universality. This is demonstrated notably well by the D_h data shown in Figure 1. Yet, as demonstrated by Figure 2 and, e.g. Sterk et al. (2013), the SL parameterizations are of fundamental importance for the accuracy of NWP systems.

Good quality observations are therefore needed for an adequate validation of the SL models. The contradictory observational evidence shown in Figure 1 underscores the need for future efforts to acquire high-quality data for near-surface boundary layers.

Because statistical methods appear problematic when applied to estimate the performance of the MOST functions in the cases under consideration, visual inspections should also be used. In the case at hand, such an inspection reveals that not only the QNSE-based functions follow the tendencies of stability dependence for the Deacon numbers, but the associated curves coincide with the main cluster of observations quite closely quantitatively. Furthermore, the results from the WRF experiments point to the ameliorating effect that QNSE MOST functions may have on the warm biases in the potential temperature profile.

The Deacon numbers present a stiff test bed for validating models of MOST similarity functions. One may question, however, the importance of replicating the Deacon numbers versus ϕ_m and ϕ_h themselves. Recall that the latter functions determine vertical gradients of the mean profiles and advective terms in the Reynolds equations and production – destruction terms in turbulence equations. The diffusion terms in both the Reynolds and mean potential temperature equations depend on the curvatures of the U and Θ profiles and are equally important. One concludes, therefore, that the tests based upon the Deacon numbers are as important as the validation of ϕ_m and ϕ_h themselves. Much work remains to be done, however, in order to bring the observational tools needed for validation of the near-surface characteristics to the precision desired.

Acknowledgements

Partial support of this research by the ARO grant W911NF-09-1-0018 is gratefully appreciated. The participation of EMT was also partially supported by Paul L. Getting Memorial Fellowship in Marine Science and Elsie and William Knight Fellowship in Marine Science. We thank the CASES-99 Program and the UK Met Office for the data used in this paper. The authors acknowledge Don P. Chambers for his assistance.

References

Andreas E. 2002. Parameterizing scalar transfer over snow and ice: a review. *Journal of Hydrometeorology* **3**: 417–432.

Andreas E, Hicks B. 2002. Comments on "A critical test of the validity of Monin-Obukhov similarity during convective conditions". *Journal of the Atmospheric Sciences* **59**: 2605–2607.

Businger J, Wyngaard J, Izumi Y, Bradley E. 1971. Flux-profile relationships in the atmospheric surface layer. *Journal of the Atmospheric Sciences* **28**: 181–189.

Cheng Y, Brutsaert W. 2005. Flux-profile relationships for wind speed and temperature in the stable atmospheric boundary layer. *Boundary-Layer Meteorology* **114**: 519–538.

Foken T. 2006. 50 Years of the Monin-Obukhov similarity theory. *Boundary-Layer Meteorology* **119**: 431–447.

Galperin B, Sukoriansky S. 2010. Geophysical flows with anisotropic turbulence and dispersive waves: flows with stable stratification. *Ocean Dynamics* **60**: 1319–1337.

Grachev A, Andreas E, Fairall C, Guest P. 2007. SHEBA flux-profile relationships in the stable atmospheric boundary layer. *Monthly Weather Review* **124**: 315–333.

Guo X, Zhang H. 2007. A performance comparison between nonlinear similarity functions in bulk parameterization for very stable boundary layer. *Environmental Fluid Mechanics* **7**: 239–257.

Högström U. 1996. Review of some basic characteristics of the atmospheric surface layer. *Boundary-Layer Meteorology* **78**: 215–246.

Holtslag A, DeBruin H. 1988. Applied modeling of the nighttime surface energy balance over land. *Journal of Applied Meteorology* **27**: 689–704.

Kim J. 1999. Spurious correlation between ratios with a common divisor. *Statistics & Probability Letters* **44**: 383–386.

Klipp J, Mahrt L. 2004. Flux-gradient relationship, self-correlation and intermittency in the stable boundary layer. *Quarterly Journal of the Royal Meteorological Society* **130**: 2087–2103.

Kosovic B, Curry J. 2000. A large-eddy simulation study of a quasi-steady, stable stratified atmospheric boundary layer. *Journal of the Atmospheric Sciences* **57**: 1057–1068.

Lin YL, Rarley R, Orville H. 1983. Bulk parameterization of the snow field in a cloud model. *Journal of Applied Meteorology* **22**: 1065–1092.

Louis J. 1979. A parametric model of vertical eddy fluxes in the atmosphere. *Boundary-Layer Meteorology* **17**: 187–202.

Luhar A, Hurley P, Rayner K. 2009. Modelling near-surface lowwinds over land under stable conditions: sensitivity tests, flux-gradient relationships, and stability parameters. *Boundary-Layer Meteorology* **130**: 249–274.

Mahrt L. 1999. Stratified atmospheric boundary layers. *Boundary-Layer Meteorology* **90**: 375–396.

Mahrt L. 2010. Variability and maintenance of turbulence in the very stable boundary layer. *Boundary-Layer Meteorology* **135**: 1–18, doi: 10.1007/s10546-009-9463-6.

Mahrt L. 2014. Stably stratified atmospheric boundary layers. *Annual Review of Fluid Mechanics* **46**: 23–45.

Monin A, Obukhov A. 1954. Basic laws of turbulent mixing in the atmospheric surface layer. *Trudy Geofizicheskogo Instituta Akademiya Nauk SSSR* **24**: 163–187.

Monin A, Yaglom A. 1965. Statistical fluid mechanics. In *Mechanics of Turbulence*, Vol. 1(Translated from the Russian by Scripta Technica, Inc). MIT Press: Cambridge, MA.

Poulos GS, Blumen W, Fritts DC, Lundquist JK, Sun J, Burns SP, Nappo C, Banta R, Newsom R, Cuxart J, Terredellas E, Balsley B, Jensen M. 2002. CASES-99: a comprehensive investigation of the stable nocturnal boundary layer. *Bulletin of the American Meterological Society* **83**: 555–581.

Skamarock W, Klemp J. 2007. A time-split nonhydrostatic atmospheric model for research and NWP applications. *Journal of Computational Physics* **135**: 3465–3485.

Skamarock W, Klemp J, Dudhia J, Gill D, Barker D, Duda M, Huang XY, Wang W, Powers J. 2008. A description of the advanced research wrf version 3. NCAR Technical note NCAR/TN475+STR. NCAR: Boulder, CO.

Sterk H, Steeneveld G, Holtslag A. 2013. The role of snow-surface coupling, radiation, and turbulent mixing in modeling a stable boundary layer over arctic sea ice. *Journal of Geophysical Research* **118**: 1199–1217.

Stoll R, Porte-Agel F. 2008. Large-eddy simulation of the stable atmospheric boundary using dynamical models with different averaging schemes. *Boundary-Layer Meteorology* **126**: 1–28.

Sukoriansky S, Galperin B. 2013. An analytical theory of the buoyancy-kolmogorov subrange transition in turbulent flows with stable stratification. *Philosophical Transactions of the Royal Society A: Mathematical, Physical and Engineering Sciences* **371**: 1982.

Sukoriansky S, Galperin B, Perov V. 2005. Application of a new spectral theory of stably stratified turbulence to the atmospheric boundary layer over sea ice. *Boundary-Layer Meteorology* **117**: 231–257.

Tennekes H, Lumley J. 1972. *A First Course in Turbulence*. MIT Press: Cambridge, MA.

Viswanadham Y. 1979. Relation of Richardson number to the curvature of the wind profile. *Boundary-Layer Meteorology* **17**: 537–544.

Zilitinkevich S, Esau I. 2007. Similarity theory and calculation of turbulent fluxes at the surface for the stably stratified atmospheric boundary layer. *Boundary-Layer Meteorology* **125**: 193–205.

Zilitinkevich S, Elperin T, Kleeorin N, Rogachevskii I, Esau I. 2013. A hierarchy of energy- and flux-budget (EFB) turbulence closure models for stably-stratified geophysical flows. *Boundary-Layer Meteorology* **146**: 341–373.

21

Observations of fire-induced turbulence regimes during low-intensity wildland fires in forested environments: implications for smoke dispersion

Warren E. Heilman,[1]* Craig B. Clements,[2] Daisuke Seto,[2] Xindi Bian,[1] Kenneth L. Clark,[3] Nicholas S. Skowronski[4] and John L. Hom[5]

[1] USDA Forest Service, Northern Research Station, Lansing, MI, USA
[2] San Jose State University, San Jose, CA, USA
[3] USDA Forest Service, Northern Research Station, New Lisbon, NJ, USA
[4] USDA Forest Service, Northern Research Station, Morgantown, WV, USA
[5] USDA Forest Service, Northern Research Station, Newtown Square, PA, USA

*Correspondence to:
Dr Warren E. Heilman, USDA
Forest Service, Northern
Research Station, 3101
Technology Blvd., Suite F, Lansing,
MI, USA.
E-mail: wheilman@fs.fed.us

Abstract

Low-intensity wildland fires occurring beneath forest canopies can result in particularly adverse local air-quality conditions. Ambient and fire-induced turbulent circulations play a substantial role in the transport and dispersion of smoke during these fire events. Recent *in situ* measurements of fire–atmosphere interactions during low-intensity wildland fires have provided new insight into the structure of fire-induced turbulence regimes and how forest overstory vegetation can affect the horizontal and vertical dispersion of smoke. In this paper, we provide a summary of the key turbulence observations made during two low-intensity wildland fire events that occurred in the New Jersey Pine Barrens.

Keywords: forest canopy; low-intensity wildland fires; smoke dispersion; turbulence

1. Introduction

Atmospheric interactions with wildland fires play an important role in fire behavior and the transport and dispersion of wildland fire smoke. The release of heat and moisture from fuel combustion during wildland fires alters the local thermal structure of the lower atmospheric boundary layer and induces turbulent circulations. These turbulent circulations, in combination with the ambient mean flow, can affect fire behavior and the transport and dispersion of smoke (Ward and Hardy, 1991; Clements *et al.*, 2008; Sun *et al.*, 2009). The presence of forest overstory vegetation can further complicate local turbulence regimes through its effect on ambient and fire-induced circulations within the fire environment (Kiefer *et al.*, 2014). A more complete understanding of the local atmospheric turbulence dynamics that occur during wildland fires, many of which occur in forested environments, is needed to build the scientific foundation upon which new and improved predictive tools for fire behavior and local smoke dispersion that more completely account for atmospheric turbulence effects can be developed.

Fortunately, recent advances in atmospheric turbulence monitoring techniques within harsh wildland fire environments have provided new opportunities for measuring and analyzing turbulence regimes in the vicinity of wildland fires, thus advancing our understanding of fire–atmosphere interactions (e.g. Clements *et al.*, 2007; Seto and Clements, 2011; Seto *et al.*, 2013). Building upon these previous wildland fire experiments, this observational study focuses specifically on the effects of forest overstory vegetation on fire-induced atmospheric turbulence regimes during low-intensity wildland fires. Low-intensity fires (maximum vertical turbulent heat fluxes on the order of $150\,kW\,m^2$ or less above the flaming region) in forested environments, including prescribed fires used for fuels management, can lead to particularly adverse local air-quality conditions (Achtemeier, 2006) because smoke from these fires may linger for relatively long periods of time within forest vegetation layers and lead to human health and local roadway safety concerns. How smoke from low-intensity fires disperses within forested environments is governed to a large extent by local ambient and fire- and forest overstory-induced turbulent circulations that are present.

In this paper, we present an overview of two prescribed fire experiments conducted in forested environments for the purpose of improving our understanding of the local atmospheric turbulence dynamics that occur during daytime low-intensity surface fires beneath forest overstory vegetation. Measurements of turbulence regimes before, during, and after fire-front passage (FFP) through *in situ* overstory towers are analyzed and the implications for local smoke dispersion in forested environments are discussed.

2. Experimental design

The two experimental sites for this study were located in the New Jersey Pinelands National Reserve (PNR), an area containing some of the most volatile fire-cycle vegetation in the eastern United States (Hom, 2014). The PNR is surrounded by wildland–urban-interface areas and by some of the densest population centers in the U.S. Parts of the surrounding area have been designated as non-attainment areas for particulate matter ($PM_{2.5}$) and ozone by the U.S. Environmental Protection Agency (EPA) (http://www.epa.gov/oar/oaqps/greenbk/rnstate.html).

The first fire experiment (E1) was conducted on 20 March 2011 in a 107-ha burn block (block center: 39.8726°N, 74.5013°W). Vegetation in the block consisted of Pitch pine (*Pinus rigida* Mill.) and mixed oak (*Quercus* spp.) overstory (~15–18 m height), with blueberry (*Vaccinium* spp.), huckleberry (*Gaylussacia* spp.), and scrub oaks in the understory. Relative maxima in plant area density occurred near the surface (~0.13 $m^2\ m^{-3}$) and at about 9 m above the surface within the forest overstory canopy (~0.08 $m^2\ m^{-3}$). The litter layer on the forest floor consisted of pine needles, shrub foliage, and woody fuels ranging in diameter from 0.6 to 7.6 cm. The second burn experiment (E2) was conducted on 6 March 2012 in a 97-ha burn block (block center: 39.9141°N, 74.6033°W). Vegetation in the E2 burn block consisted of mixed oak and scattered Pitch and Shortleaf (*P. echinata* Mill.) pines in the overstory (~20–23 m height), and primarily blueberry and huckleberry in the understory. The overall plant area density in the E2 burn block was less than that in the E1 burn block, with values less than 0.01 $m^2\ m^{-3}$ just above the surface and maximum density values ~0.1 $m^2\ m^{-3}$ at 9 m above the surface. The litter layer in the E2 burn block consisted of oak and shrub foliage, some pine needles, and 0.6 to 7.6 cm diameter woody fuels. Both burn blocks were characterized by sandy soils and were relatively flat.

A network of instrumented 3-, 10-, 20-, and 30-m towers and surface monitoring sites was established within and in the vicinity of the E1 and E2 burn blocks (Figure 1). Instrumentation mounted at multiple levels on the towers provided measurements of the three-dimensional wind speed components (U, V, W), temperature (T), relative humidity (RH), net radiation (R_n), atmospheric pressure (p), radiative heat fluxes, and carbon monoxide (CO) and carbon dioxide (CO_2) concentrations. Instrument sampling frequencies were 0.5 Hz on the 3-m towers and 10 Hz on the 10-, 20-, and 30-m towers, respectively. The high-frequency (10 Hz)

component wind-speed measurements, carried out only on the 10-, 20-, and 30-m towers within the burn blocks and on the 10-m control towers outside the burn blocks (see Figure 1), were accomplished via sonic anemometers oriented with their horizontal axes aligned in the east–west and north–south (true north) directions. The same meteorological monitoring strategy (i.e. instrumentation, monitoring levels, sampling frequency) was used for the E1 and E2 experiments.

Using drip torches, the New Jersey Forest Fire Service (NJFFS) initiated surface backing fires along the western and eastern perimeters of the E1 and E2 burn blocks, respectively, in accordance with the observed ambient wind directions. Initial ignitions occurred at 1355 UTC (E1: 0955 EDT) and 1430 UTC (E2: 0930 EST) near the southwestern (E1) and southeastern (E2) portions of the burn blocks and continued along the western (E1) and eastern (E2) burn block perimeters. Ambient near-surface temperatures and relative humidity values ranged from ~2 to 10 °C and ~30 to 70% during the E1 experiment and from ~1 to 8 °C and ~15 to 36% during the E2 experiment. Under light northeasterly to southeasterly ambient winds (<2.5 m s^{-1}) during the E1 experiment, the E1 fire line generally spread northeastward (spread rate ≈ 1.50 m min^{-1}) through the burn block until reaching the northeastern portion of the burn block around 2100 EDT (~11-h burn experiment). For the E2 experiment, subsequent fire-line ignitions along north–south oriented plow lines spaced ~200 m apart in the interior of the burn block following the initial fire-line ignition produced a more complicated burn pattern, with multiple fire lines generally spreading westward (spread rate ≈ 0.33 m min^{-1}) through the burn block against light (<3 m s^{-1}) northwesterly to southwesterly ambient winds. Active burning for the E2 experiment was completed by 1800 EST (~8.5-h burn experiment). Burning was generally confined to surface fuels, and fire-line widths were ~1–2 m for both experiments. The amount of time required for the E1 and E2 fire lines to pass through each tower location was ~1.3 and 3 min, respectively, although the effects of the fire lines on atmospheric conditions at the towers lasted much longer (~1 h).

Data collected during the experiments were subjected to a despiking and filtering routine to remove erroneous data and data values exceeding 6 standard deviations from running 1-h means. Sonic anemometer data were tilt-corrected (Wilczak *et al.*, 2001) to minimize vertical wind speed errors associated with sonic anemometers not mounted exactly level on the network towers.

The despiked and tilt-corrected 10 Hz sonic anemometer wind speed (U, V, W) and temperature (T) data were divided into 1-h block averaging periods over which mean velocities and temperatures ($\overline{U}$, $\overline{V}$, $\overline{W}$, $\overline{T}$) were computed, with perturbation velocities ($u' = U - \overline{U}$, $v' = V - \overline{V}$, $w' = W - \overline{W}$) and temperatures ($t' = T - \overline{T}$) then computed at each 0.1 s. One-hour block averaging periods were adopted for this study based on the recommendation of Sun *et al.* (2006) for eddy flux measurements over forests. 'Fire

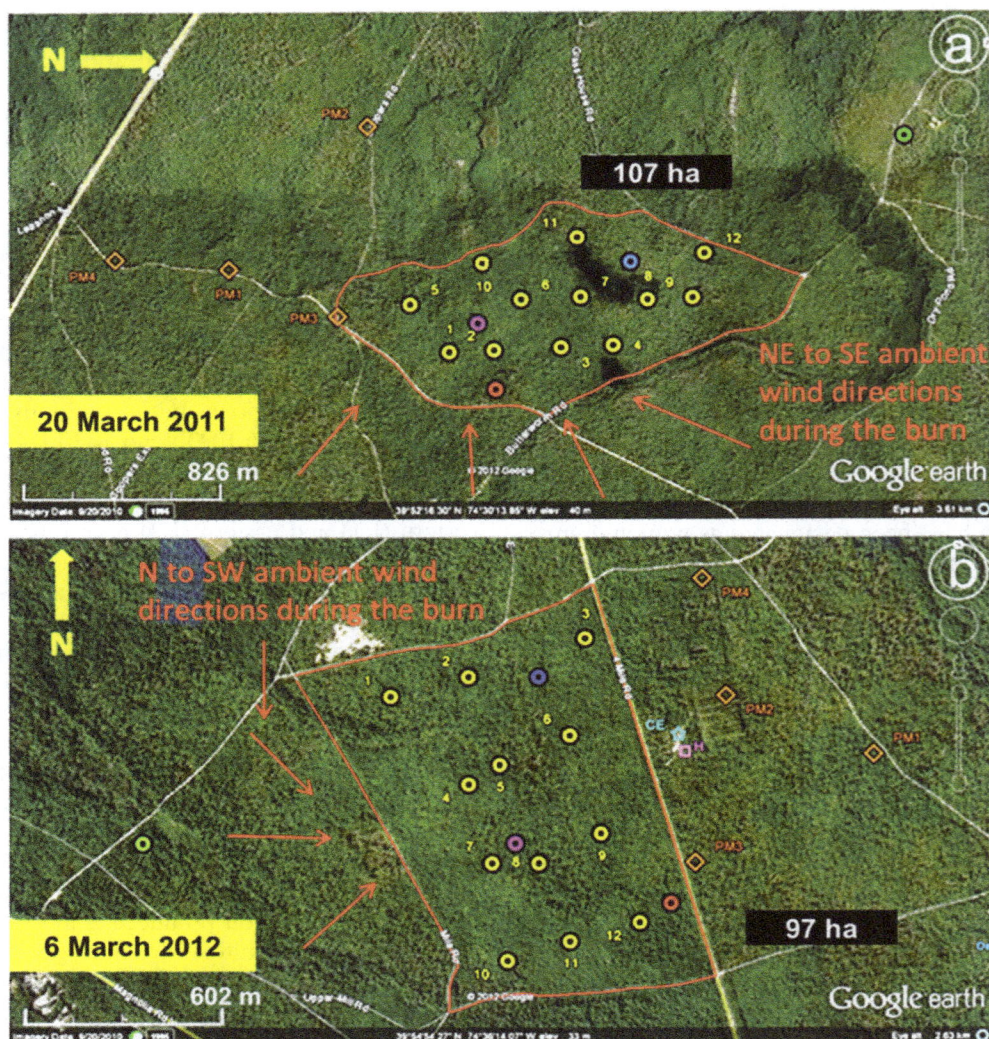

Figure 1. Locations of towers and surface monitoring stations within and in the vicinity of the burn blocks (outlined in red) for the (a) E1 and (b) E2 prescribed fire experiments conducted on 20 March 2011 and 6 March 2012, respectively in the New Jersey Pine Barrens; 3-m towers: numbered yellow circles; 10-m tower: blue circle; 20-m tower: purple circle; 30-m tower: red circle; 10-m control tower: green circle; PM$_{2.5}$ monitors: brown diamonds; ceilometer: blue star; remote helicopter: pink square; SODAR: orange square.

periods' during which FFP occurred at the tower locations were delineated for each tower, with the duration of the periods determined by subjective analysis of the temperature time series obtained from the tower sonic anemometer and thermocouple measurements. Following the methodology of Seto *et al.* (2013), perturbation velocities and temperatures during 'fire periods' were computed by subtracting the mean velocities and temperatures associated with the 1-h period prior to the onset of the 'fire period' from the measured 10 Hz 'fire period' velocities and temperatures. Although one can certainly compute a mean velocity and temperature associated with the fire-induced circulations and heating during the 'fire periods' and then compute corresponding velocity and temperature perturbations from those means, the Seto *et al.* (2013) methodology was adopted so that the computed perturbation velocities and temperatures during the

'fire periods' could provide a better representation of the true fire-induced turbulence and departures from the ambient state that were present. The computed perturbation velocities and temperatures formed the basis for spatial, temporal, and spectral analyses of the turbulence regimes that were present during the experiments.

A complete description of the two fire experiments carried out in this study, including a listing of the instrumentation, measurement strategies, and data analysis techniques, can be found in Heilman *et al.* (2013).

3. Results and discussion

For the analyses of turbulence regimes in the vicinity of the spreading fire lines through the E1 and E2 burn blocks, we focused on observations at the 20-m towers

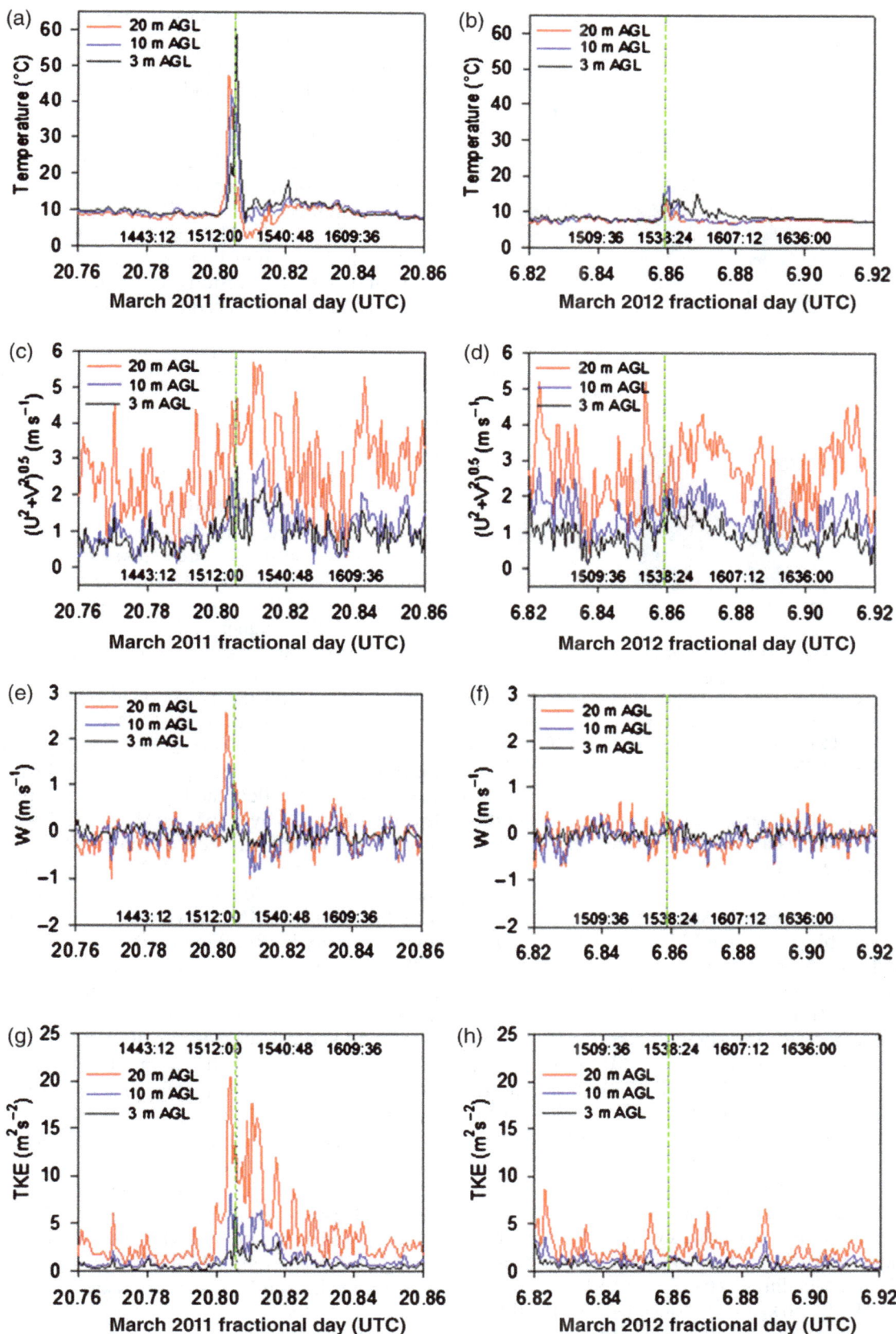

Figure 2. Observed 1-min averaged (a, b) thermocouple temperatures (°C), (c, d) horizontal [$(U^2 + V^2)^{0.5}$] wind speeds (m s^{-1}), (e, f) vertical (W) wind speeds (m s^{-1}), and (g, h) turbulent kinetic energy (TKE) (m^2 s^{-2}) at three levels on the 20 m towers before, during, and after the E1 (left column) and E2 (right column) fire lines passed the towers. Vertical dashed lines indicate times of fire-front passage (E1: 1520 EDT; E2: 1537 EST). Time stamps (hhmm:ss) in EDT (left column) and EST (right column) are shown above the lower axes or below the upper axes.

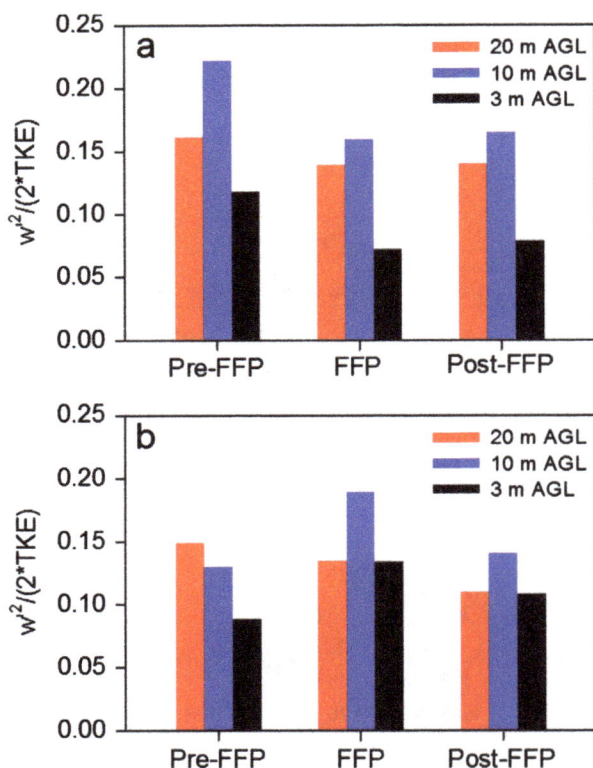

Figure 3. Observed turbulence anisotropy, as quantified by average values of $\overline{w'^2}/(2*TKE)$, during the pre-FFP (E1: 1435–1505 EDT; E2: 1452–1522 EST), FFP (E1: 1505–1535 EDT; E2: 1522–1552 EST), and post-FFP (E1: 1535–1605 EDT; E2: 1552–1622 EST) periods at three levels on the 20 m towers located in the interior of the (a) E1 and (b) E2 burn blocks.

primarily because of the availability of sonic anemometer and temperature data both within and near the top of the overstory vegetation layer from those towers. The 20-m towers were also located well-within the boundary of the burn blocks, where observed fire-line spread was less variable and where FFP occurred during daytime conditions for both experiments.

The effects of the fire lines on thermal (T) and kinematic (U, V, W) fields at the 20-m tower locations are shown in Figure 2. Temperature time series (1-min averages) before, during, and after FFP for the E1 and E2 experiments (Figure. 2(a) and (b)) indicate the E1 fire line had a more pronounced local impact on the thermal regime than the E2 fire line. The different temperature responses are consistent with the different maximum vertical turbulent heat fluxes (1-min averages) observed at 3 m AGL above the E1 and E2 fire lines (E1: 23.0 kW m^{-2}; E2: 3.2 kW m^{-2}). Fire-intensity differences were due in part to differences in pre-fire surface fuel loadings (E1: 1.478 ± 0.388 kg m^{-2}; E2: 1.104 ± 0.246 kg m^{-2}) (mean $\pm$ 1 SD), differences in average surface fuel moisture contents (E1: $21.9 \pm 9.8\%$; E2: $49.5 \pm 19.6\%$), and differences in fuel type and arrangement (Heilman *et al.*, 2013). Note that maximum instantaneous (10 Hz) thermocouple temperatures above the E1 and E2 fire lines reached 145.6 and 28.9 °C, respectively.

Consistent with the observed thermal regime variations, local circulation responses to the E1 and E2 fire lines at the 20-m tower locations (Figure 2(c)–(f)) indicate the E1 fire line had a more pronounced local impact on the horizontal $[(U^2 + V^2)^{0.5}]$ and vertical (W) wind speeds than the E2 fire line. FFP through the E1 20 m tower location produced a southeasterly to southwesterly horizontal wind-direction shift and relatively strong updrafts/downdrafts (Figure 2(e)), particularly at 10 and 20 m AGL. FFP through the E2 20-m tower location had a minimal impact on the speed of the ambient horizontal westerly to southwesterly winds (Figure 2(d)) and the speed of the updrafts/downdrafts above the fire line (Figure 2(f)).

The different intensity E1 and E2 fires also generated different turbulence regimes within and near the top of the forest vegetation layers inside the burn blocks (Figure 2(g)–(h)). Turbulence at the 20-m tower locations, quantified by turbulent kinetic energy (TKE) per unit mass (equal to one-half of the sum of the 1-min averaged horizontal and vertical velocity variances ($\overline{u'^2}$, $\overline{v'^2}$, $\overline{w'^2}$) computed from the sonic anemometer component wind-speed measurements), was consistently higher at 20 m (near the canopy top) than at the 10- and 3-m heights. The higher-intensity E1 fire resulted in substantially higher TKE values (Figure 2(g)) within and just above the vegetation layer during and immediately following FFP compared to the E2 fire (Figure 2(h)), with the largest increases occurring at 20 m AGL. At the 20-m level, TKE increased from less than 5 m^2 s^{-2} well before the E1 FFP (1520 EDT) to about 20 m^2 s^{-2} 3 min prior to FFP. TKE values then fluctuated wildly and generally diminished to less than 5 m^2 s^{-2} by ~1610 EDT. At the 10- and 3-m levels, TKE values reached maxima of ~8 m^2 s^{-2} (1517 EDT) and ~7 m^2 s^{-2} (1520 EDT), respectively, and then diminished to less than 2 m^2 s^{-2} by ~1541 EDT. This fire-induced TKE behavior was absent during the lower-intensity E2 fire (Figure 2(h)).

Turbulent mixing of heat, momentum, moisture, and smoke in a particular direction during fire events depends on the distribution of energy among the horizontal and vertical components of the total TKE field. To assess the relative contributions of these components to the total TKE field, a specific measure of turbulence anisotropy, values of TKE$_w$ = $\overline{w'^2}/(2*TKE)$ were computed for both experiments. Note that TKE$_w$ $\approx$ 0.33 under isotropic conditions, whereas TKE$_w$ $\approx$ 0.14 for classical atmospheric surface layers (Panofsky and Dutton, 1984). Average observed levels of turbulence anisotropy as measured by TKE$_w$ 30 min before, 30 min during, and 30 min after FFP at the 20-m tower locations for the E1 and E2 experiments are summarized in Figure 3.

Anisotropic turbulence was prevalent during all periods, with the vertical component of TKE usually comprising less than 22% of the total TKE on average. Furthermore, anisotropy tended to be stronger at the 3-m level than at the near-canopy-top 20-m level and the mid-canopy 10-m level. Mean values

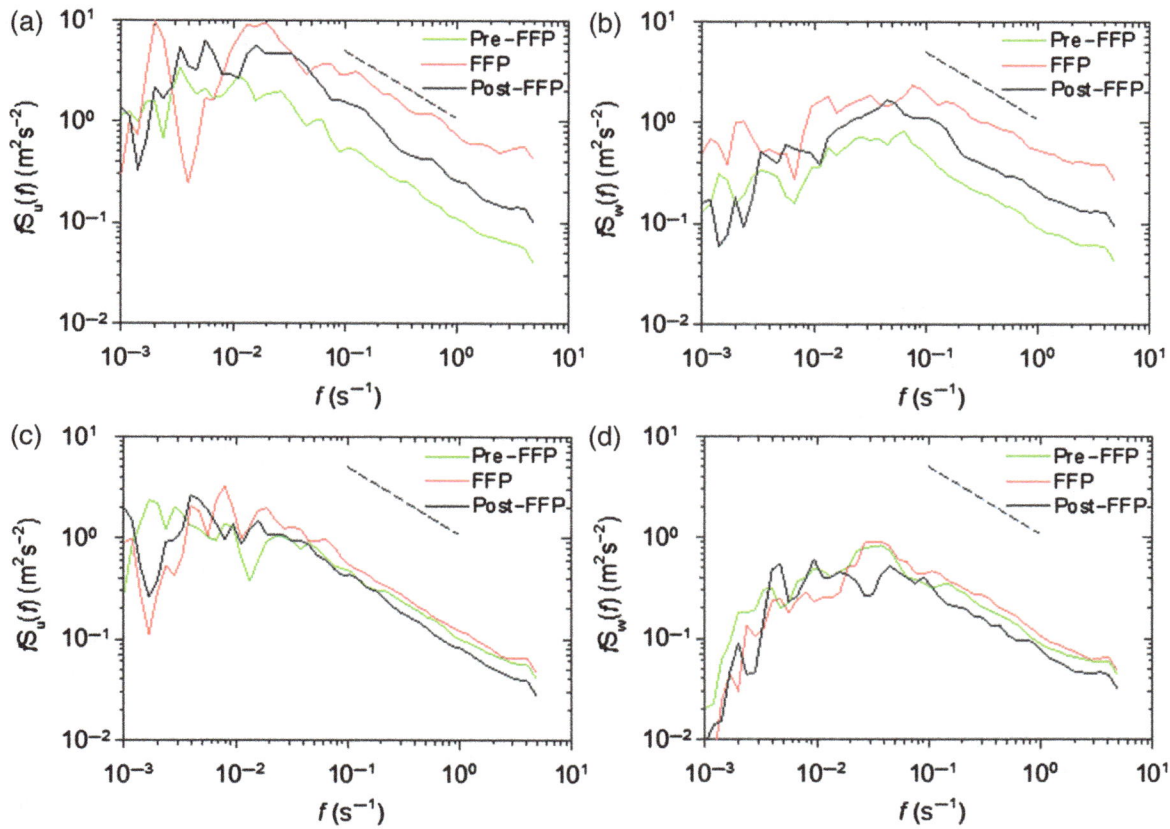

Figure 4. Frequency weighted power spectra (m^2 s^{-2}) at 20 m AGL for the (a, c) horizontal (streamwise) wind velocity [$f\,S_w(f)$] and (b, d) vertical wind velocity [$f\,S_w(f)$] as a function of spectral frequency f (s^{-1}) during the pre-FFP period (E1: 1435–1505 EDT; E2: 1452–1522 EST), the FFP period (E1: 1505–1535 EDT; E2: 1522–1552 EST), and the post-FFP period (E1: 1535–1605 EDT; E2: 1552–1622 EST) for the E1 (top row) and E2 (bottom row) fire experiments. The dashed line represents the theoretical −2/3 slope of spectral power versus frequency curves within the inertial subrange according to Kolmogorov theory.

of TKE$_w$ diminished at all levels from the pre-FFP period to the post-FFP period for the higher-intensity E1 fire (Figure 3(a)). For the lower-intensity E2 fire, mean TKE$_w$ values at the 3- and 10-m levels actually increased during the FFP period (Figure 3(b)). Note that maximum w'^2 values at each level occurred during the FFP periods for both experiments (E1: 10.72 m^2 s^{-2} at 20 m AGL, 3.71 m^2 s^{-2} at 10 m AGL, 1.25 m^2 s^{-2} at 3 m AGL; E2: 1.12 m^2 s^{-2} at 20 m AGL, 1.25 m^2 s^{-2} at 10 m AGL, 0.45 m^2 s^{-2} at 3 m AGL). This observed behavior in w'^2 and mean TKE$_w$ values suggests that even though lower-intensity fires in forested environments will probably result in lower overall fire-induced TKE and lower buoyancy-induced vertical velocity perturbations (w') compared to higher-intensity fires, the magnitudes of the vertical velocity perturbations compared to the horizontal velocity perturbations (u' and v') above the fire front may still be large enough such that turbulence fields could actually be less anisotropic than the fields associated with higher-intensity fires.

Using wavelet spectrum analyses (Torrence and Compo, 1998; Seto *et al.*, 2013), anisotropy during the E1 and E2 fires was also assessed in terms of its variation across the different spatial scales (frequencies) of turbulent eddies that contributed to the

total TKE fields. The wavelet analyses indicate that the relatively large increases in TKE during FFP for the higher-intensity E1 fire, especially near the canopy top (Figure 2(g)), were associated with energy increases in both the horizontal (streamwise) and vertical velocity perturbations primarily at mid to high frequencies ($>10^{-1}$ s^{-1}) (Figure 4(a) and (b)). Similarly, for the minor changes in TKE during FFP for the lower-intensity E2 fire (Figure 2(h)), slight energy increases in the horizontal (streamwise) and vertical velocity perturbations during FFP were again observed mainly over the mid- to high-frequency portions of the spectrum (Figure 4(c)–(d)). During the E1 post-FFP period, the horizontal and vertical velocity perturbation energies consistently exceeded the energies observed in the pre-FFP period over the mid- to high-frequency portion of the spectrum; the opposite occurred during the lower-intensity E2 fire. Vertical velocity spectra for both fire experiments exhibited peak energy values at the mid-frequency portion of the spectrum ($\sim 10^{-1}$ s^{-1}) before, during, and after FFP, whereas the horizontal (streamwise) spectra exhibited peak energy values at low frequencies ($\sim 10^{-3}$–10^{-2} s^{-1}). Within the inertial subrange portion of the frequency spectrum, the energy curves exhibited slopes similar to the −2/3 slope suggested by Kolmogorov theory (Kolmogorov, 1941).

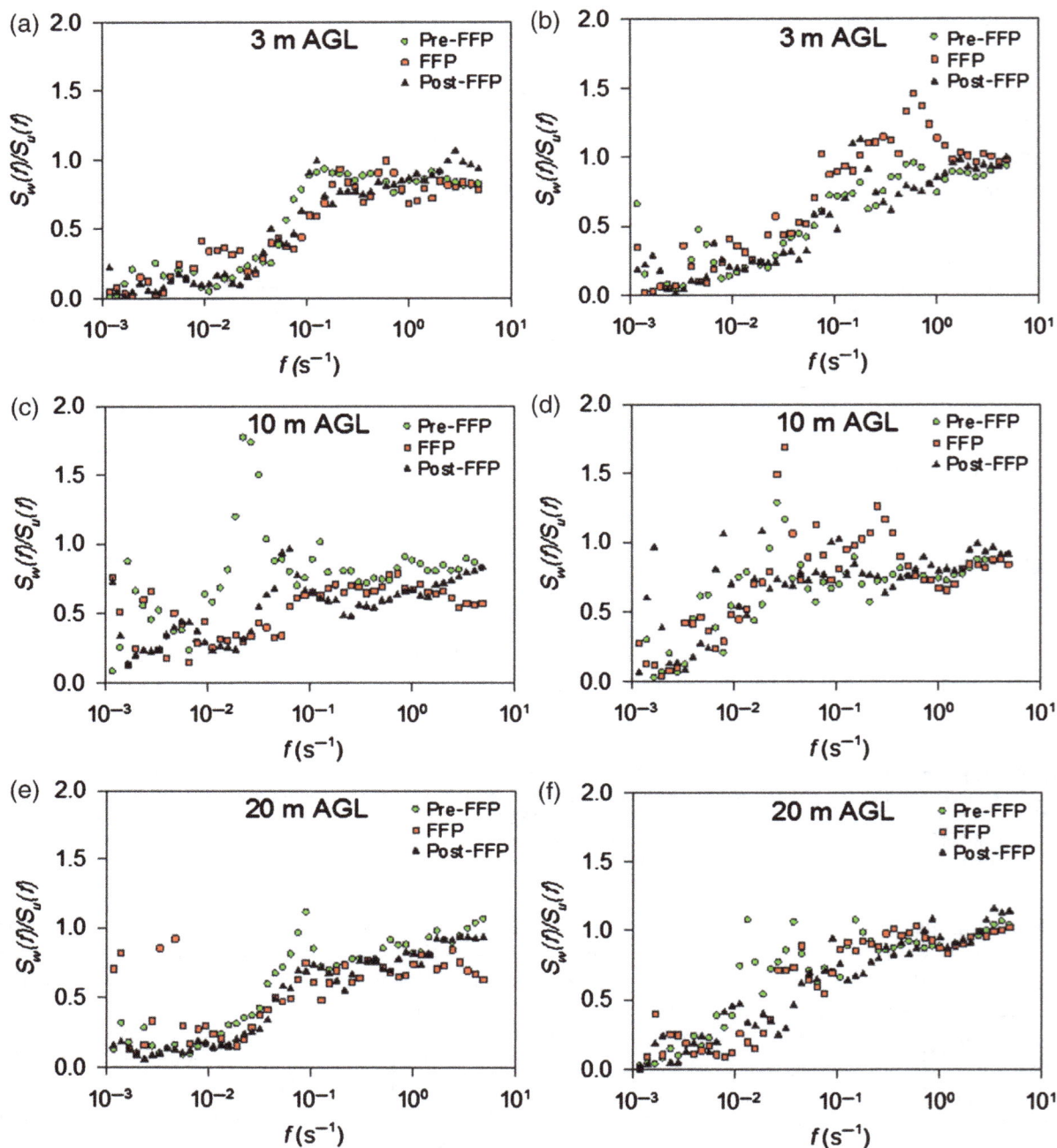

Figure 5. Ratios of the vertical to horizontal (streamwise) power spectra as a function of spectral frequency f (s^{-1}) at (a, b) 3 m, (c, d) 10 m, and (e, f) 20 m AGL during the pre-FFP period (E1: 1435–1505 EDT; E2: 1452–1522 EST), the FFP period (E1: 1505–1535 EDT; E2: 1522–1552 EST), and the post-FFP period (E1: 1535–1605 EDT; E2: 1552–1622 EST) for the E1 (left column) and E2 (right column) fire experiments.

Vertical to horizontal (streamwise) spectra ratios (Figure 5) reveal that low-frequency (large-eddy) turbulent circulations that occurred within and near the top of the vegetation layers during both experiments were more anisotropic than the high-frequency (small-eddy) turbulent circulations, with horizontal (streamwise) turbulence dominating vertical turbulence over most of the low-frequency portion of the spectrum. At higher frequencies, the vertical to horizontal power spectra ratios for both experiments were generally closer to a value of 1 as opposed to the isotropic 4/3 value as

predicted by the Kolmogorov (1941) inertial subrange law. This result is consistent with Biltoft (2001), who also provided observational evidence of spectral ratios approaching a value of 1 in the inertial subrange. The dominance of horizontal turbulence over vertical turbulence was prevalent over most of the frequency spectrum regardless of whether a surface fire was present or not (note pre- and post-FFP periods versus FFP periods in Figure 5). As noted in Figure 3, the most anisotropic conditions were generally observed near the surface and canopy top, while turbulence tended to be a bit less anisotropic at the mid-canopy 10 m level. The power spectra ratios shown in Figure 5 indicate the tendency toward more isotropic conditions at the mid-canopy level occurred primarily over the low- to mid-frequency range of the spectrum ($10^{-3}-10^{-1}$ s^{-1}), which corresponds to eddy sizes greater than $\sim 10-30$ m under the observed $1-3$ m s^{-1} mean wind speeds within and near the canopy top.

4. Summary and conclusions

Atmospheric turbulence plays an important role in the evolution of smoke plumes during wildland fire events. Turbulence regimes observed during our prescribed fire experiments suggest the presence of forest overstory vegetation during low-intensity surface fires could be an important factor in the local dispersion of smoke from those fires. Depending on actual fire intensity, increases in fire-induced TKE can be much larger at or near the canopy top than at levels just above the surface fire. Under those circumstances, the turbulent mixing or diffusion of smoke as it exits the top of the canopy could be much more substantial than the mixing occurring near the surface and within the vegetation layer. The observations also suggest that turbulence within forest vegetation layers is more anisotropic near the surface and near the canopy top than at mid-canopy levels, with the horizontal component of TKE dominating the vertical component primarily at large eddy sizes (low frequencies). While the presence of a low-intensity surface fire in a forested environment will tend to increase vertical velocity perturbations and the vertical component of TKE due to buoyancy effects, anisotropic turbulence regimes within the forest overstory vegetation layer may still persist. It follows then that horizontal turbulent mixing of smoke from low-intensity surface fires may dominate vertical turbulent mixing processes, particularly near the surface and canopy top.

More research is needed to compare results from this study with turbulence observations during fires of varying intensity in forests characterized by different canopy structure and under different ambient atmospheric conditions. It is through these observational turbulence studies under different environmental conditions that we can develop a better understanding of turbulence regimes that develop during wildland fire events and set the scientific foundation for developing operational air-quality predictive tools that more completely account for forest overstory and fire-intensity impacts on local smoke dispersion.

Acknowledgements

Research support was provided by the U.S. Joint Fire Science Program (Project # 09-1-04-1) and the USDA Forest Service (Research Cost Reimbursable Agreement # 13-CR-11242306-073). We thank the New Jersey Forest Fire Service for managing and conducting the prescribed fires for our experiments. We would also like to thank the anonymous reviewers of this paper for their constructive comments and suggested edits.

References

Achtemeier GL. 2006. Measurements of moisture in smoldering smoke and implications for fog. *International Journal of Wildland Fire* **15**: 517–525.

Biltoft CA. 2001. Some thoughts on local isotropy and the 4/3 lateral to longitudinal velocity spectrum ratio. *Boundary-Layer Meteorology* **100**: 393–404.

Clements CB, Zhong S, Goodrick S, Li J, Potter BE, Bian X, Heilman WE, Charney JJ, Perna R, Jang M, Lee D, Patel M, Street S, Aumann G. 2007. Observing the dynamics of wildland grass fires. *Bulletin of the American Meteorological Society* **88**: 1369–1382.

Clements CB, Zhong S, Bian X, Heilman WE, Byun DW. 2008. First observations of turbulence generated by grass fires. *Journal of Geophysical Research* **113**: D22102, doi: 10.1029/2008JD010014.

Heilman WE, Zhong S, Hom JL, Charney JJ, Kiefer MT, Clark KL, Skowronski N, Bohrer G, Lu W, Liu Y, Kremens R, Bian X, Gallagher M, Patterson M, Nikolic J, Chatziefstratiou T, Stegall C, Forbus K. 2013. Development of modeling tools for predicting smoke dispersion from low-intensity fires. Final Report, U.S. Joint Fire Science Program, Project 09-1-04-1 [Online]. http://www.firescience.gov/projects/09-1-04-1/project/09-1-04-1_final_report.pdf (accessed 25 March 2015).

Hom JL. 2014. Fire research in the New Jersey Pine Barrens. In *Remote Sensing Modeling and Applications to Wildland Fires*, Qu JJ, Sommers W, Yang R, Riebau A, Kafatos M (eds). Springer and Tsinghua University Press: Beijing; 181–191.

Kiefer MT, Heilman WE, Zhong S, Charney JJ, Bian X, Skowronski NS, Hom JL, Clark KL, Patterson M, Gallagher MR. 2014. Multiscale simulation of a prescribed fire event in the New Jersey Pine Barrens using ARPS-CANOPY. *Journal of Applied Meteorology and Climatology* **53**: 793–812.

Kolmogorov AN. 1941. The local structure of turbulence in incompressible viscous fluid for very large Reynolds numbers. *Doklady Akademii Nauk SSSR* **30**: 299–303.

Panofsky HA, Dutton JA. 1984. *Atmospheric Turbulence*. John Wiley and Sons: New York, NY, 397 pp.

Seto D, Clements CB. 2011. Fire whirl evolution observed during a valley wind-sea breeze reversal. *Journal of Combustion* **2011**: 569475, doi: 10.1155/2011/569475.

Seto D, Clements CB, Heilman WE. 2013. Turbulence spectra measured during fire front passage. *Agricultural and Forest Meteorology* **169**: 195–210.

Sun XM, Zhu ZL, Wen XF, Yuan GF, Yu GR. 2006. The impact of averaging period on eddy fluxes observed at ChinaFLUX sites. *Agricultural and Forest Meteorology* **137**: 188–193.

Sun R, Krueger SK, Jenkins MA, Zulauf MA, Charney JJ. 2009. The importance of fire–atmosphere coupling and boundary-layer turbulence to wildfire spread. *International Journal of Wildland Fire* **18**: 50–60.

Torrence C, Compo GP. 1998. A practical guide to wavelet analysis. *Bulletin of the American Meteorological Society* **79**: 61–78.

Ward DE, Hardy CC. 1991. Smoke emissions from wildland fires. *Environment International* **17**: 117–134.

Wilczak JM, Oncley SP, Stage SA. 2001. Sonic anemometer tilt correction algorithms. *Boundary-Layer Meteorology* **99**: 127–150.

22

Assessment of the variability of pollutants concentration over the metropolitan area of São Paulo, Brazil, using the wavelet transform

M. Zeri,[1]* V. S. B. Carvalho,[2] G. Cunha-Zeri,[1] J. F. Oliveira-Júnior,[3] G. B. Lyra[3] and E. D. Freitas[4]

[1] Brazilian Center for Monitoring and Early Warnings of Natural Disasters (CEMADEN), São José dos Campos, Brazil
[2] Instituto de Recursos Naturais, Universidade Federal de Itajubá, Brazil
[3] Departamento de Ciências Ambientais, Instituto de Florestas, Universidade Federal Rural do Rio de Janeiro, Seropédica, Brazil
[4] Instituto de Astronomia, Geofísica e Ciências Atmosféricas, Universidade de São Paulo, Brazil

*Correspondence to:
M. Zeri, Brazilian Center for
Monitoring and Early Warnings
of Natural Disasters
(CEMADEN), Estrada Doutor
Altino Bondesan, 500 − Eugênio
de Melo, 12247−016, São José
dos Campos, São Paulo, Brazil.
E-mail:
marcelo.zeri@cemaden.gov.br

Abstract

The objective of this work was to investigate the mean and variability of a dataset of pollutant concentrations from measurements taken over the metropolitan area of São Paulo city, Brazil. Wavelet analysis was applied to the time series of pollutant concentrations, revealing the strongest harmonics influencing the signals. A mode of variability of 4−8 days was significant until the middle of the last decade and is likely associated with the approach and passage of meteorological systems. A dataset representing the number of frontal systems moving across the state helped to explain the interannual variability during wintertime. Years with fewer frontal systems had higher levels of pollutants in several locations. Weather events such as inversions and the passage of frontal systems influence the concentration of pollutants. Public policies on air quality should focus not only on reducing the long-term exposure of city-dwellers to the negative effects of pollutants but also account for the possible short-term effects of the weather on air quality.

Keywords: air pollution; wavelet analysis; meteorological systems; data analysis; industrial activity

1. Introduction

Air pollution is a common problem in cities around the world, particularly in metropolitan areas (Sharma *et al.*, 1983; deLeon *et al.*, 1996; Schwartz, 1996; Samet *et al.*, 2000; de Miranda *et al.*, 2002; Godoy *et al.*, 2009). The effects of air pollution – such as particulate matter with diameter $\leq 10\,\mu m$ (PM_{10}), sulfur dioxide (SO_2), carbon monoxide (CO), or ozone (O_3) – increase hospitalizations due to respiratory problems, lung cancer trends, acid rain, and the black dust covering building's façades (Fajersztajn *et al.*, 2013). The effects of air pollution over the metropolitan area of São Paulo (MASP), Brazil, are already linked to impacts on human health (Gonçalves *et al.*, 2005; Cançado *et al.*, 2006), as well as to feedbacks with atmospheric and climatic conditions including composition of aerosols (Castanho and Artaxo, 2001; Bourotte *et al.*, 2005), the urban heat island (UHI) effect, local circulation patterns (Silva Dias and Machado, 1997; Freitas *et al.*, 2007), and mesoscale circulations induced by topography and sea/land breezes (Oliveira *et al.*, 2003). In this work, a dataset of pollutants concentration was analyzed using statistical inference (means and variances) as well as wavelet analysis, to detect the most important modes of variability influencing the levels of pollution. In general, the concentrations have a daily cycle associated with traffic of vehicles or industrial activity.

There is also an annual trend, with a maximum in winter; the cold air makes the atmospheric surface layer shallow and not well mixed, increasing the concentrations. In addition, atmospheric inversions are sometimes observed in winter, when the vertical profile of air temperature makes it impossible to an air parcel to rise due to buoyancy. Here, we used daily averages, thus the daily cycle was filtered out. The wavelets helped to identify harmonics of several days, which contributed to modulate the concentrations beyond the daily cycle.

Observed variability in air pollutant concentrations was analyzed in the context of technological changes facing Brazil regarding air quality policies and the use of ethanol as a substitute to gasoline (Goldemberg, 2007). Brazilian federal agencies, such as the Federal Environmental Council – CONAMA, and state agencies continue to study environmental quality, establish pollutant concentration standards, and operate pollution monitoring networks. These agencies also enforce the use of new technologies to help reduce pollutant emissions, such as catalytic converters which chemically catalyze reactions to oxidize or reduce toxic pollutants into less toxic products and are required by CONAMA to be installed in all new cars from 1997 onwards. Additional examples that public policies decisions have on pollutants concentrations can be found in Andrade *et al.* (2015) and Carvalho *et al.* (2015). The oil crisis in the 1970s also had a significant impact on technology in

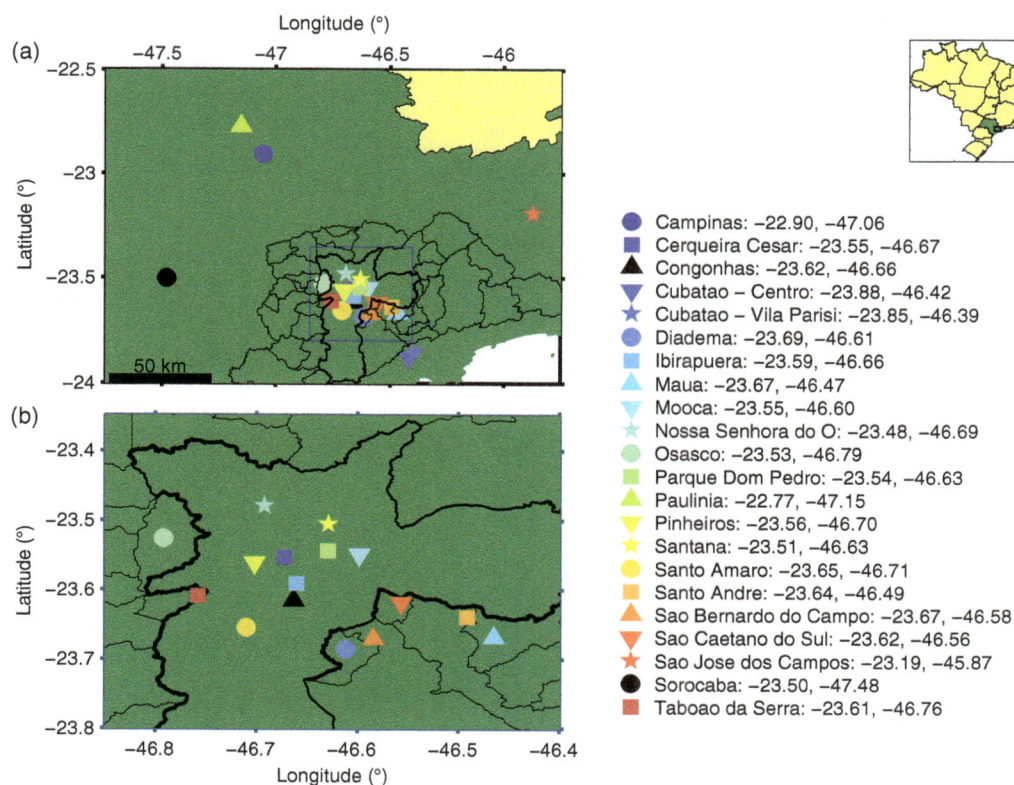

Figure 1. Map of stations and coordinates (latitude and longitude, °) within the metropolitan region of São Paulo (a). The São Paulo city limits are highlighted in bold in the zoomed in detail (b).

Brazil. The rising prices of gasoline compelled the federal government to implement an ethanol fuel industry for road transportation guaranteeing a supply of ethanol from sugarcane coupled with the auto industry building vehicles to run on the new fuel (Goldemberg, 2007). Currently, ethanol-only or flex fuel cars represent nearly 60% of the total vehicle fleet and contribute to 80% of new licensed vehicle in Brazil.

In this work, we report on the variability of pollutants in the MASP using time series of pollutants concentrations measured over 22 stations. The objective here was not a comprehensive study of annual or daily cycles or causes and effects of public policies, which was described in other studies (Salvo and Geiger, 2014; Andrade *et al.*, 2015; Carvalho *et al.*, 2015), but to complement the analysis in those studies by applying wavelet decomposition to the time series of pollutants concentrations.

2. Site and data

The MASP is located in São Paulo state (Figure 1), southeastern Brazil. The territorial area of MASP is of 7944 km^2, while the urbanized area covers 2139 km^2, including many cities which are adjacent to the border of São Paulo city (thick black line in bottom panel). For this study, cities outside the limits of MASP were also included, such as São José dos Campos, Cubatão, Sorocaba, Paulínia, and Campinas. The MASP is surrounded

by topographical features with altitudes up to 1100 m above sea level, such as the hills of *Serra do Mar*, to the south, and *Serra da Cantareira*, to the north. The climate is tropical wet with rainy summers and dry winters (Bourotte *et al.*, 2005).

During summer, the region is influenced by South Atlantic convergence zone (SACZ) and mesoscale convective systems (MCS), which typically cause thunderstorms before sunset and nighttime fog (Silva Dias and Machado, 1997; Castanho and Artaxo, 2001; Freitas *et al.*, 2007). During winter, the reduction in rainfall and the occurrence of thermal inversions in the atmospheric boundary layer (ABL) contribute to higher concentrations of SO_2, CO, and PM_{10} (Angevine *et al.*, 1998; Martins *et al.*, 2004; Barbaro *et al.*, 2014). The dispersion of pollutants in this region is influenced by sea breeze and by valley-mountain circulations, due to the proximity to both the coast and the mountain range that runs parallel to the Atlantic Ocean known as *Serra do Mar* (Silva Dias and Machado, 1997; Oliveira *et al.*, 2003; Carvalho *et al.*, 2012). In addition, circulations associated with the UHI effect contribute to the general dispersion of pollutants (Freitas *et al.*, 2007). Similar to other metropolitan areas of the world, emissions by industrial activity and vehicle traffic are the main source of anthropogenic pollutants in the MASP (Castanho and Artaxo, 2001).

Pollutant concentration datasets were obtained from the São Paulo Environmental Agency (CETESB) from stations primarily located within the urban perimeter.

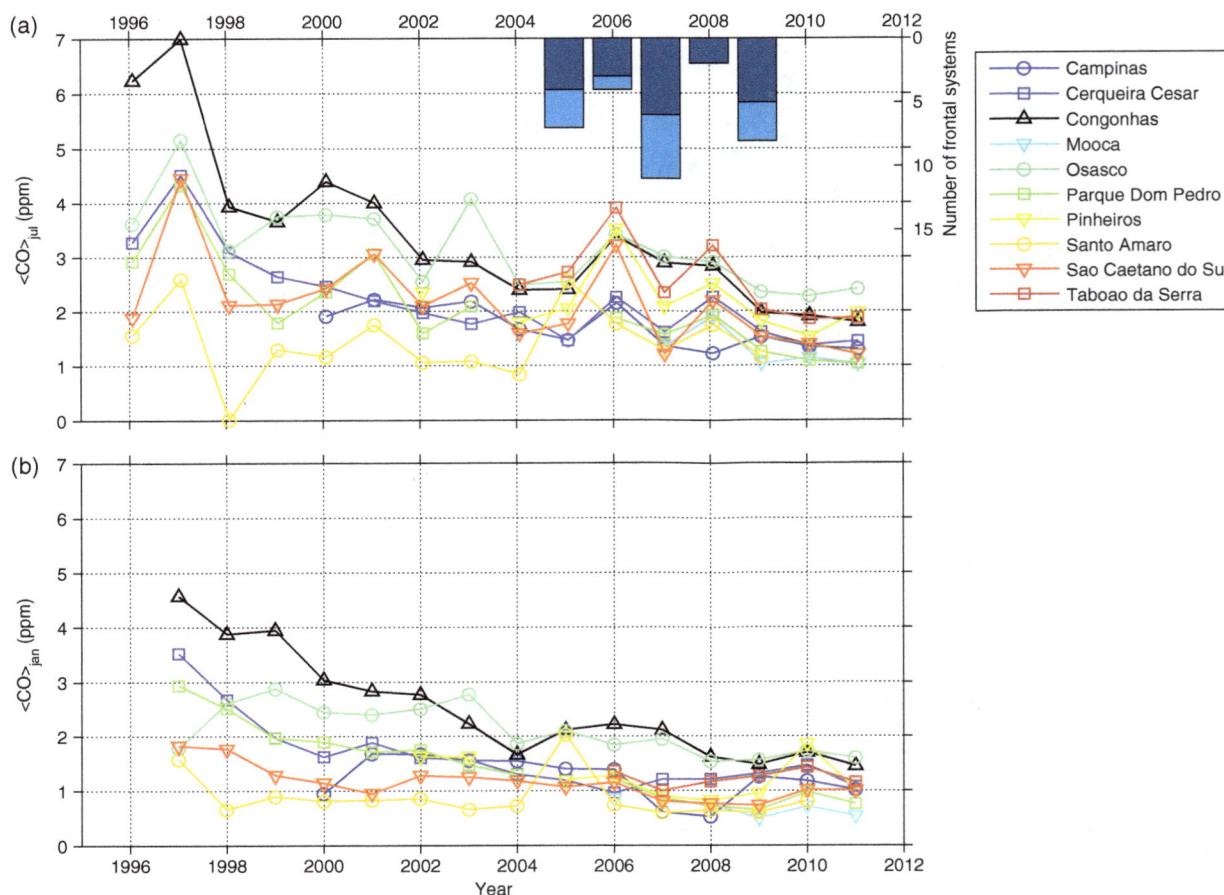

Figure 2. Evolution of monthly averages of CO (ppm) from 1996 to 2011. (a) July and (b) January. Bars denote the number of frontal systems reaching the coast (deep blue) and the countryside (light blue).

Most of the air quality stations used in this study are influenced by vehicle or/and industrial emissions with exception of the Ibirapuera station (Figures S2(e) and S5, Supporting Information), which is located in a park. Stations located outside the MASP, such as Campinas, Sorocaba, and São José dos Campos, had historically lower concentrations for some pollutants. Information regarding the number of licensed vehicles and the size of the ethanol fleet were obtained from the Brazilian Sugarcane Industry Association (UNICA).

The dataset of pollutants concentration from each station included hourly records on PM_{10}, O_3, CO, and SO_2, averaged in this work over 8- or 24-h windows, as recommend by World Health Organization (WHO) standards. In recent years, many studies have addressed both the variability, annual cycle, daily and weekday patterns (Carvalho *et al.*, 2015), and relationships of ethanol prices and public policies on this variability (Salvo and Geiger, 2014; Andrade *et al.*, 2015). In this work, we chose to work only with CO and SO_2, because they presented the highest change from 1996 to 2012 and also had continuous time series of concentrations which were suitable for wavelet analysis. Results on O_3 and PM_{10} are shown in Figures S1–S12 of Supporting Information.

To help explain some peaks in mean monthly concentrations from year to year, data on the number of frontal systems reaching the coast and the countryside were obtained from Climanálise (Climanálise, 2005). These data were plotted together with the monthly concentrations in Figures 2 and 3.

3. Methodology

Data analysis consisted of statistical inferences on time series of concentrations, such as averages, and also correlations with time, to give support to observed trends in pollutant concentrations. In general, wintertime is characterized by less mixing of pollutants, due to colder temperatures and the proximity of the South Atlantic Subtropical Anticyclone. To better identify interannual trends in concentrations, averages were calculated for both July (winter) and January (summer). This procedure enhances the long-term trends because annual maxima (and minima) are compared together. Because of the nature of the annual cycle, with a strong peak during winter but low concentration during summer, yearly averages tend to weaken the maxima, masking the effects of frontal systems on air pollution. It should be noted that not all pollutants were measured at all stations.

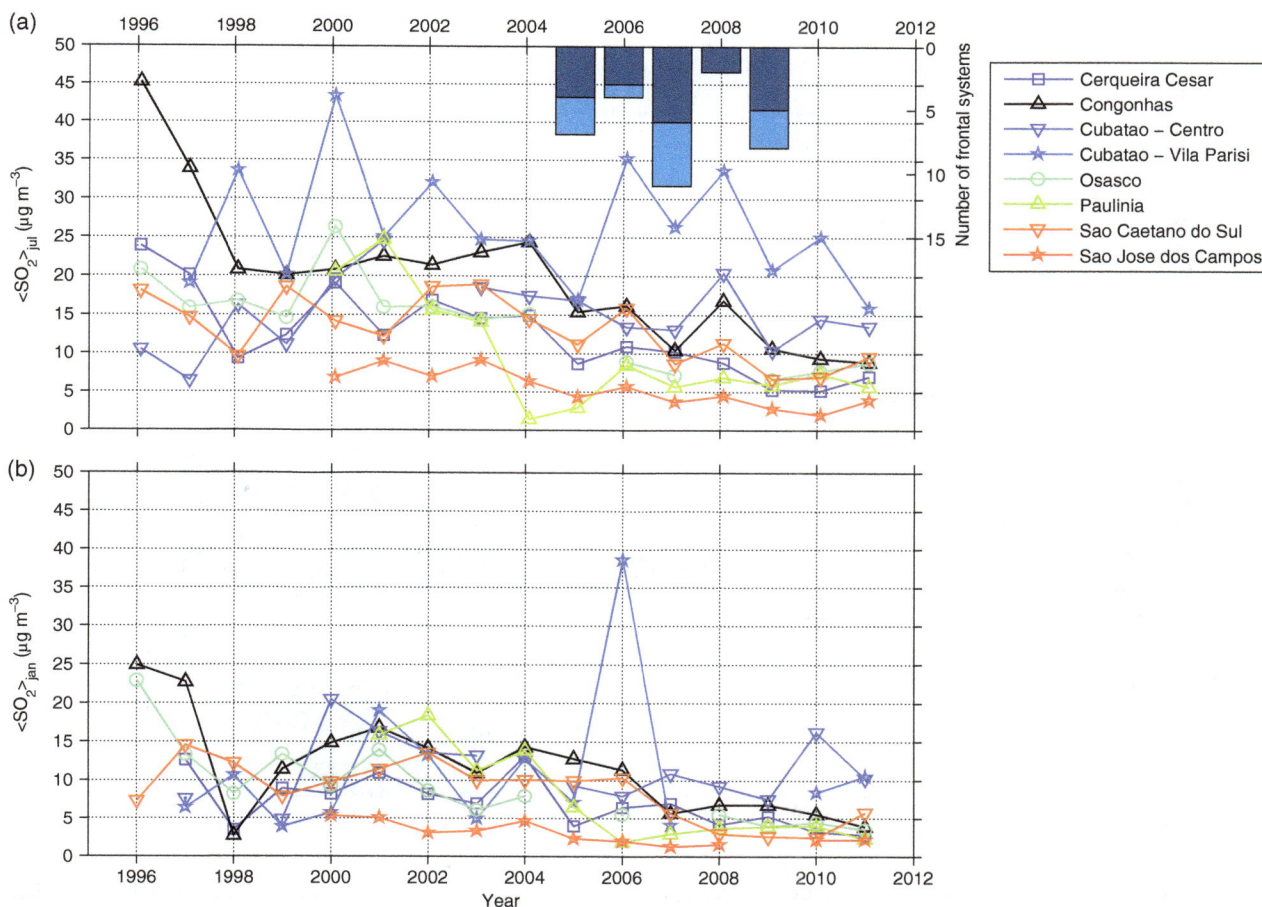

Figure 3. Evolution of monthly averages of SO$_2$ (μg m^{-3}) from 1996 to 2011. (a) July and (b) January. Bars denote the number of frontal systems reaching the coast (deep blue) and the countryside (light blue).

The full dataset was presented in Figures S1–S4 together with standards for each pollutant following the WHO or CONAMA. The standards used were 9 ppm for CO (maximum 8-h moving average), 150 μg m^{-3} for PM$_{10}$ (24-h mean), 100 μg m^{-3} for O$_3$ (maximum 8-h moving average), and 20 μg m^{-3} for SO$_2$, (24-h mean). The standards for PM$_{10}$ and CO were established by CONAMA.

The time series of pollutant concentrations were analyzed using wavelet analysis, a technique that enables the most important frequencies influencing the variability of a signal to be inferred (Daubechies, 1992; Torrence and Compo, 1998). Although Fourier analysis can also be used to identify the most important harmonics in time series, wavelets make it possible to locate in time the influence of harmonics which are not stationary. In recent years, wavelet analysis has been used in many studies of geophysical data, such as river levels, turbulence over plant canopies, and pollutant concentrations (Collineau and Brunet, 1993; Sá *et al.*, 1998; Terradellas *et al.*, 2005; Zeri *et al.*, 2011). The wavelet decomposition works similar to a spectrum, separating the harmonics in a signal while assigning a 'wavelet power' to them, which is proportional to the overall variance. The most important harmonics will be the ones with high wavelet power. Mathematically,

the wavelet power is calculated from the convolution of a function (the wavelet mother) with portions of the signal. The wavelet mother chosen for this study was the Morlet (wavenumber 6), because it was shown to be appropriate to identify the variability of climatological data (Torrence and Compo, 1998). Wavelet analysis requires continuous time series. For this reason, gaps in the data were filled using linear interpolation. Because the wavelet power is calculated locally, the influence of interpolated gaps is easily identified in the scalograms. The wavelet power shown in the scalograms of Figures 5 and 6 (and Figures S1–S12) was calculated as the squared modulus of the wavelet coefficients, having units of signal variance.

4. Results, discussion, and conclusions

The concentration of some pollutants over the MASP is well below the standards (WHO) while others have been continuously surpassing the safe limits. The concentration of CO reached more than the limit of 9 ppm only in the beginning of the period analyzed here (1996–1997), except for two stations that are influenced heavily by heavy vehicle traffic (Congonhas and Cerqueira César). From 2004 to 2011, only 2 or 3 events of CO higher than

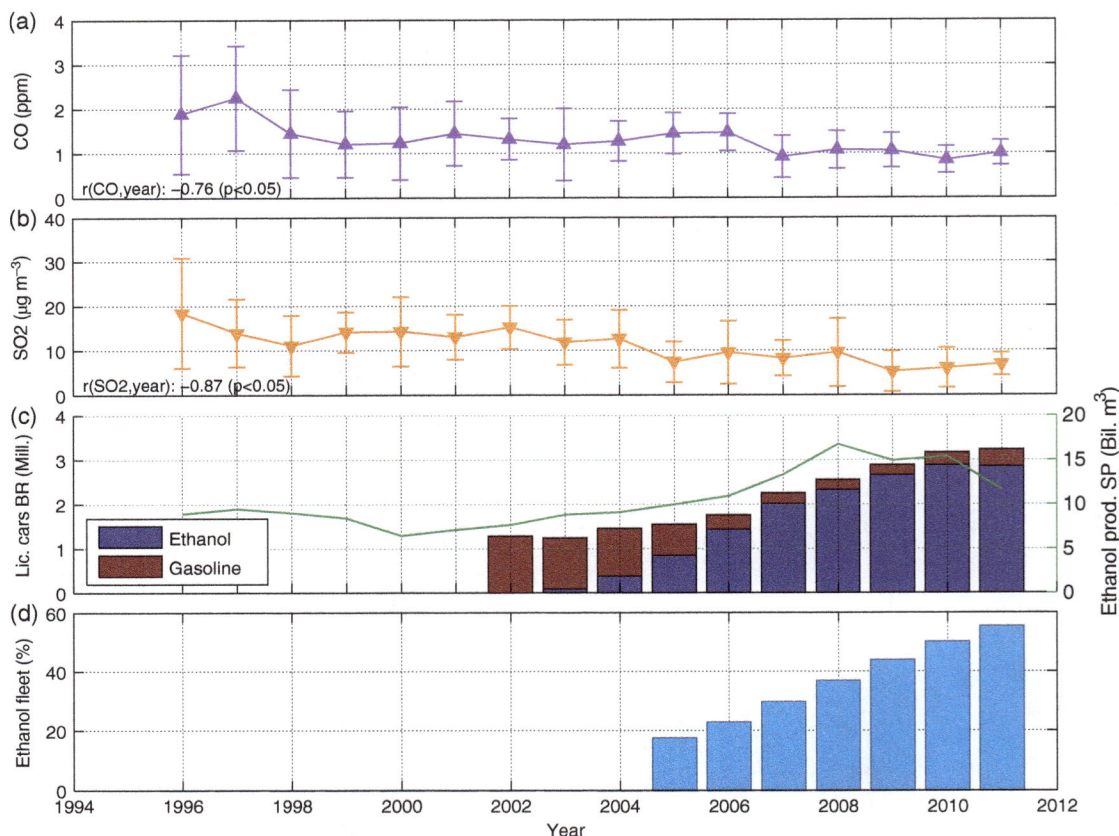

Figure 4. Evolution of average July concentration for CO (ppm) (a), and SO_2 ($\mu g\,m^{-3}$) (b). (c) Licensed cars in Brazil for gasoline and ethanol (left axis); evolution of ethanol production in São Paulo state (right axis). (d) Proportion of the fleet associated with ethanol cars.

9 ppm were observed (Figure S1). For PM_{10} (Figure S2), even wintertime peaks have been below the safe limit of $150\,\mu g\,m^{-3}$. The only station that deviated from these results is Cubatão (Figure S2(d)), where average concentrations are rarely below $70\,\mu g\,m^{-3}$. Cubatão has intense industrial activity associated with its chemical and petrochemical complex, and as a result it has the highest levels of pollution in the dataset. This is also reflected in the concentrations of SO_2 (Figure S3), with both stations in Cubatão frequently showing SO_2 above the $20\,\mu g\,m^{-3}$ limit. Finally, O_3 has been increasing in several stations and frequently reaching over the limit of $100\,\mu g\,m^{-3}$, creating concerns associated with this damaging pollutant (Figure S4).

The monthly averaged concentrations of CO and SO_2 (Figures 2 and 3) presented strong interannual variability, with peaks following minima and vice versa. The wintertime concentrations of SO_2 (Figure 3(a)) peaked in several cities in 2006 and later in 2008 with different amplitudes. These peaks and valleys during wintertime were also observed for other pollutants, including PM_{10} (Figure S5) and O_3 (Figure S6). The number of frontal systems reaching the coast and the countryside (bars) helps to explain this variability. In general, a frontal system brings rainfall for several days, washing out pollutants from the air. Indeed, July of 2006 and 2008 had the lowest number of frontal systems and the highest

values of mean monthly pollution concentration. The monthly concentrations of PM_{10} decreased for January (Figure S5(b)) and are approximately constant when averaged for July (wintertime), responding strongly only to months with lower rainfall (lower number of frontal systems, Figure S5(a)).

Overall, a decrease in CO and SO_2 concentrations from 1996 to 2011 is evident for most of the stations. However, the decreasing trend is stronger for July (Figures 2(a) and 3(a)). For some locations (Osasco, Cerqueira César), CO decreased by about 50% from 1996 to 2011 while the station near the airport of Congonhas, a site strongly influenced by vehicles emissions, decreased from 4.7 to 1.5 ppm in the period, a change of almost 70%. Similar changes were observed for SO_2 (Figure 3), with largest reductions observed in winter compared to summer. For PM_{10} and O_3 (Figures S5 and S6), a decreasing trend was observed only for summer in PM_{10} (from 1996 to 2002). This trend could be associated with public policies enforced to reduce vehicular emissions (Carvalho *et al.*, 2015). On the other hand, the summertime concentration of O_3 has been increasing (Figure S6) since 2007, which could be associated with the increasing fleet of vehicles using ethanol, producing more precursors to O_3 formation, particularly aldehydes (Salvo and Geiger, 2014).

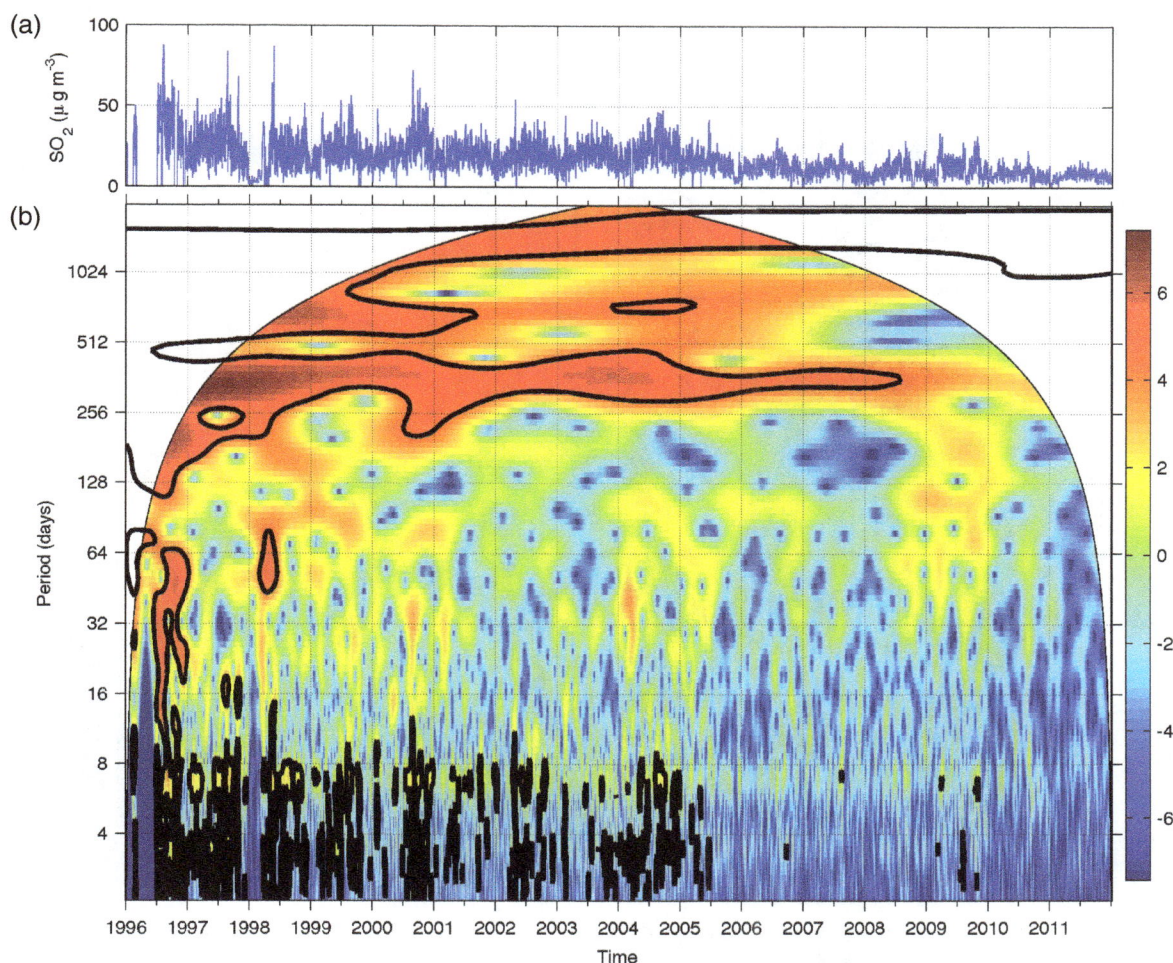

Figure 5. (a) Time series of the concentration of SO_2 ($\mu g\,m^{-3}$) for the station Congonhas and (b) wavelet decomposition for the time series of SO_2. The color scale is proportional to the series variance while bold contours enclose regions where the wavelet power is statistically significant, when compared to a random noise.

The wintertime concentrations for CO and SO_2 were averaged for all locations for each year (Figure 4), revealing three significant results: (1) three phases are evident for CO: a sharp decrease in average concentration from 1996 to 1998, followed by a steady phase, and another decrease by 2006–2007, (2) both average and spatial variability (indicated by lower error bars) in CO among stations separated by 100 s of kilometers were lower at the end of the period, (3) in the 2000s, SO_2 concentration decreased from 2004 to 2005 and later from 2008 to 2009. The different phases observed in the two topmost panels (Figure 4) were likely associated with the changes in policies of air pollution (until ~2005) and later by the widespread adoption of ethanol cars, discussed in more details by Andrade *et al.* (2015) and Carvalho *et al.* (2015). Here, we present some data of ethanol use (Figure 4(b) and (c)) to give context to the changes observed in the time series of pollutants. The decrease in concentrations of CO and SO_2 after 2004 was likely influenced by the increasing number ethanol fueled cars, which pollute less CO and SO_2, licensed each year (data for whole country), coupled with an increase in ethanol production in the state of

São Paulo – which is responsible for more than 60% of ethanol production in Brazil (Goldemberg, 2007). Overall, the ethanol fleet increased from 20% in 2005 to 60% in 2011. The true causation, however, of the reductions observed in the data goes beyond the scope of this paper. To accomplish this, a detailed analysis of sources of pollutants over temporal and spatial scales is required.

The wavelet decomposition revealed the strongest and statistically significant harmonics or temporal scales from 1996 to 2011 (SO_2 measured at Congonhas airport, Figure 5, and at São Caetano do Sul, Figure 6. Other examples were included in Figures S1–S12.). As expected, the annual cycle is strong and significant, between 256 and 512 days of duration. Because the series are composed of daily averages, the influence of the diurnal cycle is not shown. However, the contours between 4 and 8 days are present from the start of the period until ~2006–2007. This indicates that the concentrations are likely modulated in those scales by precipitation, horizontal advection, stability of the ABL, the solar radiation at the surface, or other effects. The results in Figures S7–S12 show

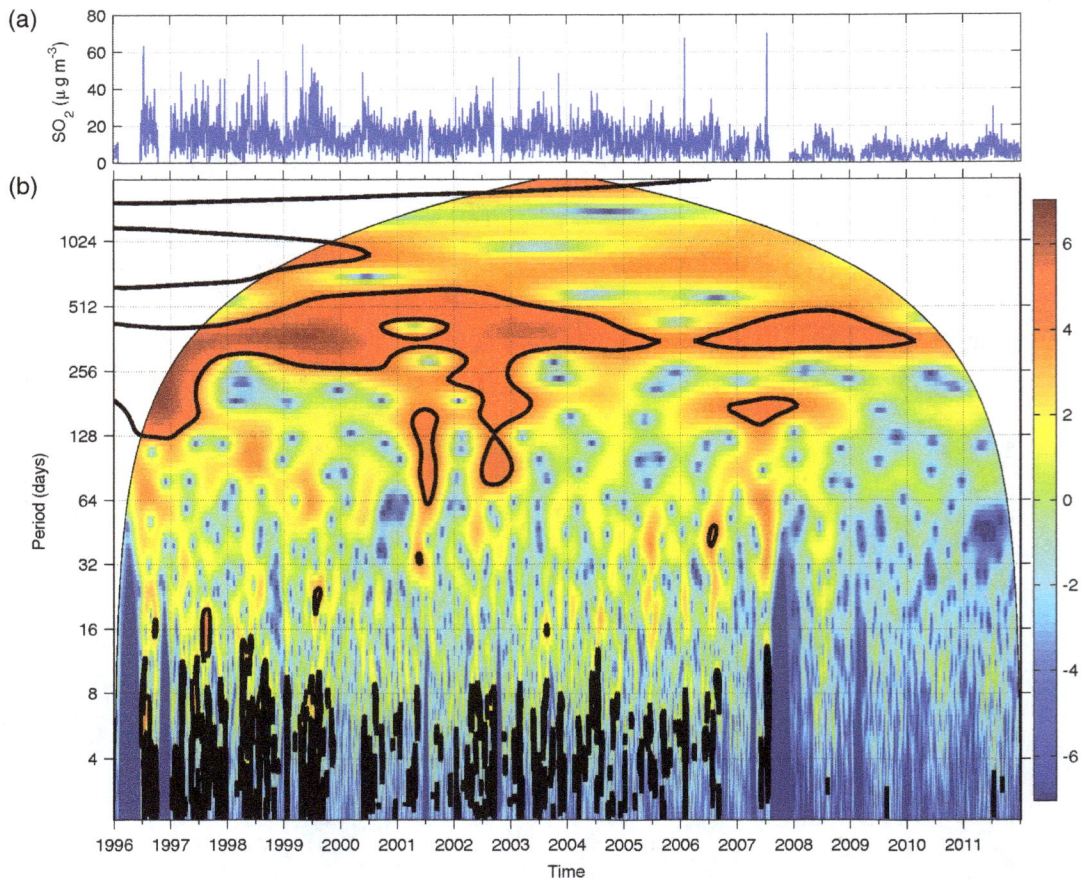

Figure 6. (a) Time series of the concentration of SO$_2$ (μg m^{-3}) for the station of São Caetano do Sul and (b) wavelet decomposition for the time series of SO$_2$. The color scale is proportional to the series variance while bold contours enclose regions where the wavelet power is statistically significant, when compared to a random noise.

a mixture of trends for some pollutants and cities. While the 4−8 days harmonic became less prevalent in Figures S7−S9, it was still strong and frequent for PM$_{10}$ (Figures S10−S12), modulating the high variance observed until 2011. The same result was found in another study for the city of Rio de Janeiro (Zeri *et al.*, 2011). This variability was associated with the passage of frontal systems, which have a similar duration when they occur in the southeastern region of Brazil. While the dataset for Rio de Janeiro limited to 2 years, the longer time series of pollutants presented here made it possible to register a sudden change in the influence of this harmonic of 4−8 days of duration, becoming non-statistically significant after 2006−2007. The disappearance of this harmonic suggests a much weaker influence of the meteorological systems that reach the region, at least in time scales of several days, beyond the daily cycles. The daily cycle would still be influenced by the evolution of air temperature, humidity and wind speed, cycles influence by solar energy as well as by anthropogenic factors, such as traffic and industrial activity. In addition, the temporal resolution of this analysis (daily averages) makes it impossible to infer on short-lived spikes in concentrations (hours), beyond the acceptable limits determined by air quality agencies.

The patterns identified in the wavelet analysis are a result of the modulation of the variance by the harmonics. When the variance is reduced, the modulation loses significance and the patterns disappear from the wavelet plots. The reduction in both daily averages and variances of SO$_2$ and CO concentrations is obvious from the plots in Figures S1 and S4, with the exception of SO$_2$ in Cubatão panels S4(c) and S4(d). The sources of pollutants differ between locations, causing great differences in averages and variances between stations separated by a few kilometers. For example, São Caetano do Sul and Congonhas are separated by ~15 km, but the variance observed in the signal of SO$_2$ reduced from 2005 to 2006, in Congonhas, and later from 2007 to 2008, in São Caetano do Sul. As a result, the modulation with periods of 4−8 days identified by the wavelet lasted longer in São Caetano do Sul. Thus, the variance in the signals is a result of local sources and not induced by the harmonics. Finally, the harmonic of 4−8 days could be associated with low-pressure systems, as evidenced by the number of frontal systems reaching the state during July in Figures 2 and 3, or high pressure systems, leading to lower temperatures and reduced turbulent mixing, increasing concentrations of pollutants. A more detailed analysis using continuous meteorological data near the stations should make a clear distinction between both influences.

Low levels of pollutants should be a target of public policies so that the MASP, and cities in general, become more resilient to the influence of meteorological systems on air quality. The consequence of ideal public policies is a reduction in the vulnerability of city dwellers to the presence of meteorological systems that disturb the lower atmosphere for days, trapping pollutants during cold inversions or spreading – by turbulence – dust deposited over the ground. Wavelet analysis is a helpful tool that identifies the harmonics in time series of pollutants. Future work using finer temporal resolutions (hours) should explore other harmonics associated with human or industrial activity, such as patterns of traffic of cars or industrial activity.

Acknowledgements

The authors are grateful to CETESB for sharing the dataset of pollutants; data can be accessed at http://ar.cetesb.sp.gov.br/qualar/. The authors acknowledge the helpful comments from three anonymous reviewers as well as the assistance of C. J. Bernacchi with the language.

Supporting information

The following supporting information is available:

Figure S1. Time series of concentrations of CO (ppm). The standard of 9 ppm (maximum 8-h moving average) is marked with the dashed line.

Figure S2. Time series of concentrations of PM_{10} ($\mu g\,m^{-3}$). The standard of 150 $\mu g\,m^{-3}$ (24-h mean) is marked with the dashed line.

Figure S3. Time series of concentrations of O_3 ($\mu g\,m^{-3}$). The standard of 100 $\mu g\,m^{-3}$ (maximum 8-h moving average) is marked with the dashed line.

Figure S4. Time series of concentrations of SO_2 ($\mu g\,m^{-3}$). The standard of 20 $\mu g\,m^{-3}$ (24-h mean) is marked with the dashed line.

Figure S5. Evolution of monthly averages of PM_{10} ($\mu g\,m^{-3}$) from 1996 to 2011. Top: July and bottom: January. Bars denote the number of frontal systems reaching the coast (deep blue) and the countryside (light blue).

Figure S6. Evolution of monthly averages of O_3 ($\mu g\,m^{-3}$) from 1996 to 2011. Top: July and bottom: January. Bars denote the number of frontal systems reaching the coast (deep blue) and the countryside (light blue).

Figure S7. Top: time series of the concentration of CO (ppm) for the station Congonhas; bottom: wavelet decomposition for the time series. The color scale is proportional to the series variance while bold contours enclose regions where the wavelet power is statistically significant, when compared to a random noise.

Figure S8. Top: time series of the concentration of CO (ppm) for the station São Caetano do Sul; bottom: wavelet decomposition for the time series. The color scale is proportional to the series variance while bold contours enclose regions where the wavelet power is statistically significant, when compared to a random noise.

Figure S9. Top: time series of the concentration of PM_{10} ($\mu g\,m^{-3}$) for the station Congonhas; bottom: wavelet decomposition for the time series. The color scale is proportional to the series variance while bold contours enclose regions where the wavelet power is statistically significant, when compared to a random noise.

Figure S10. Top: time series of the concentration of PM_{10} ($\mu g\,m^{-3}$) for the station São Caetano do Sul; bottom: wavelet decomposition for the time series. The color scale is proportional to the series variance while bold contours enclose regions where the wavelet power is statistically significant, when compared to a random noise.

Figure S11. Top: time series of the concentration of PM_{10} ($\mu g\,m^{-3}$) for the station Cubatão; bottom: wavelet decomposition for the time series. The color scale is proportional to the series variance while bold contours enclose regions where the wavelet power is statistically significant, when compared to a random noise.

Figure S12. Top: time series of the concentration of SO_2 ($\mu g\,m^{-3}$) for the station Cubatão; bottom: wavelet decomposition for the time series. The color scale is proportional to the series variance while bold contours enclose regions where the wavelet power is statistically significant, when compared to a random noise.

References

Andrade MF, Ynoue RY, Freitas ED, Todesco E, Vara Vela A, Ibarra S, Martins LD, Martins JA, Carvalho VSB. 2015. Air quality forecasting system for Southeastern Brazil. *Frontiers in Environmental Science* **3**: 9, doi: 10.3389/fenvs.2015.00009

Angevine WM, Grimsdell AW, Hartten LM, Delany AC. 1998. The flatland boundary layer experiments. *Bulletin of the American Meteorological Society* **79**: 419–431, doi: 10.1175/1520-0477(1998)079<0419:TFBLE>2.0.CO;2.

Barbaro E, de Arellano JV-G, Ouwersloot HG, Schröter JS, Donovan DP, Krol MC. 2014. Aerosols in the convective boundary layer: shortwave radiation effects on the coupled land-atmosphere system. *Journal of Geophysical Research, [Atmospheres]* **119**: 5845–5863, doi: 10.1002/2013JD021237.

Bourotte C, Forti M-C, Taniguchi S, Bícego MC, Lotufo PA. 2005. A wintertime study of PAHs in fine and coarse aerosols in São Paulo city, Brazil. *Atmospheric Environment* **39**: 3799–3811, doi: 10.1016/j.atmosenv.2005.02.054.

Cançado JED, Saldiva PHN, Pereira LAA, Lara LBLS, Artaxo P, Martinelli LA, Arbex MA, Zanobetti A, Braga ALF. 2006. The impact of sugar cane-burning emissions on the respiratory system of children and the elderly. *Environmental Health Perspectives* **114**: 725–729.

Carvalho VSB, Freitas ED, Mazzoli CR, Andrade MF. 2012. Avaliação da influência de condições meteorológicas na ocorrência e manutenção de um episódio prolongado com altas concentrações de ozônio sobre a região metropolitana de São Paulo. *Revista Brasileira de Meteorologia* **27**: 463–474, doi: 10.1590/S0102-77862012000400009.

Carvalho VSB, Freitas ED, Martins LD, Martins JA, Mazzoli CR, Andrade MF. 2015. Air quality status and trends over the metropolitan area of São Paulo, Brazil as a result of emission control policies. *Environmental Science & Policy* **47**: 68–79, doi: 10.1016/j.envsci.2014.11.001.

Castanho ADA, Artaxo P. 2001. Wintertime and summertime São Paulo aerosol source apportionment study. *Atmospheric Environment* **35**: 4889–4902, doi: 10.1016/s1352-2310(01)00357-0.

Climanálise. 2005. Produtos Climanálise INPE/CPTEC. http://www.cptec.inpe.br/products/climanalise/ (accessed 3 March 2011).

Collineau S, Brunet Y. 1993. Detection of turbulent coherent motions in a forest canopy. Part 1. wavelet analysis. *Boundary-Layer Meteorology* **65**: 357–379.

Daubechies I. 1992. *Ten Lectures on Wavelets*, Vol. **61**. Society for Industrial and Applied Mathematics: Philadelphia, PA, 377 pp.

Fajersztajn L, Veras M, Barrozo LV, Saldiva P. 2013. Air pollution: a potentially modifiable risk factor for lung cancer. *Nature Reviews Cancer* **13**: 674–678, doi: 10.1038/nrc3572.

Freitas E, Rozoff C, Cotton W, Dias PS. 2007. Interactions of an urban heat island and sea-breeze circulations during winter over the metropolitan area of São Paulo, Brazil. *Boundary-Layer Meteorology* **122**: 43–65, doi: 10.1007/s10546-006-9091-3

Godoy MLDP, Godoy JM, Roldão LA, Soluri DS, Donagemma RA. 2009. Coarse and fine aerosol source apportionment in Rio de Janeiro, Brazil. *Atmospheric Environment* **43**: 2366–2374, doi: 10.1016/j.atmosenv.2008.12.046.

Goldemberg J. 2007. Ethanol for a sustainable energy future. *Science* **315**: 808–810.

Gonçalves FLT, Carvalho LMV, Conde FC, Latorre M, Saldiva PHN, Braga ALF. 2005. The effects of air pollution and meteorological parameters on respiratory morbidity during the summer in Sao Paulo City. *Environment International* **31**: 343–349.

deLeon AP, Anderson HR, Bland JM, Strachan DP, Bower J. 1996. Effects of air pollution on daily hospital admissions for respiratory disease in London between 1987–88 and 1991–92. *Journal of Epidemiology and Community Health* **50**: S63–S70.

Martins MHRB, Anazia R, Guardani MLG, Lacava CIV, Romano J, Silva SR. 2004. Evolution of air quality in the Sao Paulo Metropolitan Area and its relation with public policies. *International Journal of Environment and Pollution* **22**: 430–440, doi: 10.1504/IJEP.2004.005679

de Miranda RM, de Fátima Andrade M, Worobiec A, Grieken RV. 2002. Characterisation of aerosol particles in the São Paulo Metropolitan Area. *Atmospheric Environment* **36**: 345–352, doi: 10.1016/s1352-2310(01)00363-6

Oliveira AP, Bornstein RD, Soares J. 2003. Annual and diurnal wind patterns in the city of São Paulo. *Water, Air and Soil Pollution: Focus* **3**: 3–15, doi: 10.1023/a:1026090103764

Sá LDA, Sambatti SBM, Galvao GP. 1998. Applying the Morlet wavelet in a study of variability of the level of Paraguay River at Ladario, MS. *Pesquisa Agropecuária Brasileira* **33**: 1775–1785.

Salvo A, Geiger FM. 2014. Reduction in local ozone levels in urban Sao Paulo due to a shift from ethanol to gasoline use. *Nature Geoscience* **7**: 450–458, doi: 10.1038/Ngeo2144

Samet JM, Dominici F, Curriero FC, Coursac I, Zeger SL. 2000. Fine particulate air pollution and mortality in 20 US Cities, 1987–1994. *New England Journal of Medicine* **343**: 1742–1749, doi: 10.1056/Nejm200012143432401

Schwartz J. 1996. Air pollution and hospital admissions for respiratory disease. *Epidemiology* **7**: 20–28.

Sharma VP, Arora HC, Gupta RK. 1983. Atmospheric pollution studies at Kanpur – suspended particulate matter. *Atmospheric Environment* **17**: 1307–1313, doi: 10.1016/0004-6981(83)90405-5

Silva Dias MF, Machado AJ. 1997. The role of local circulations in summertime convective development and nocturnal fog in São Paulo, Brazil. *Boundary-Layer Meteorology* **82**: 135–157, doi: 10.1023/A:1000241602661

Terradellas E, Soler MR, Ferreres E, Bravo M. 2005. Analysis of oscillations in the stable atmospheric boundary layer using wavelet methods. *Boundary-Layer Meteorology* **114**: 489–518, doi: 10.1007/S10546-004-1293-Y

Torrence C, Compo GP. 1998. A practical guide to wavelet analysis. *Bulletin of the American Meteorological Society* **79**: 61–78, doi: 10.1175/1520-0477(1998)079<0061:Apgtwa>2.0.Co;2

Zeri M, Oliveira JF, Lyra GB. 2011. Spatiotemporal analysis of particulate matter, sulfur dioxide and carbon monoxide concentrations over the city of Rio de Janeiro, Brazil. *Meteorology and Atmospheric Physics* **113**: 139–152, doi: 10.1007/S00703-011-0153-9

23

Improvement of land surface temperature simulation over the Tibetan Plateau and the associated impact on circulation in East Asia

Haifeng Zhuo,[1,2] Yimin Liu[1]* and Jiming Jin[3,4]

[1] State Key Laboratory of Numerical Modeling for Atmospheric Sciences and Geophysical Fluid Dynamics (LASG), Institute of Atmospheric Physics, Chinese Academy of Sciences, Beijing, China
[2] University of Chinese Academy of Sciences, Beijing, China
[3] College of Water Resources and Architectural Engineering, Northwest A&F University, Yangling, China
[4] Department of Watershed Sciences and Plants, Soils and Climate, Utah State University, Logan, UT, USA

*Correspondence to:
Y. Liu, No. 40, Huayanli, Institute of Atmospheric Physics, Chinese Academy of Sciences, Chaoyang District, Beijing 100029, China.
E-mail: lym@lasg.iap.ac.cn

Abstract

A new sensible-heat (*SH*) parameterization was used in the Weather Research and Forecasting model to improve the vertical heat transfer simulation in arid regions. With this new scheme, the simulated *SH* decreased, resulting in a 2 °C reduction of the cold bias in the land surface temperature over the Tibetan Plateau. The weakened *SH* led to anticyclonic circulation and cyclonical flow changes in the northern East China due to Rossby wave propagation. The summer-rainfall in East Asian is marginally improved. It is suggested that the *SH* simulation over the Plateau is a land–atmosphere coupling issue that is important to the dynamical downscaling in East Asian.

Keywords: Tibetan Plateau; land surface temperature; East Asian summer monsoon; WRF; CLM

1. Introduction

Owing to the unique geography of the Tibetan Plateau (TP) and land–atmosphere interaction characteristics, all of the global climate models underestimate surface air temperature in the TP, and most overestimate its precipitation (Hao *et al.*, 2013). Most of the reanalysis datasets also tend to underestimate temperature in the TP (You *et al.*, 2010; Wang and Zeng, 2012). Land surfaces form the lower boundary conditions of the atmosphere, which essentially dominate local energy and water transfer, and can affect local, regional, and global climate processes. Proper description of land surface processes in earth system models is a popular research topic.

Arid and semi-arid regions constitute approximately one-fourth of the total global land area, and account for approximately 40% of the land area in China (Hu and Zhang, 2001). They are mostly distributed in Northwest China and in the TP. Because vegetation coverage and rainfall are extremely rare in these regions, the land–atmosphere interaction process is dominated by heat transfer processes. However in the arid and semi-arid regions, surface temperature and energy are poorly simulated.

The unique solar radiation and energy transfer processes in the TP differ significantly from those in regions of lower altitude (Smith and Shi, 1992; Zhang *et al.*, 2014). The development of the turbulent transfer theory in recent years has allowed scientists to conduct numerous land surface parameterization field experiments in arid and semi-arid regions (Zeng *et al.*, 1998; Yang *et al.*, 2008; Chen *et al.*, 2010). In such studies, the surface energy transfer process was investigated, and several new parameterization schemes of land surface models were developed (Zeng *et al.*, 2012; Wang *et al.*, 2014).

Without considering vegetation, the emissivity ε of bare soil in Community Land Model (CLM) is set to a constant at approximately 0.96. Therefore, in arid and semi-arid regions, the main factors affecting the calculation of the land–atmosphere energy balance are surface albedo α, aerodynamic roughness length z_{0m}, thermal roughness length z_{0h}, soil heat capacity c_v, and soil thermal conductivity K_{soil}. After conducting more than 10 years of observation at three stations in alpine and desert areas including Gobi, Huang *et al.* (2005) assigned a surface albedo value of 0.255 ± 0.021 to the Dunhuang area of Gobi. Zhou *et al.* (2012) assigned a momentum roughness length z_{0m} of 0.61 ± 0.02 mm in Dunhuang, and an average thermal roughness length z_{0h} of approximately 0.05 mm during the daytime. Chen *et al.* (2009) used these parameters to evaluate the Biosphere–Atmosphere Transfer Scheme (BATS) and showed that the cold bias in land surface temperature (*LST*) simulation has been improved in the daytime, and its diurnal cycle has been accurately simulated. In addition, some studies have shown that vegetation degeneration in northwest arid areas not only affects local precipitation (Li and Xue, 2010) but also drive an anticyclone anomalies in the upper troposphere in the spring, and maintain to summer, at last affect the

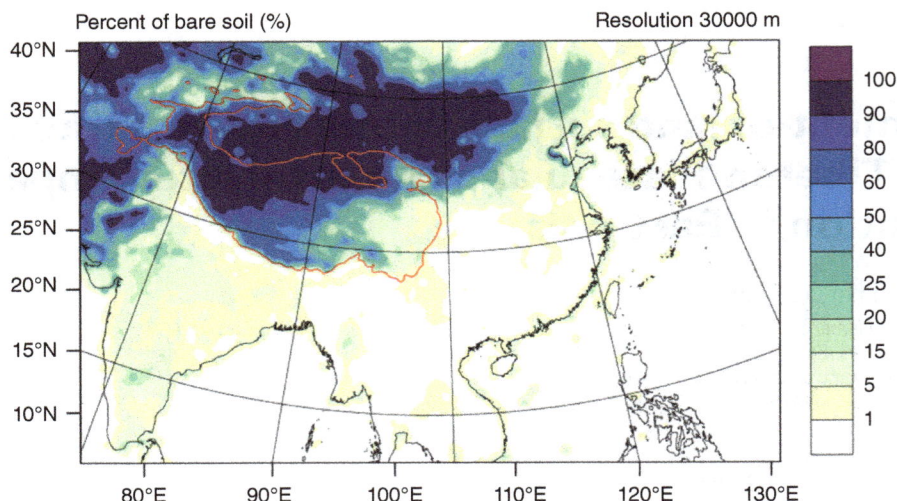

Figure 1. Model domain and the distribution of bare soil in the Community Land Model. The red solid line area indicates elevation >3000 m.

precipitation in the northeast part of TP through "silk road" wave train (Zhou and Huang, 2010; Huang *et al.*, 2011).

Moreover, many studies have derived surface sub-layer turbulent heat transfer parameterization for incorporating z_{0h} in past decades (Sheppard, 1958; Brutsaert, 1982; Zeng and Dickinson, 1998; Kanda *et al.*, 2007; Yang *et al.*, 2008). However, some problems remain (Hogue *et al.*, 2005; Jiménez *et al.*, 2011; Decker *et al.*, 2012). Many land surface models such as Noah land surface model (Noah LSM) and CLM overestimate the *SH* and thus underestimate *LST* during the dry season in arid areas (Hogue *et al.*, 2005). Chen *et al.* (2010) evaluated the parameterization of z_{0h} when simulating *LST* with six different schemes of the Noah LSM. They found that original Noah significantly underestimated *LST* and overestimated *SH* in the daytime. The offline run with revised parameterization schemes can give better results of both *LST* and *SH* owing to better simulation of the diurnal variation in z_{0h}.

Recently, through both theoretical analyses and data-model comparison, Zeng *et al.* (2012) used *in situ* observation and offline experiments at several stations in arid and semi-arid regions, including the Arizona desert and the TP, to further revise the coefficients of z_{0h} both in Noah and CLM. They also set a constraint of minimum friction velocity and take a prescribed soil texture. This measure significantly reduced the underestimation of *LST* in the daytime. Their new scheme has been successfully applied to the National Centers for Environmental Prediction Global Forecast System (NCEP-GFS) assimilation (Zheng *et al.*, 2012). However, it is unknown whether these improvements can be effectively applied in the land–atmosphere coupled model. Simulation is then applied over East Asia. These analyses will be used to examine the extent to which the new *SH* parameterization improves *LST* simulation. Moreover, we discuss the impacts of

land–atmosphere interaction on diabatic heating and local and downstream circulation.

This study helps us deeply understand the attribution of the land surface process of large-scale topography. Moreover, the discovery of direct and indirect effects of the land surface heat transfer process from TP to the downstream regions can be applied to improve the climate prediction of East Asia effectively.

2. Experiment design

2.1. Model and data

The Weather Research and Forecasting (WRF) model is currently the most widely used mesoscale model. Its latest WRFV3.6.1 (Skamarock *et al.*, 2008) coupled with CLM4 (Oleson *et al.*, 2013) was used in this study. The related physics includes MYNN surface layer and Mellor-Yamada Nakanishi and Niino Level 3 PBL. According to the model domain shown in Figure 1, the percentage of bare soil has gradually increased from Southeast China, with rich vegetation, to the Northwest China, which is sparsely vegetated. Particularly in the western plateau, most parts of the TP are sparsely vegetated.

We conducted two experiments: a control experiment (CTL) and a sensitivity experiment (SEN). Both used the same basic settings except that the new parameterization of *SH* was used in the SEN experiment. The central point of the model domain is (30°N, 103°E). The horizontal resolution was 30 km with 150 grid points from south to north and 240 grid points from west to east. Integration was from September 1, 2003, to August 31, 2010, and was continuous with the same initial condition at 00Z UTC. The forcing data used in this study were 6-hourly NCEP climate forecast system reanalysis (CFSR) products (Saha *et al.*, 2010). In addition, we used the daily site observation data of the Chinese Meteorological Administration (CMA) from 2003 to 2010;

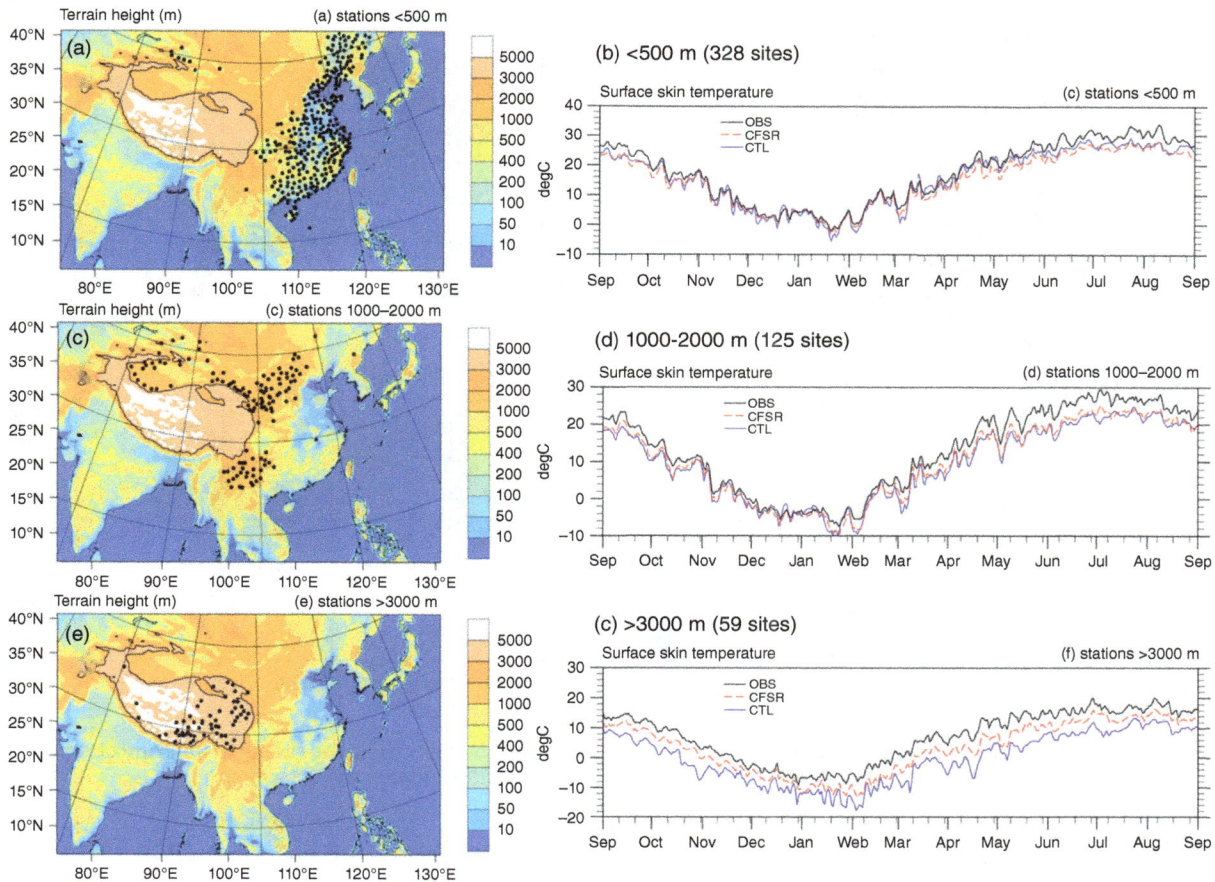

Figure 2. (a), (c), (e) Site distribution and corresponding land surface temperature time series of observation (black line), Climate Forecast System Reanalysis data (red dashed line), and Weather Research and Forecasting simulation (blue line) for (b) 0–500 m, (d) 1000–2000 m, and (f) >3000 m recorded September to August for 1-year integration.

the *LST* observation of the Moderate Resolution Imaging Spectroradiometer (MODIS) which recorded four times daily; and the Tropical Rainfall Measuring Mission (TRMM)-3B42 dataset.

2.2. Evaluation of the WRF model

In the present study, we compare the CMA station data against the corresponding model location in the CTL (Figure 2). The sites used for comparison are distributed in three different regions: elevation below 500 m (328 sites), at 1000–2000 m (125 sites), and higher than 3000 m (59 sites). The percentage of bare soil is larger at the stations located at higher elevations. In Figure 2, the left-hand side shows the terrain height and locations of the CMA sites, and the right-hand side shows averaged *LST*. The values of the CTL were obtained from the grid point in closest proximity to the site location. Because each site's *LST* is based on actual terrain elevation and the model's elevation may not be the same, we adjusted the temperature by using temperature vertical lapse rate of ~6 °C km^{-1}. The results show that the cold bias of *LST* gradually increased with an increase in altitude. The cold bias was obviously greater for the sites in the TP, with a root−mean−square deviation (RMSD) of ~7.6 °C and elevation higher than 3000 m,

than that in the eastern plain, with an RMSD of ~2.4 °C and elevation below 500 m. The original CLM model underestimates the *LST* in arid and semi-arid regions.

2.3. Model configuration with new scheme

Zeng and Dickinson (1998) previously reported a relationship between z_{0h} and z_{0m}. Recently, they revised the constants "a" and "b" from 0.13/0.45 to 0.36/0.5 and constrained the minimum friction velocity u_{*min} to 0.07 m in bare soil (Zeng *et al.*, 2012). The three parameters used in the *SH* of non-vegetated surfaces are z_{0h}, z_{0m}, and minimum friction velocity u_{*min}:

$$\ln\left(\frac{z_{0m}}{z_{0h}}\right) = a\left(\frac{u_* z_{0m}}{\nu}\right)^b \tag{1}$$

$$u_{*min} = 0.07\frac{\rho_0}{\rho}\left(\frac{z_{0m}}{z_{0g}}\right)^{0.18} \tag{2}$$

where $\nu = 1.5 \times 10^{-5}$ m^2 s^{-1} refers to the molecular viscosity. The air density at sea level ρ_0 is 1.22 kg m^{-3}, and the roughness length of bare soil is 0.01 m. It is shown that the underestimation of *LST* is significantly improved in the daytime (Zeng *et al.*, 2012; Wang *et al.*, 2014).

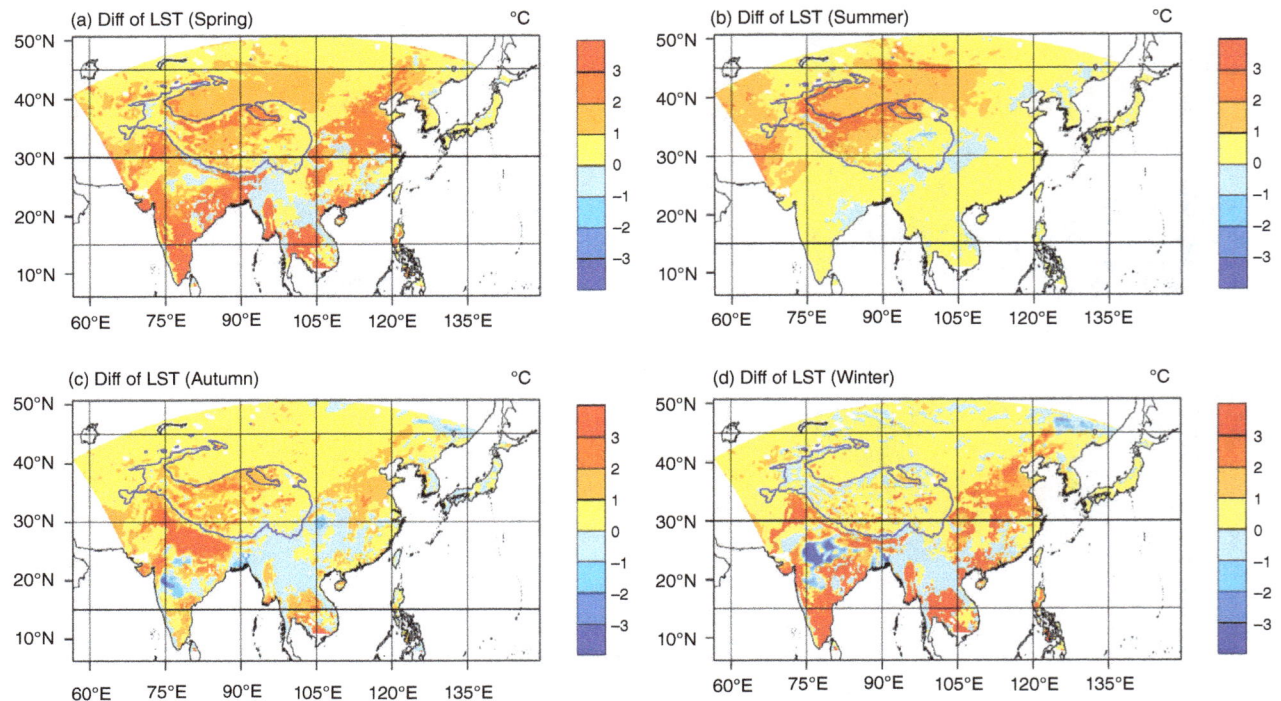

Figure 3. Differences in seasonal mean land surface temperature determined by subtracting the results of the control experiment from those of the sensitivity experiment (2003–2010): (a) spring, (b) summer, (c) autumn, and (d) winter.

3. Results

3.1. *LST* simulation

When a two-way feedback of land–atmosphere interaction was considered, the *LST* showed an obvious improvement in all of the arid and semi-arid regions, particularly in the TP and the Taklamakan Desert, but showed variations among seasons. The TP has the largest impacts in summer because the thermal forcing in the weak wind plays a more important role than the heating in strong wind in other seasons (Wu *et al.*, 2007). The surface sensible heating over the Plateau pumps moisture from oceans so contributes the formation and variation of the Asian monsoon (Wu *et al.*, 2012a, 2012b; Liu *et al.*, 2102). The maximum *LST* increase, 3 °C, occurred in summer over the western TP; otherwise, the *LST* increase in the Taklamakan Desert was smaller than that in the TP. The changes in other regions indicate responses to the interaction in arid regions (Figure 3).

Figure 4 shows the differences in *SH* between SEN and CTL. The *SH* decreased in arid and semi-arid regions when using the new scheme. Of the four seasons, the *SH* decrease was most obvious in summer. Moreover, the surface wind changed with the *SH*; wind divergence was noted near the TP. The impact on East Asia was strong in summer. The cyclonic wind deviation was located in Northeast China, although the anti-cyclonic wind deviation occurred in the South China. Because thermal forcing had the strongest impacts on circulation in summer, the following analyses and discussions focus on summer.

3.2. Impact on the circulation and the East Asia climate in summer

To determine the causes of the cyclonic deviation of surface wind in summer, we analyzed the thermal heating effects of the TP on the atmospheric circulation. In summer, the cumulus convection is strong at the south slope and southern edge of the plateau; the maximum diabatic heating is $10\,K\,day^{-1}$ at the south slope (Figure 5(a)). The updraft airflow is located at both sides of the TP due to the *SH* air pump effect (Wu *et al.*, 2007). The flow on the south slope is considerably stronger owing to moisture physics feedback. However, the pumping effect was weakened after the application of the new scheme. The upward movement and the diabatic heating also became weaker on both north and south of the TP (Figure 5(b)).

The new scheme also affects the summer circulation and rainfall in East China. Figure 5(c) shows a comparison of CTL precipitation and TRMM observation, and Figure 5(d) shows the difference in precipitation and 700 hPa wind between SEN and CTL. The East Asia Summer Monsoon rain belt is underestimated by approximately $4\,mm\,day^{-1}$ in East–Central China and is overestimated at the same rate in the eastern TP and the southeastern coast of China (Figure 5(c)). The new scheme improves the precipitation in most parts of East Asia. The underestimated rainfall in Northeast and East–Central China showed marginal improvement, as did the rainfall overestimated in the eastern TP, the Hetao region in Mongolia, and the southeast coast of China in summer. This result occurred because the new scheme obviously reduces the *SH* in the western and

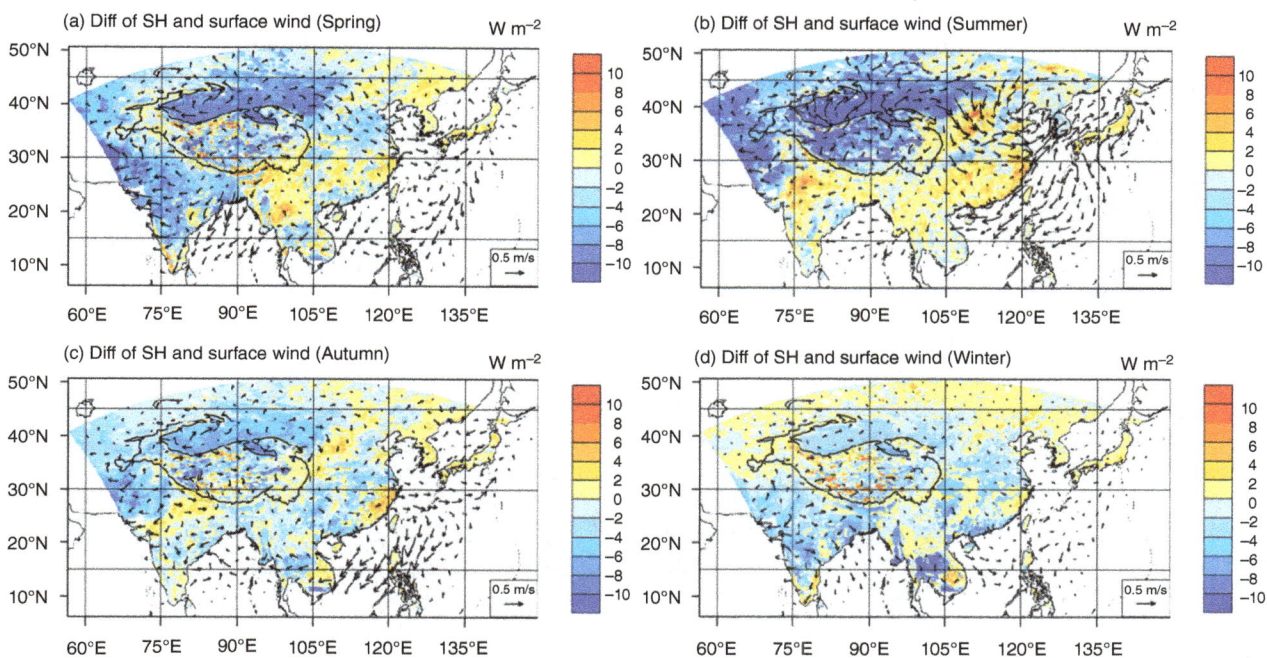

Figure 4. Comparison of seasonal mean surface sensible heat (shading, units: wm⁻²) and surface wind (vector, units: m s⁻¹), determined by subtracting the results of the control experiment from those of the sensitivity experiment (2003–2010): (a) spring, (b) summer, (c) autumn, and (d) winter.

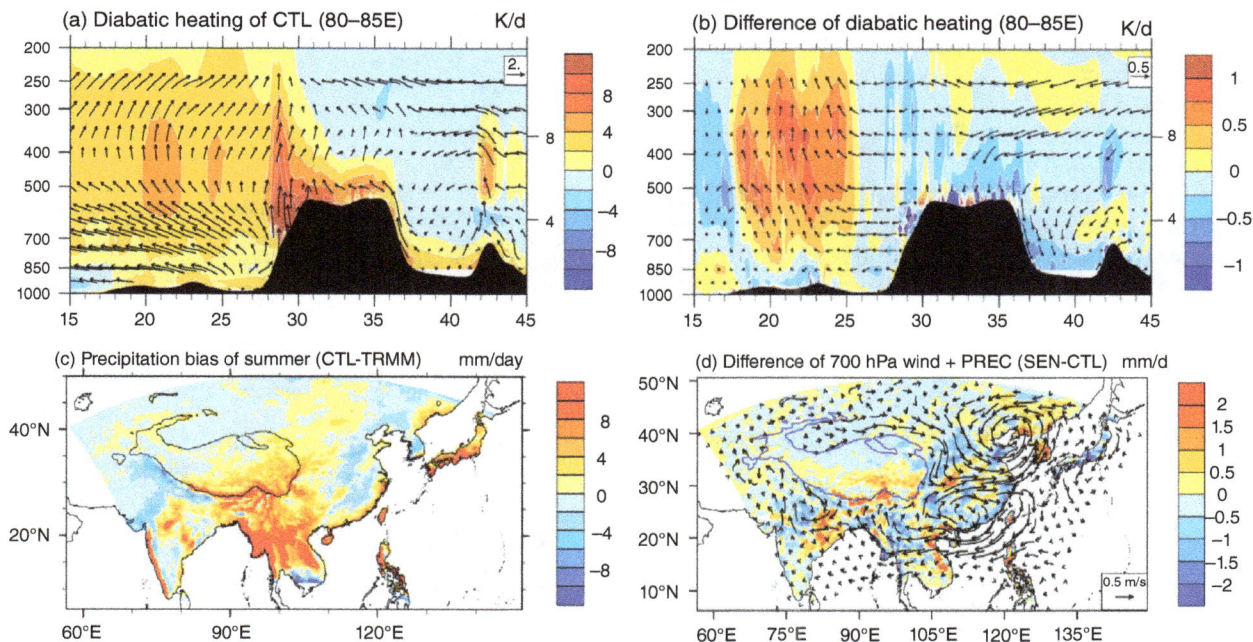

Figure 5. Diabatic heating, wind, and precipitation in summer. (a) 80°−85°E mean diabatic heating and wind in the control experiment (CTL); (b) 80°−85°E mean difference of diabatic heating and wind determined by subtracting the results of CTL from those of the sensitivity experiment (SEN); (c) precipitation bias of CTL and the Tropical Rainfall Measuring Mission (TRMM); (d) difference in precipitation and 700 hPa wind (SEN minus CTL).

central TP, where the ground surface is dominated by bare soil. Thus, the *SH* air pump effect of the TP is weakened, and the diabatic heating and the updraft airflow are reduced in both north and south sides of the TP. The weakened surface heating resulted in divergence and anticyclonic circulation close to the surface near the TP and cyclonical flow change in the northern part of East China due to Rossby wave propagation. The southeasterly (westerly) winds of the cyclone led to convergence over Northeast (East–Central) China and contributed to the *in situ* rainfall increase. A compensatory anticyclone developed in South China, and its northwestern flow reduced the rainfall along the southeast coast.

In summary, the new *SH* scheme shows improvement in *LST* simulations in the TP and precipitation simulations in parts of East China in the summer monsoon season. On the other hand, soil moisture (Hong *et al.*, 2009; Moufouma-Okia and Rowell, 2010; Jaeger and Seneviratne, 2011) also play a role in the simulation in the regional models.

4. Summary

This study investigated the influence of a new *SH* parameterization scheme for bare soil in the land–atmosphere coupling system of WRF. The results show that the overestimation of the *SH* and underestimation of *LST* over arid and semi-arid regions are improved by using the new scheme. The new scheme significantly improves the *LST* simulation over most of the western and central TP. The average cold bias of the *LST* was reduced by approximately 2 °C in the TP. The decrease in *SH* affects the vertical circulation near the TP. The *SH* air pump effect in the TP and the updraft motion of airflow around the plateau were also weakened. The diabatic heating over both the southern and northern slopes of the TP was also reduced.

Moreover, the decrease in *SH* affects the circulation and precipitation in East Asia. The new scheme led to divergence with the development of an anticyclone over the TP and a cyclone and anticyclone over Northeast and South China, respectively. The rain belt simulations were thus improved. The cyclonic circulation in Northeast China led to more precipitation in Northeastern and Central China. The anticyclonic circulation in South China decreased the overestimated rainfall in the southeastern coast. It is suggested that the SH simulation over the Plateau is a land–atmosphere coupled issue that is important to climate dynamical downscaling, as well as the climate prediction and weather forecasting in East Asian. The influence of the new scheme on global climate models deserves future investigation.

Acknowledgements

This work is supported by the National Natural Science Foundation of China (Grant No. 91437219) and the Third Tibetan Plateau Scientific Experiment (Grant No. GYHY201406001).

References

Brutsaert WH. 1982. *Evaporation into the Atmosphere: Theory, History, and Applications*. D. Reidel Publishing Co. Dordrecht. Holland.

Chen W, Zhu D, Liu H, Sun SF. 2009. Land–air interaction over arid/semi-arid areas in China and its impact on the East Asian summer monsoon. Part I: calibration of the land surface model (BATS) using multicriteria methods. *Advances in Atmospheric Sciences* **26**(6): 1088–1098.

Chen YY, Yang K, Zhou D, Qin J, Guo X. 2010. Improving the Noah land surface model in arid regions with an appropriate parameterization of the thermal roughness length. *Journal of Hydrometeorology* **11**: 995–1006.

Decker M, Brunke M, Wang Z, Sakaguchi K, Zeng XB, Bosilovich MG. 2012. Evaluation of the reanalysis products from GSFC, NCEP, and ECMWF using flux tower observations. *Journal of Climate* **25**: 1916–1944.

Hao ZC, Ju Q, Jiang WJ, Zhu CJ. 2013. Characteristics and scenarios projection of climate change on the Tibetan Plateau. *The Scientific World Journal* **2013**(7): 1903–1912. Article ID 129793, doi: 10.1155/2013/129793.

Hogue TS, Bastidas L, Gupta H, Sorooshian S, Mitchell K, Emmerich W. 2005. Evaluation and transferability of the Noah land surface model in semiarid environments. *Journal of Hydrometeorology* **6**(1): 68–84.

Hong S, Lakshmi V, Small EE, Chen F, Tewari M, Manning KW. 2009. Effects of vegetation and soil moisture on the simulated land surface processes from the coupled WRF/Noah model. *Journal of Geophysical Research, [Atmospheres]* **114**(D18): 3151–3157.

Hu Y, Zhang Q. 2001. Some issues of arid environment dynamics (in Chinese). *Advances in Earth Sciences* **1**: 18–23.

Huang RH, Wei GA, Zhang Q, Gao XQ. 2005. The preliminary scientific achievements of the field experiment on Air–Land Interaction in the Arid Area of Northwest China (NWC-ALAIEX). In *Proceedings of the 4th CTWF International Workshop on the Land Surface Models and their Applications*, 15–18 November, Zhuhai, China.

Huang RH, Chen W, Zhang Q. 2011. *Land–Atmosphere Interaction over Arid Region of Northwest China and Its Impact on East Asian Climate Variability*. China Meteorological Press: Beijing (in Chinese).

Jaeger EB, Seneviratne SI. 2011. Impact of soil moisture–atmosphere coupling on european climate extremes and trends in a regional climate model. *Climate Dynamics* **36**(9–10): 1919–1939.

Jiménez C, Prigent C, Mueller B, Seneviratne SI, McCabe MF, Wood EF, Rossow WB, Balsamo G, Betts AK, Dirmeyer PA, Fisher JB, Jung M, Kanamitsu M, Reichle RH, Reichstein M, Rodell M, Sheffield J, Tu K, Wang K. 2011. Global intercomparison of 12 land surface heat flux estimates. *Journal of Geophysical Research* **116**: D02102, doi: 10.1029/2010JD014545.

Kanda M, Kanega M, Kawai T, Moriwaki R, Sugawara H. 2007. Roughness lengths for momentum and heat derived from outdoor urban scale models. *Journal of Applied Meteorology and Climatology* **46**: 1067–1079.

Li Q, Xue Y. 2010. Simulated impacts of land cover change on summer climate in the Tibetan Plateau. *Environmental Research Letters* **5**: 015102, doi: 10.1088/1748–9326/5/1/015102.

Liu YM, Wu GX, Hong JL, Dong BW, Duan AM, Bao Q, Zhou LJ. 2012. Revisiting Asian monsoon formation and change associated with Tibetan plateau forcing: ii. change. *Climate Dynamics* **39**(5): 1183–1195.

Moufouma-Okia W, Rowell DP. 2010. Impact of soil moisture initialisation and lateral boundary conditions on regional climate model simulations of the west African monsoon. *Climate Dynamics* **35**(1): 213–229.

Oleson KW, Dai YJ, Bonan G, Bosilovich M, Dickinson R, Dirmeyer P, Hoffman F, Houser P, Levis S, Niu GY, Thornton P, Vertenstein M, Yang ZL, Zeng XB. 2013. Technical Description of the Community Land Model (CLM), NCAR Technical Note. NCAR/TN-503+STR, National Center for Atmospheric Research, Boulder, CO.

Saha S, Moorthi S, Pan HL, Wu XR, Wang JD, Nadiga S, Tripp P, Kistler R, Woollen J, Behringer D, Liu HX, Stokes D, Grumbine R, Gayno G, Wang J, Hou YT, Chuang HY, Juang HMH, Sela J, Iredell M, Treadon R, Kleist D, Delst P van, Keyser D, Derber J, Ek M. 2010. The NCEP climate forecast system reanalysis. *Bulletin of the American Meteorological Society* **91**(8): 1015–1057.

Sheppard PA. 1958. Transfer across the earth's surface and through the air above. *Quarterly Journal of the Royal Meteorological Society* **84**: 205–224.

Shi L, Smith EA. 1992. Surface forcing of the infrared cooling profile over the Tibetan Plateau. Part II: cooling-rate variation over large-scale plateau domain during summer monsoon transition. *Journal of Atmospheric Science* **49**: 823–844.

Skamarock W, Klemp J, Dudhia J, Gill D, Barker D, Duda M, Huang X, Wang W, Powers J. 2008. A description of the advanced research

WRF Version 3, National Center for Atmospheric Research Technical Note. NCAR/TN-475+STR: 113 p.

Smith EA, Shi L. 1992. Surface forcing of the infrared cooling profile over the Tibetan Plateau. Part I: influence of relative longwave radiative heating at high altitude. *Journal of Atmospheric Science* **49**: 805–822.

Wang AH, Zeng XB. 2012. Evaluation of multi-reanalysis products with in situ observations over the Tibetan Plateau. *Journal of Geophysical Research* **117**: D05102, doi: 10.1029/2011JD016553.

Wang AH, Barlage M, Zeng XB, Draper CS. 2014. Comparison of land skin temperature from a land model, remote sensing, and in situ measurement. *Journal of Geophysical Research* **119**: 3093–3106.

Wu GX, Liu YM, Zhang Q, Duan AM, Wang T, Wan RJ, Liu X, Li W, Wang ZZ, Liang XY. 2007. The influence of mechanical and thermal forcing by the Tibetan Plateau on Asian Climate. *Journal of Hydrometeorology* **8**: 770–789.

Wu GX, Liu YM, Dong BW, Liang XY, Duan AM, Bao Q, Yu JJ. 2012a. Revisiting Asian monsoon formation and change associated with Tibetan plateau forcing: I. Formation. *Climate Dynamics* **39**(5): 1169–1181.

Wu GX, Liu YM, He B, Bao Q, Duan AM, Jin FF. 2012b. Thermal controls on the Asian summer monsoon. *Scientific Reports* **2**(5): 404.

Yang K, Koike T, Ishikawa H, Kim J, Li X, Liu HZ, Liu SM, Ma YM, Wang JM. 2008. Turbulent flux transfer over bare-soil surfaces: characteristics and parameterization. *Journal of Applied Meteorology and Climatology* **47**: 276–290.

You QL, Kang SC, Pepin N, Flügel WA, Yan YP, Behrawan H, Huang J. 2010. Relationship between temperature trend magnitude, elevation and mean temperature in the Tibetan plateau from homogenized surface stations and reanalysis data. *Global & Planetary Change* **71**(1): 124–133.

Zeng XB, Dickinson RE. 1998. Effect of surface sublayer on surface skin temperature and fluxes. *Journal of Climate* **11**: 537–550.

Zeng XB, Wang ZZ, Wang AH. 2012. Surface skin temperature and the interplay between sensible and ground heat fluxes over arid regions. *Journal of Hydrometeorology* **13**(4): 1359.

Zhang BQ, Wu PT, Zhao X, Gao X. 2014. Spatiotemporal analysis of climate variability (1971–2010) in spring and summer on the loess plateau, china. *Hydrological Processes* **28**(4): 1689–1702.

Zheng W, Wei H, Wang Z, Zeng XB, Meng J, Ek M, Mitchell K, Derber J. 2012. Improvement of daytime land surface skin temperature over arid regions in the NCEP GFS model and its impact on satellite data assimilation. *Journal of Geophysical Research* **117**: D06117, doi: 10.1029/2011JD015901.

Zhou L, Huang RH. 2010. Interdecadal variability of summer rainfall in Northwest China and its possible causes. *International Journal of Climatology* **30**: 549–557.

Zhou D, Huang G, Ma YM. 2012. Summer heat transfer over a Gobi underlying surface in the arid region of Northwest China (in Chinese). *Transactions of Atmospheric Sciences* **35**(5): 541–549.

24

Impact of Turkish ground-based GPS-PW data assimilation on regional forecast: 8–9 March 2011 heavy snow case

Seyda Tilev-Tanriover[1]* and Abdullah Kahraman[2,3]

[1] Department of Meteorological Engineering, Istanbul Technical University, Istanbul, Turkey
[2] Turkish State Meteorological Service, ITU Met-Office, Istanbul, Turkey
[3] Graduate School of Science, Engineering and Technology, Istanbul Technical University, Istanbul, Turkey

*Correspondence to:
S. Tilev-Tanriover, Department of Meteorological Engineering, Istanbul Technical University, Istanbul, Turkey.
E-mail: tanriovers@itu.edu.tr

Abstract

A ground-based Global Positioning System (GPS) network which consists of 147 uniformly distributed stations has been implemented in Turkey in May 2009. Precipitable water (PW) data estimated using this network is available since October 2010. Three experiments for a specific heavy snow case have been conducted to assess the impact of assimilation of this new data on short range weather forecasts. Verifications with surface observations indicate some reduction in the biases of basic variables for whole domain, and improvement in precipitation forecasts. Upper air variables of all simulations are compared with radiosonde observations, and some improvement in dew point temperatures through troposphere is observed with cycling mode assimilation.

Keywords: data assimilation; GPS-PW; ground-based GPS; numerical weather prediction; heavy snow

1. Introduction

Water vapour, with its highly fluctuating spatial and temporal distribution, is the most critical component of the atmosphere. It is the key ingredient of deep moist convection; small scale variability of moisture can strongly influence the existence, location and severity of the storms. Its amount is also of primary importance in frontal and orographic rain processes. One other fact about water vapour is being one of the major greenhouse gases significantly affecting radiative transfer within the atmosphere. Furthermore, latent heat and moisture advection due to global atmospheric circulation contribute to meridional energy balance of the earth (Bevis *et al.*, 1992). From planetary to microscale, phase change of water has a fundamental role in meteorological processes.

Surface observations of humidity is exceptionally important over areas with complex topography, such as Anatolia, including several mountain ranges and plateaus, surrounded by Black Sea on the north, Aegean Sea on the west, and Mediterranean sea to the south. Conventional observations from meteorological stations inside the cities do not always represent the distribution of moisture content appropriately, and radiosondes from eight locations in Turkey are simply not enough to capture mesoscale variability of water vapour. Moreover, they are available two times a day, and are very expensive. An innovative solution for observation of the water content of the atmosphere is GPS-derived precipitable water data.

GPS networks are easy and reasonable to operate once they are installed. Furthermore, their temporal and spatial resolution is drastically high. Thus, currently GPS networks are more commonly used for atmospheric observations (Bevis *et al.*, 1992).

Time integration of an atmospheric model is an initial value problem, so that success of a numerical weather prediction depends not only on realistic representation of the atmosphere, but also on another fundamental component, namely accurate determination of the initial conditions (Kalnay, 2003). Ground-based GPS networks are powerful sources of humidity information for fine-scale regional models as following studies have shown. Various parameters such as zenith total delay (ZTD), integrated water vapour (IWV) and precipitable water (PW) from GPS networks are assimilated with alternative techniques at different regions of the world in the last decade: One of the earliest is Higgins's (2001) study, assimilating GPS ZTD data with Met Office's 3D variational assimilation system. He showed that better forecasts are possible with these new data in winter cases, with an expectation of the positive impact to be greater in particular during summer weather conditions. In Germany, IWV product from German GPS ground station network is assimilated by Tomassini *et al.* (2002). Meanwhile, Falvey and Beavan's (2002) GPS-PW assimilation for New Zealand resulted in a significant improvement in total rainfall on the upwind side of the Southern Alps and a positive impact on humidity profiles of the lee side. Guerova *et al.* (2004) used IWV data

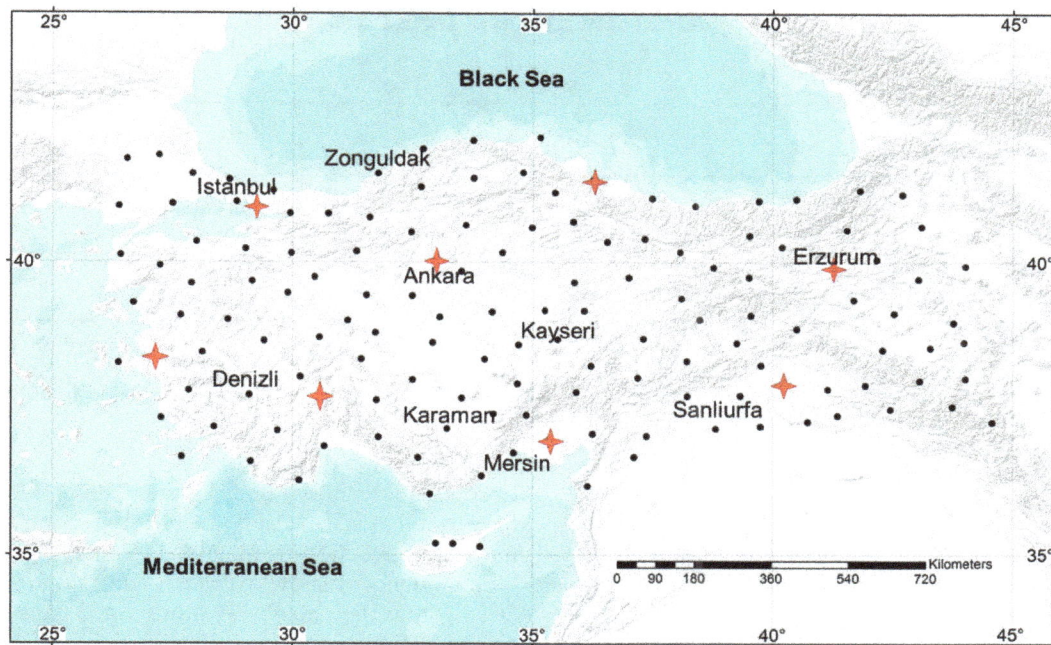

Figure 1. Locations of TUSAGA stations (black dots) and radiosonde stations (red stars).

from 80 European GPS sites with a continuous data assimilation scheme based on the nudging technique at MeteoSwiss. Their operational assimilation into high-resolution aLpine model domain (7 km of grid intervals) resulted in better model performance based on bias precipitation scores. Beyond, ZTD measurements from European network are assimilated into MM5 model using three-dimensional variational approach by Cucurull *et al.* (2004). Their results indicated that cycling mode assimilation performed better, in terms of root mean square errors (RMSEs) of basic variables, especially for specific humidity. Faccani *et al.*'s (2005) study of ZTD data assimilation into a 9-km horizontal grid interval model domain also improved skills of operational forecasts in terms of RMS and mean error scores calculated with precipitation observations in Basilicata region of Italy, in particular during the seasonal transition from winter to spring. A different study of ZTD data assimilation using Three-dimensional variational data assimilation (3DVAR) into AROME model with a domain of 2.5-km horizontal grid intervals was conducted by Boniface *et al.* (2009). They experienced a more advanced accuracy of Mediterranean heavy rain simulations with assimilation of these data. Besides, a 4DVAR assimilation experiment by Shoji *et al.* (2011) was focused on the tropical cyclone Nargis in 2008. Their implementation of PW modified the fields of pressure and wind, and increased precipitation rates. As a common point, all these studies encourage usage of these new observations by assimilating into mesoscale atmospheric models.

Main objective of this study is investigation of the impact of a newly available data from Turkish ground-based GPS network in short range mesoscale forecasts.

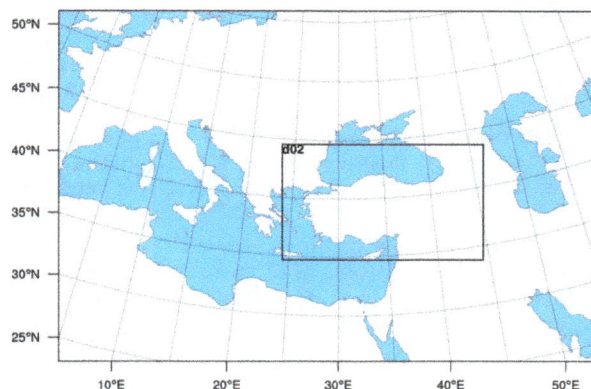

Figure 2. Domains used for the simulations. Coarse domain has 24 km and nested domain has 8 km horizontal grid intervals.

The scale of the impact, either positive or negative, is needed to be addressed. One other point is to see if different assimilation procedures give different results. For this purpose, three preliminary assimilation experiments using 3DVAR technique have been carried out. Assimilated data is introduced in Section 2 and meteorological conditions for the case are shortly explained in Section 3. Detailed information on model configuration and experimental design is given in Sections 4 and 5, respectively. In the last section, results of the experiments are discussed.

2. GPS-PW data

A ground-based GPS network in Turkey (CORS-TR/TUSAGA-Aktif) was established in 2009, by Istanbul Kultur University in association with

EXPERIMENTAL DESIGN

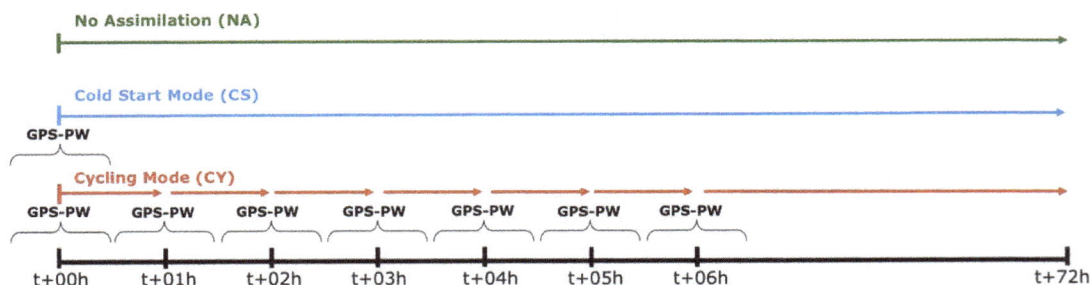

Figure 3. Experimental design for three runs. Time windows for assimilated data are 1 hour long, starting from 30th minute of the previous hour and ending at 30th minute of the ongoing hour.

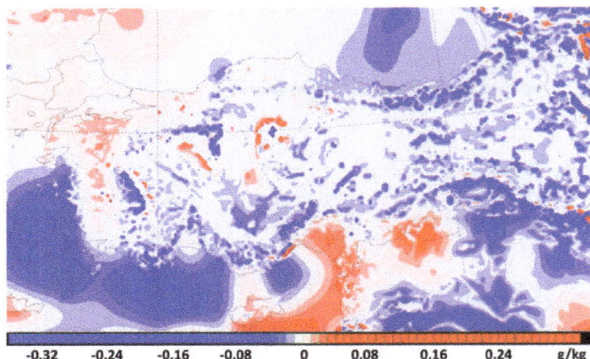

Figure 4. Water vapour mixing ratio difference of 850 hPa for the initial time of the simulations with and without assimilation. Positive values refer to higher values of mixing ratio with assimilation.

the General Directorate of Land Registration and Cadastre of Turkey and the General Command of Mapping of Turkey (Yildirim *et al*, 2011). The network consists of approximately 147 uniformly distributed stations. The main purpose of the project was serving data for geodetic, terrestrial mapping and cadaster applications. General Command of Mapping and Turkish State Meteorological Service have been estimating GPS-based PW values by using TUSAGA-Aktif system since 23 October 2010 according to a protocol signed between them on 21 July 2009 (Erkan *et al.*, 2010). The PW data which are obtained using GAMIT software and Saastamoinen model (Aysezen *et al.*, 2009) has approximately 80 km horizontal and a very high temporal resolution (can be reduced to 1 min) were validated by radiosonde and surface observations. Figure 1 shows the locations of the GPS (black dots) and radiosonde (red stars) stations.

3. A brief description of heavy snow case over Turkey on 8 and 9 March 2011

On 7 March 2011, a sharp through over the Balkans with a cold tongue is tilted and a cut-off low is formed as a strong high-pressure system over Eastern

Europe approached the region. The cold-core low moved inlands of Anatolia, with a contribution of moisture advection from surrounding seas, resulted in high amounts of snow over the country until 10 March 2011. Snow height observations for 3 days exceeded 45 cm in some regions.

4. Model configuration

Two domains with 24 and 8 km horizontal resolutions and 35 vertical levels are set with two-way nesting option for the numerical simulations (Figure 2). WRF-ARW 3.4 (Skamarock *et al.*, 2008) is used, and the physical options are chosen according to sensitivity tests performed in previous studies for the region (e.g. Kahraman and Tanriover, 2009), namely Kain–Fritsch (Kain, 2004) as cumulus parameterization, WSM 6-class (Hong and Lim, 2006) as microphysics scheme, RRTMG (Mlawer *et al.*, 1997) as radiative model, Mellor–Yamada–Janjic (Mellor and Yamada, 1982; Janjic, 1990, 1996, 2002) as planetary boundary layer scheme and NOAH (Chen and Dudhia, 2001) as land surface model. Simulation periods are between 1200 UTC 7 March 2011 and 1200 UTC 10 March 2011 for all runs. National Centers for Environmental Prediction-Global Forecast System (NCEP GFS) analysis with $1.0°$ horizontal resolution is used as initial and boundary conditions. Surface observations as well as radiosondes are used for verification.

5. Data assimilation experiments

Weather Research and Forecasting Model's Community Variational/Ensemble Data Assimilation System (WRFDA) is used for assimilation processes (Barker *et al.*, 2012). Three experiments for the heavy snow case have been conducted: first is a control run without assimilation (NA), second is a cold start run with 3DVAR PW assimilation (CS), and the last run with the cycling mode 3DVAR assimilation of PW data (CY). No other data than GPS-PW is assimilated to the model. Figure 3 shows the experimental design of

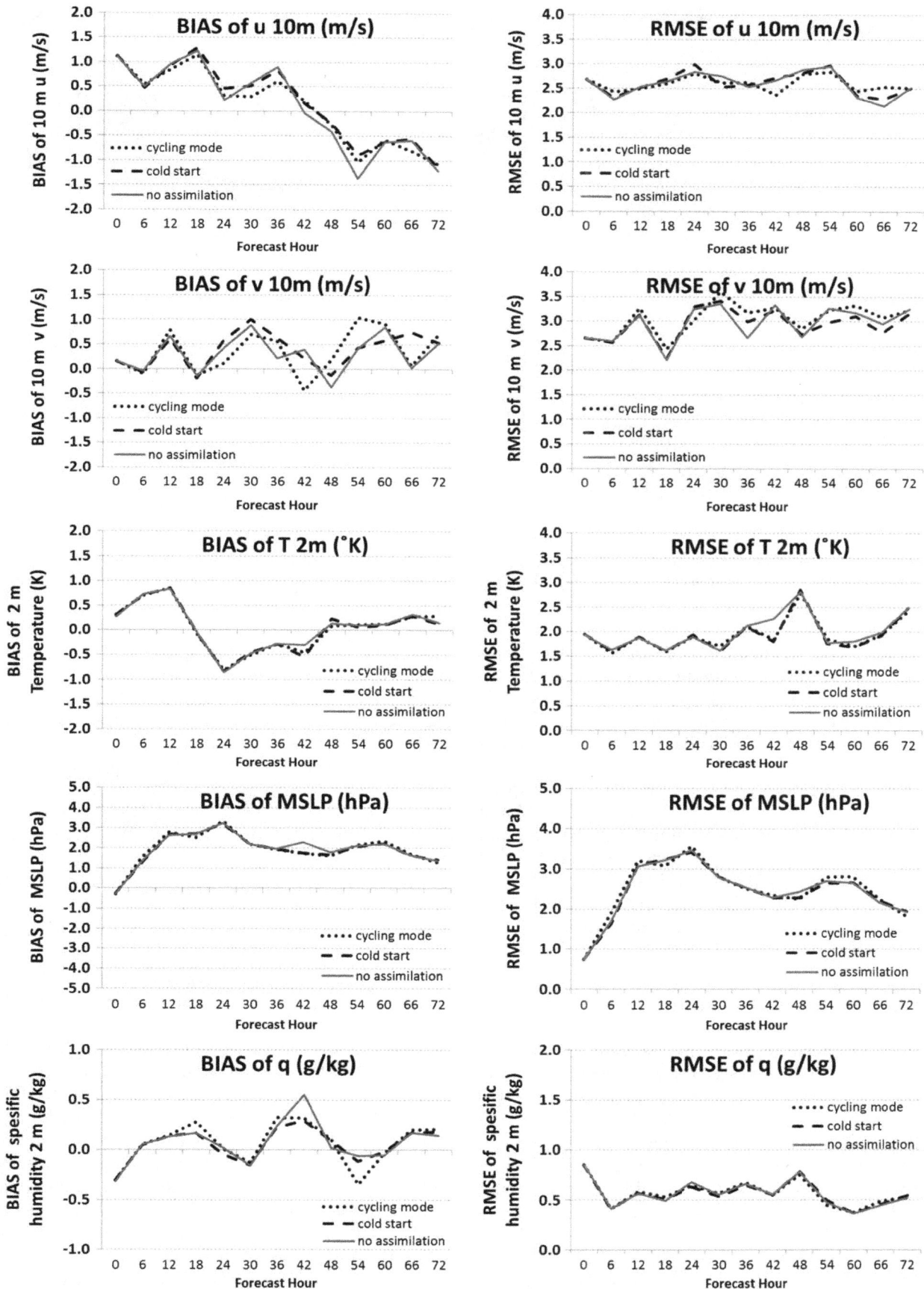

Figure 5. Mean error (bias) and RMSE statistics of horizontal wind components (u, v), temperature (T), MSLP and specific humidity (q) computed using synoptic observations with 6 h intervals during the simulation period for the second domain.

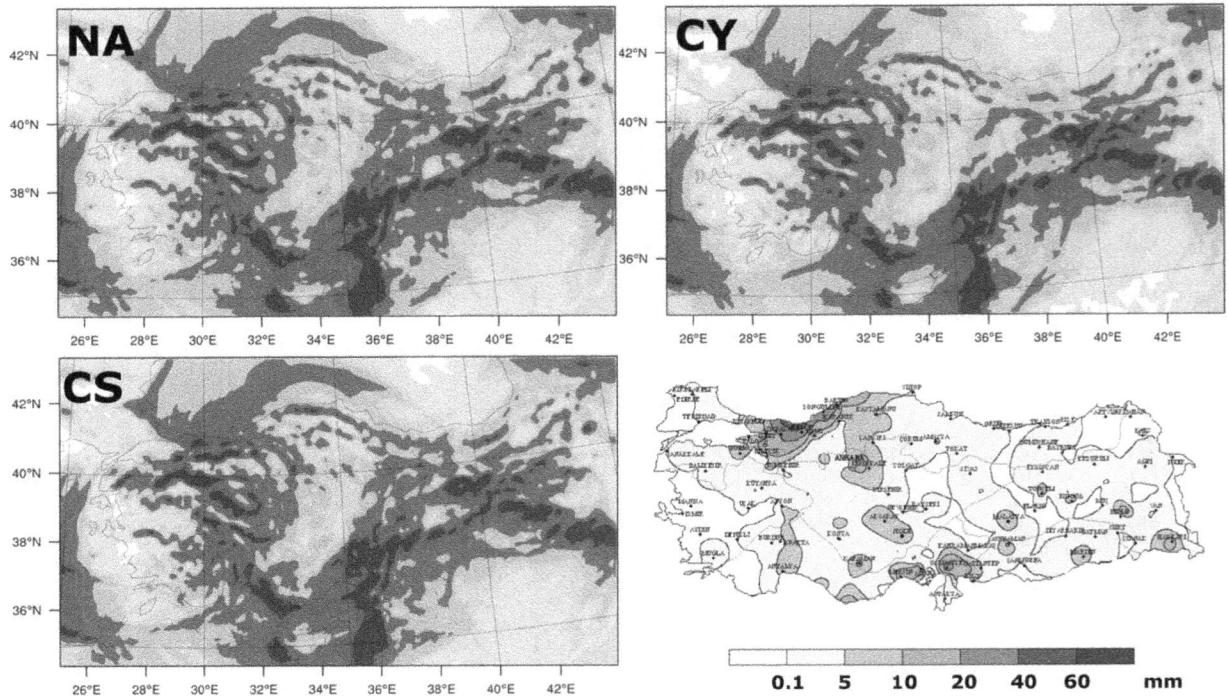

Figure 6. The 24-h total precipitation maps for and 0600 UTC 10 March 2011. Top left is the control run with no GPS-PW assimilation (NA), lower left is the cold start run (CS), top right is the cycling mode run (CY). Turkish State Meteorological Service (TSMS) observations for the same period are given at the lower right (subfigure from TSMS).

the study. After the first run initialized with GFS analysis, the cold start run experiment is performed, with assimilation applied to 1200 UTC 7 March 2011 analysis time with a ±30-min time window. Water vapour mixing ratio fields of 850 hPa for these two experiments reveal that up to 0.3 g kg^{-1} differences occur regionally (Figure 4). For the cycling mode experiment, assimilation is applied to 1200 UTC 7 March 2011 and the next 6 h, with 1 h interval and ±30-min time windows. In this experiment, free model run is performed from 1800 UTC 7 March 2011 to 1200 UTC 10 March 2011. Consequently, three 72-h forecasts are acquired with three different ways.

6. Results and discussions

Outputs from the experiments are verified with surface and upper air observations. Mean error (bias) and RMSE statistics of horizontal wind components (u, v), temperature (T), mean sea level pressure (MSLP) and specific humidity (q) are computed using synoptic observations with 6 h intervals during the simulation period for the domain with 8 km grid intervals (Figure 5). Biases of u have a variation approximately between -1.3 and $+1.3$ m s^{-1}, those of v between -0.5 and $+1.0$ m s^{-1}, of T between -0.8 and $+0.8$ K, of MSLP between 0 and 3 hPa and of q between -0.3 and $+0.3$ g kg^{-1}. All three simulations are considered to be successful in terms of a mesoscale simulation of a snowstorm with respect to order of bias values.

RMSEs of MSLP as well as q are almost equal for all simulations. Decrease at RMSE of T near 42nd forecast hour occurs with both type of assimilations. The results indicate that success of forecast skills of three experiments fluctuate throughout the simulation period. Nevertheless, statistics of all experiments generally have similar values for surface variables.

Simulations are verified with radiosonde observations at eight locations in Turkey. Average correlation of dew point temperature for 850, 700, 500 and 300 hPa levels for whole simulation period is 0.975 for the run without assimilation, 0.974 for the cold start run and 0.977 for the cycling mode run. Average RMSE scores of the same field are 6.65, 6.75 and 6.58 K, respectively. These results denote a positive impact of cycling mode assimilation compared to other simulations.

Daily precipitation field from the experiments show that the simulations are generally able to capture the overall patterns (Figure 6). Here, it must be noted that Turkey is a mountainous country, and there are limited number of stations over mountains where simulated precipitation amounts have peaks (e.g. Taurus Mountains surrounding the Southeast Anatolia Region, some parts of mid-west Anatolia). All three simulations have similar distribution of precipitation in most parts of the domain, but regional differences exist. Over west part of Black Sea, area covered with 20–40 mm precipitation associated with the cycling mode assimilation experiment is approximately half of that for other experiments. Nonetheless, it is unknown if this

Table I. Average of daily precipitation observations at Zonguldak, Istanbul Sariyer, Erzurum, Ankara, Kayseri, Denizli, Karaman, Sanliurfa and Mersin compared to simulation results.

	Average of observations	NA	CY
Daily precipitation (mm)	11.69	8.56	10.42
BIAS (mm)		−3.13	−1.27
Normalized BIAS (%)		−26.75	−10.86

distinction over the sea is of positive or negative in terms of being more close to the real precipitation distribution.

A comparison of daily precipitation forecasts and observations for nine synoptic stations, namely Zonguldak, Istanbul Sariyer, Erzurum, Ankara, Kayseri, Denizli, Karaman, Sanliurfa and Mersin shows that average of normalized BIAS is −26.75 for NA and −10.86 for CY (Table I). For example, on 10 March 2011 12.2 mm of precipitation is observed in Ankara, 5.6 mm is simulated without assimilation of GPS-PW data, whereas 15.5 mm is simulated with cycling mode assimilation. However, some stations like Sanliurfa resulted in worsening of precipitation values with cycling mode assimilation (4.8 mm observations, 10.4 mm NA, 13.3 mm CY). Consequently, both simulations have underestimated the daily precipitation in average, but 16% lower normalized BIAS in point-based verification is obtained using cycling mode assimilation.

To conclude, it is suggested that cycling mode 3DVAR assimilation of ground-based GPS-PW observations has resulted in considerable improvement in forecast skills according to verification with vertical distribution of dew point temperature and daily precipitation observations. On the other hand, the results of cold start mode assimilation indicate limited effects in the simulation with respect to verification statistics and visual comparison of precipitation distribution. Regional differences in spatial patterns of precipitation are controversial, and one should be cautious to regard them as a success of assimilation, because of limited observations. Owing to the small spatial scale of the regional impacts, model outputs should be verified with finer resolution observations such as radar or satellite products, mesonetworks, etc. It is suggested that more experiments must be conducted to investigate forecast improvements for different weather types, even with a different verification approach like feature-based quality measures. Operational assimilation of these new data can be beneficial for regional forecasts.

Acknowledgements

We are thankful to Turkish State Meteorological Service and General Command of Mapping for the data. We also want to express our gratitude to Dr Altug Aksoy and to anonymous reviewers for their valuable advises. Numerical simulations are performed at Meteorological Modelling and Analysis Laboratory of Istanbul Technical University, Meteorological Engineering Department.

References

Aysezen MŞ, Cingöz A, Aktuð B, Lenk O. 2009. *Continuously Operating Reference Stations and Datum Transformation Project (CORS-TR)*, Retrieved June 24, 2013. http://academic.cankaya.edu.tr/agorur/MTS1-2/MTS2/Bildiriler/89-95.pdf.

Barker D, Huang X-Y, Liu Z, Auligne T, Zhang X, Rugg S, Ajjaji R, Bourgeois A, Bray J, Chen Y, Demirtas M, Guo Y-R, Henderson T, Huang W, Lin H-C, Michalakes J, Rizvi S, Zhang X. 2012. The weather research and forecasting model's community variational/ensemble data assimilation system: WRFDA. *Bulletin of the American Meteorological Society* **93**: 831–843.

Bevis M, Businger S, Herring TA, Rocken C, Anthes RA, Ware RH. 1992. GPS meteorology: remote sensing of atmospheric water vapor using the global positioning system. *Journal of Geophysical Research* **97**(D14): 15787–15801.

Boniface K, Ducrocq V, Jaubert G, Yan X, Brousseau P, Masson F, Champollion C, Chéry J, Doerflinger E. 2009. Impact of high-resolution data assimilation of GPS zenith delay on Mediterranean heavy rainfall forecasting. *Annales Geophysicae* **27**: 2739–2753.

Chen F, Dudhia J. 2001. Coupling an advanced land-surface/hydrology model with the Penn State/NCAR MM5 modeling system. Part I: Model Implementation and Sensitivity. *Monthly Weather Review* **129**: 569–585.

Cucurull L, Vandenberghe F, Barker D, Vilaclara E, Rius A. 2004. Three-dimensional variational data assimilation of ground-based GPS ZTD and meteorological observations during the 14 December 2001 storm event over the western Mediterranean Sea. *Monthly Weather Review* **132**: 749–763.

Erkan Y, Aktuğ B, Lenk O, Parmaksız E, Mert İ, Bacanlı H. 2010. TUSAGA-Aktif Sistemi ve Atmosferik Calismalara Ait On Sonuclar. Uluslararasi Katilimli 1.Meteoroloji Sempozyumu, Ankara, Turkey, 28 May 2010.

Faccani C, Ferretti R, Pacione R, Paolucci T, Vespe F, Cucurull L. 2005. Impact of a high density GPS network on the operational forecast. *Advances in Geosciences* **2**: 73–79.

Falvey M, Beavan J. 2002. The impact of GPS precipitable water assimilation on mesoscale model retrievals of orographic rainfall during SALPEX'96. *Monthly Weather Review* **130**: 2874–2888.

Guerova G, Bettems J-M, Brockmann E, Matzler CH. 2004. Assimilation of the GPS-derived integrated water vapour (IWV) in the MeteoSwiss numerical weather prediction model – a first experiment. *Physics and Chemistry of the Earth* **29**: 177–186.

Higgins M. 2001. Progress in 3D-variational assimilation of total zenith delay at the Met Office. *Physics and Chemistry of the Earth* **26**(6–8): 445–449.

Hong S-Y, Lim J-O J. 2006. The WRF single-moment 6-class microphysics scheme (WSM6). *Journal of the Korean Meteorological Society* **42**: 129–151.

Janjic ZI. 1990. The step-mountain coordinate: physical package. *Monthly Weather Review* **118**: 1429–1443.

Janjic ZI. 1996. The surface layer in the NCEP Eta Model. In Eleventh Conference on Numerical Weather Prediction, Norfolk, VA, 19–23 August 1996. American Meteorological Society, Boston, MA; 354–355.

Janjic ZI. 2002. Nonsingular Implementation of the Mellor–Yamada Level 2.5 Scheme in the NCEP Meso model. NCEP Office Note **No. 437**, 61 pp.

Kahraman A, Tanriover S. 2009. Sensitivity and predictability analysis of Advanced Research WRF Model (WRF-ARW) in Eastern Mediterranean Region. In European Geosciences Union, General Assembly 2009, Vienna, Austria. EGU, EGU2009-8801.

Kain JS. 2004. The Kain-Fritsch convective parameterization: an update. *Journal of Applied Meteorology* **43**: 170–181.

Kalnay E. 2003. *Atmospheric Modeling, Data Assimilation and Predictability*. Cambridge University Press: Cambridge, UK; 341 pp.

Mellor GL, Yamada T. 1982. Development of a turbulence closure model for geophysical fluid problems. *Reviews of Geophysics and Space Physics* **20**: 851–875.

Mlawer EJ, Taubman SJ, Brown PD, Iacono MJ, Clough SA. 1997. Radiative transfer for inhomogeneous atmosphere: RRTM, a validated correlated-k model for the longwave. *Journal of Geophysical Research* **102**(D14): 16663–16682.

Shoji Y, Kunii M, Saito K. 2011. Mesoscale data assimilation of Myanmar cyclone Nargis Part II: assimilation of GPS-derived precipitable water vapor. *Journal of the Meteorological Society of Japan* **89**(1): 67–88.

Skamarock WC, Klemp JB, Dudhia J, Gill DO, Barker DM, Duda MG, Huang X-Y, Wang W, Powers JG. 2008. A description of the advanced research WRF version 3. NCAR Tech. Note NCAR/TN-4751STR, 113pp. http://www.mmm.ucar.edu/wrf/users/docs/arw_v3.pdf.

Tomassini M, Gendt G, Dick G, Ramatschi M, Schraff C. 2002. Monitoring of integrated water vapour from ground-based GPS observations and their assimilation in a limited-area NWP model. *Physics and Chemistry of the Earth* **27**: 341–346.

Yildirim O, Salgin O, Bakici S. 2011. *The Turkish CORS Network (TUSAGA-Aktif)*, Retrieved June 24, 2013. http://www.fig.net/pub/fig2011/papers/ts03g/ts03g_yildirim_algin_et_al_5244.pdf.

25

Changes to radiosonde reports and their processing for numerical weather prediction

Bruce Ingleby[1,2*] and David Edwards[1]

[1] Met Office, Exeter EX1 3PB, UK
[2] ECMWF, Reading RG2 9AX, UK

*Correspondence to:
B. Ingleby, ECMWF, Shinfield
Park, Reading, RG2 9AX, UK.
E-mail: bruce.ingleby@ecmwf.int

This article is published with the permission of the Controller of HMSO and the Queen's Printer for Scotland.

Abstract

In the biggest change to radiosonde reporting for many years binary BUFR reports are replacing alphanumeric TEMP code. The binary reports can provide much higher vertical resolution and the time and position of each level, but changes to observation processing at forecast centres are needed to use the new reports and in particular the level dependent positions. Using data vertically averaged over model layers, observation minus background $(O - B)$ statistics from both formats compare closely (with a temperature offset of about 0.1°C due to rounding in the older format). Correct treatment of balloon drift improves the $O - B$ statistics at upper-levels.

Keywords: radiosonde; balloon drift; data assimilation; temperature; humidity; wind

1. Introduction

Despite their relative sparsity radiosonde reports are still important for Numerical Weather Prediction (NWP) because they provide complete vertical profiles through much of the atmosphere and act as reference data for aircraft and satellite reports. As part of a World Meteorological Organization (WMO) migration from alphanumeric codes, binary format radiosonde data are now reported from some countries at higher vertical resolution, higher precision and with extra metadata. There is also a trend for some radiosondes not to use a pressure sensor. This article describes work at the Met Office to use the binary data, starting with UK reports, and to improve the processing. Balloon drift (displacement in the horizontal and time) is included in the interpolation of background (short range forecast) fields to observation locations.

2. Radiosonde data

Figure 1(a) shows the locations of the main UK radiosonde stations and the downwind drift of the balloons – the typical Westerlies are apparent. Ascents from the manned stations, Camborne and Lerwick, reach around 30 and 34 km at 00 and 12 UTC respectively, ascents from four Autosonde stations (usually only at 00 UTC) reach around 24 km using smaller balloons. (Larkhill is an artillery station and reports irregularly.) The stations all use Vaisala RS92-SGP radiosondes with DigiCORA version 3.62 processing. The balloons are launched 45 min before the nominal report time, they take about 50 min to reach 100 hPa

and some take about 2 h to balloon burst. The binary reports include data every 2 s – about 5000 levels for the higher ascents. Figure 1(b) shows the average horizontal drift at each level.

The Vaisala RS92 is currently the most widely used radiosonde in the world, deployed in much of Europe, South America, the Middle East, Australia, Canada and elsewhere. (In early 2014 out of 742 stations reporting globally, 278 were using RS92, 48 of these were Autosondes). It has been the subject of various detailed studies (e.g. Steinbrecht *et al.*, 2008) and is used in the GCOS (Global Climate Observing System) Reference Upper Air Network (GRUAN, www.gruan.org). As documented by Nash *et al.* (2011) the RS92 copes better with cloud wetting than most other operational radiosondes due to a hydrophobic coating on the temperature sensor and the use of two humidity sensors which are alternately heated (down to −60°C) to reduce contamination problems.

2.1. TEMP coding/decoding and rounding of temperature

For decades radiosonde data has been exchanged on the WMO Global Telecommunications System (GTS) using the alphanumeric TEMP code (FM-35 in WMO, 2011). TEMP code was designed to keep telecommunications messages as short as possible whilst providing a good representation of the ascent. A subset of levels is chosen (manually or automatically) so that linear interpolation in log pressure can reconstruct the original profile to given accuracy (e.g. 1 K up to 300 hPa and 2 K at higher levels). These 'significant' levels correspond to turning points (such as temperature inversions) but also to more subtle features of the profile, Figure 2

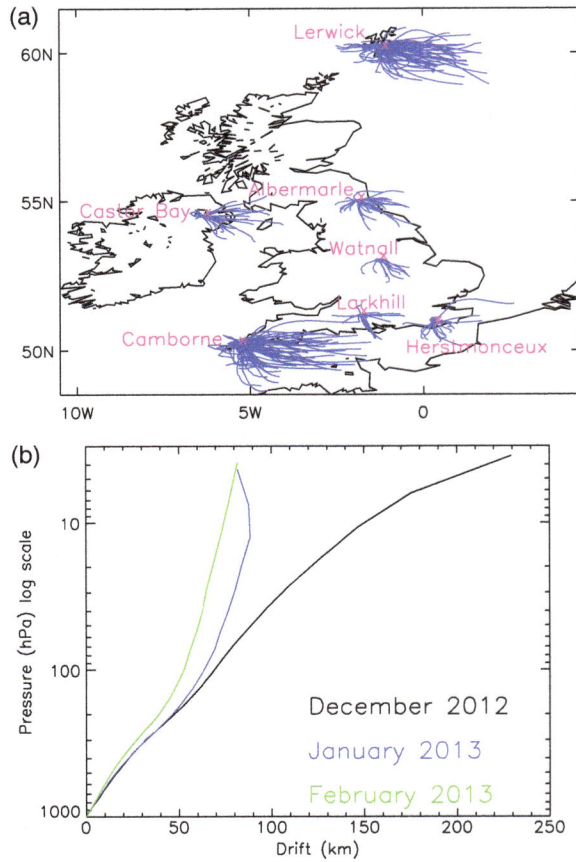

Figure 1. Radiosonde drift for UK radiosondes for December 2012–February 2013. (a) Positions for 12 UTC ascents, (b) average horizontal drift as a function of pressure for 00 and 12 UTC ascents.

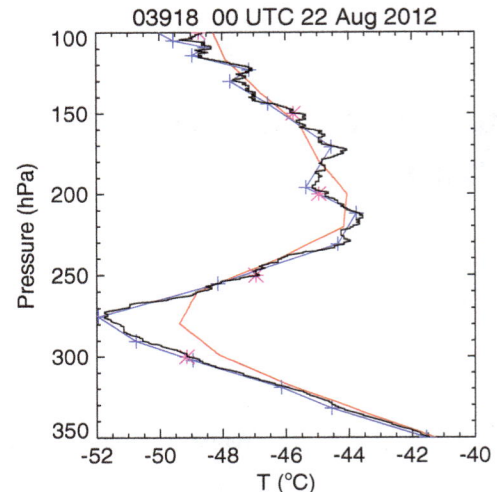

Figure 2. Part of a temperature profile from a radiosonde ascent. Black: high resolution, blue + joined by blue lines: significant levels, purple *: standard levels; red: interpolated background values.

shows an example. To ensure timeliness separate reports are sent when the ascent reaches 100 hPa and following balloon burst. There are four 'Parts' of the TEMP code: Part A (standard levels: 1000, 925, 850, 700, 500, 400, 300, 250, 200, 150 and 100 hPa plus surface, tropopause and maximum wind levels), Part B (significant levels up to 100 hPa), Part C (standard levels: 70, 50, 30, 20 and 10 hPa) and Part D (significant levels above 100 hPa). Complete ascents (all parts) typically have 60–120 levels, but there can be more than 200.

Temperatures are reported in Celsius with a precision of 0.2°C, to save space the sign is coded with the tenths figure. The code figure is even (odd) for a positive (negative) temperature (Table 3931 of WMO 2011). For example the temperatures +13.4 and +13.5°C are both coded as 134. The Met Office decodes this at face value, i.e. 13.4°C – giving an average offset of −0.05°C for temperatures measured to one decimal place ('rounding down' also applies to negative temperatures). RS92 uses two decimal places and all values between +13.40 and +13.59°C are coded as 134 (DigiCORA III processing, Marttila, 2012, Vaisala, pers. comm.; other manufacturers might differ): decoding the value as 13.4°C gives an average offset of −0.095°C. European Centre for Medium-Range Weather Forecasts (ECMWF)

decoding makes a 0.05°C allowance for rounding: values measured to one decimal place are decoded unbiased, but RS92 TEMP temperatures have a mean offset of −0.045°C. (Internally in the Met Office there is an additional −0.05°C bias for TEMP data because for data reported to one decimal place the conversion to Kelvin uses 273.1 rather than 273.15 K as 0°C.) The rounding was noticed when comparing with data from the BUFR messages (below). Small biases are more important for climate users, such as the HadAT record of standard level temperatures (Thorne et al., 2005), than NWP. HadAT and Integrated Global Radiosonde Archive (Durre et al., 2006) do not currently account for rounding of TEMP data (Thorne, 2014, pers. comm.).

2.2. BUFR (Binary Universal Form for the Representation of meteorological data)

For about 15 years WMO has been promoting and coordinating a switch to binary (BUFR) reporting for surface and radiosonde observations. It is a complex change involving many different countries (and four different groups within the Met Office). For each radiosonde ascent there should be one BUFR report when the radiosonde reaches 100 hPa, and one report containing the whole ascent following balloon burst (see templates linked from http://www.wmo.int/pages/prog/www/WMOCodes/MigrationTDCF.html also https://software.ecmwf.int/wiki/display/TCBUF/TAC+To+BUFR+Migration). Temperatures are reported in Kelvin to two decimal places and there is provision for reporting the latitude, longitude and time of each level. Unlike TEMP the launch position is included in the message (some positions are erroneous, often due to wrong conversion from degrees/minutes to decimals). In early 2014 about 45% of global radiosonde reports are available in BUFR on the GTS, for Europe the proportion is much higher and many of these are 'proper' BUFR. Some of the BUFR reports are currently reformatted from TEMP code (still as separate parts which complicates processing) – some have the temperature offset removed and others do not.

2.3. Vertical coordinate

Use of Global Positioning System (GPS) on radiosondes in place of radar tracking has increased over the last 15 years; the horizontal positions are used to derive winds. GPS heights measured by radiosonde are now so accurate that in most cases there is no need to use a pressure sensor (Nash *et al.*, 2006, 2011; Steinbrecht *et al.*, 2008). (For many years some Soviet/Russian radiosondes have computed pressures hydrostatically from radar heights. At low radar elevations the radar heights have reduced accuracy – worse than GPS). Most radiosondes are 'single use' so a balance is needed between cost and performance. Some current and near-future GPS radiosondes (including the Vaisala RS41) do not include a pressure sensor as standard. In preparation for this the BUFR template used by the Met Office includes geopotential height and not pressure (although the RS92-SGP has a pressure sensor). Before transmission geometric height, provided by GPS, is converted to geopotential height using the variation of gravity with latitude and elevation. The Met Office NWP processing (below and Supporting Information) requires pressure. Thus pressure values are calculated using the hydrostatic equation starting from station height and station pressure (an error in either of these would degrade the whole ascent).

3. Processing and results

NWP systems are a major user of radiosonde data and comparison with short range forecasts (background fields) are a powerful tool for assessing observation quality. The forecast model used in this analysis is the UKV which has 70 vertical levels (terrain following at low levels, with top at ~2.8 hPa) and 1.5 km horizontal grid spacing over the UK. Piccolo and Cullen (2012) used a similar model with a 4 km grid. The UKV uses a 3D-Var analysis (Lorenc *et al.*, 2000) with a 3 h time window. The forecast (background) values are interpolated bilinearly in the horizontal and linearly in time from 2, 3 and 4 h forecasts (valid at 11, 12 and 13 UTC for a 12 UTC analysis). Macpherson (1995) found little benefit from treating balloon drift and concluded that for simple advection without wind shear the errors due to neglecting the exact time and position of each level cancelled. The model that he used had 31 levels and 17 km grid spacing, given the improvements in model resolution and accuracy since 1995 it is appropriate to re-evaluate this aspect. Radiosonde drift has been studied by Seidel *et al.* (2011) and the effect on climate statistics by McGrath *et al.* (2006). Laroche and Sarrazin (2013) showed that taking account of radiosonde

drift (calculated from the winds and an assumed rate of ascent) gave a positive impact for global NWP.

In the Met Office processing (see Supporting Information) the different TEMP reports are combined to give complete ascents. Then, interpolating linearly in log pressure (using the definition of significant levels) to provide complete profiles, average temperature and wind components are calculated for each model layer that the ascent covers. Processing the high resolution BUFR data was straightforward – mainly allowing for the much larger numbers of levels. The data averaged onto model layers provides a useful comparison between the two sources. An option was added to use the latitude, longitude and time where the radiosonde ascent intersects the model levels (the calculation is iterated a few times because the model levels are not flat): 'slant' processing. For comparison there is also an option to use the correct latitude and longitude but the launch time: 'partial' processing.

Observation-minus-background (O − B) statistics were produced for more than a year; here we present statistics for December 2012 to February 2013, because on average the radiosondes drift more in the winter. Only cases with both TEMP and BUFR reports from a particular station are included (there were some gaps in BUFR availability).

Figure 3 shows vertical profiles of temperature O − B statistics: mean and standard deviation (SD). The black and blue lines – 'vertical' processing of TEMP and BUFR data – are very similar except for the temperature offset mentioned above. At most levels the TEMP O − B bias is smaller – perhaps because these data are currently assimilated. (There are larger biases in the upper stratosphere for several reasons). Standard level SDs (from BUFR data using vertical processing) are similar to nearby layer averaged SDs up to 300 hPa but larger at higher levels reflecting larger representativeness errors in the stratosphere (partly due to thick model layers there). At upper levels the slant processing (green line) performs best – results for the partial processing (red line) are more mixed, being worse than vertical processing at some levels.

Figure 4 shows similar plots for relative humidity (RH). Currently we use RH values interpolated to the layer mid-points (the original motivation was to retain cloud, i.e. values close to 100% RH, in the processing) and the first five lines use interpolation. The orange line is slant processing but with vertical averaging and this gives slightly better SD O − B statistics. The biases are very similar between the different options. At 200 hPa (just below the tropopause) the observed profiles are much wetter than the background. Hilton *et al.* (2012) reported an upper tropospheric dry bias in the Met Office global forecasts – this has been partially alleviated since. The smaller SD O − B values below 850 hPa reflect lower natural variability there: the mean observed RH at low levels is between 80 and 86% with an SD between 9 and 13%, at 800 hPa the mean RH is 66% and SD is 33%. The specific humidity (q) O − B SD (not shown) has a peak at about 800 hPa consistent

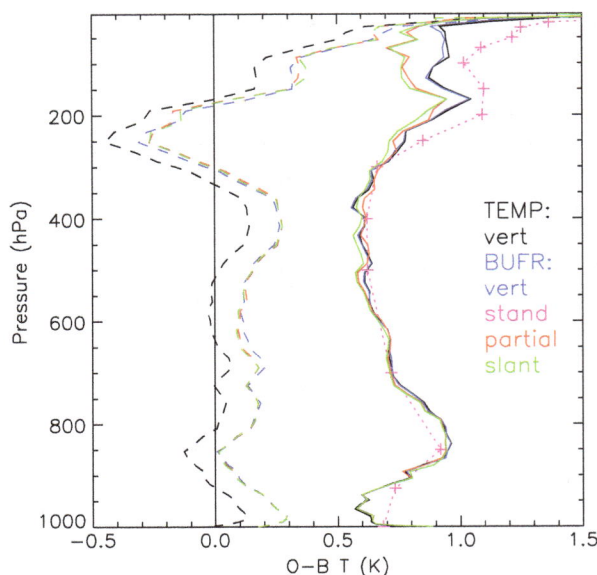

Figure 3. Mean (dashed) and SD (solid) of temperature O − B for December 2012–February 2013, 00 and 12 UTC ascents. Except for standard levels (+ symbols, SD only) the observed values are those averaged over model layers and they are plotted at the mean pressure of the layer mid-points. The lines are for vertical processing (TEMP black and BUFR blue); partial (red) and full slant (green) processing of BUFR data.

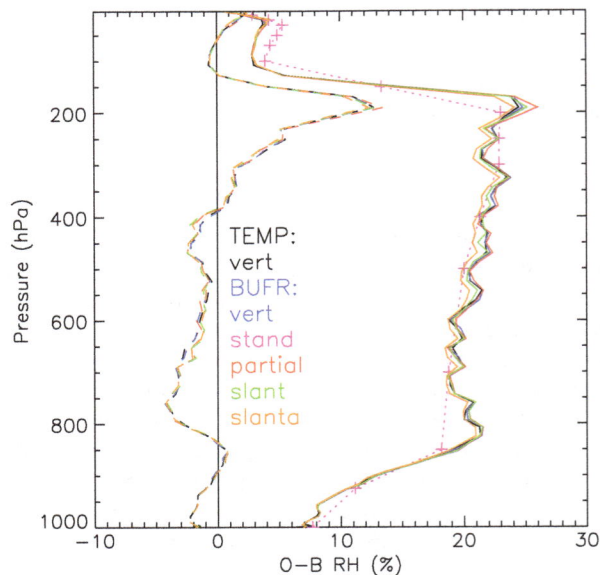

Figure 4. Mean (dashed) and SD (solid) of RH (%) O − B for December 2012–February 2013, 00 and 12 UTC ascents. Except for standard levels (+ symbols, SD only) the observed values are plotted at the mean pressure of the layer mid-points. The lines are for vertical processing (TEMP black and BUFR blue); partial (red) and full slant (green/orange) processing of BUFR data using interpolation to model layer mid-points except that orange line uses vertical averaging.

with the peak in q increments that Ingleby *et al.* (2013) found in the global assimilation system.

Figure 5 shows wind speed statistics (the vertical averaging is performed on wind components, their statistics – not shown – display similar features). As for temperature vertical processing of TEMP and BUFR gives very similar results (but above 10 hPa the extra vertical resolution of the BUFR data gives a useful improvement to the SDs), full slant processing is best at upper levels (lower plot) and partial processing gives more mixed results. At most levels the bias is quite small, at very low levels the background speeds are too strong – broadly consistent with results compared with surface stations (Ingleby, 2014), especially at night. For the first 20 s or so (∼100 m) of the ascent there can be problems with GPS lock which may give large wind errors – the BUFR O − B SDs are slightly worse than those for TEMPs at very low levels (TEMP reports would not normally include wind levels so close to the ground). Nash *et al.* (2011) show large wind differences between radiosonde types near the surface. At 200 hPa and above the vertical averaging shows a marked improvement over the standard level statistics – making more difference than the other processing options.

4. Discussion and summary

Any change in the characteristics of reported data (including the change of vertical coordinate) is of interest for NWP and climate monitoring. The comparison

of TEMP and BUFR data brought to light a small offset in the temperature values due to rounding/decoding issues with the TEMP reports – but small offsets matter in climate studies. Globally there will be a transition of several years with an uncomfortable mixture of TEMP reports, BUFR reports and TEMP data massaged into BUFR format. The BUFR template used by the UK provides extra metadata such as the radiosonde serial number and the version of the processing system used. The serial number can be used to identify minor changes to the humidity sensor, such as the mounting or the coating, which nevertheless have an impact on the radiosonde humidity bias (Jones, 2010, chapters 3 and 4). As in TEMP reports there is a generic indicator of whether a radiation correction to temperature (and humidity) has been performed, but this does not specify which version of the correction has been used. For some radiosonde systems (not RS92) the Met Office performs a temperature bias correction (updated infrequently, based on separate 00 and 12 UTC O − B statistics for 100 hPa height from individual stations). In October 2013 a test was performed of DigiCORA version 3.64.1 together with new radiation corrections and time lag corrections for humidity applied before transmission (Nash *et al.*, 2011; see also links in Appendix S2, Supporting Information). The upgrade (to be applied operationally during 2014) makes little difference to the temperatures but significant differences to upper tropospheric humidities (generally increasing them during daylight).

The vertical averaging performed in the Met Office NWP system improves the O − B statistics in the

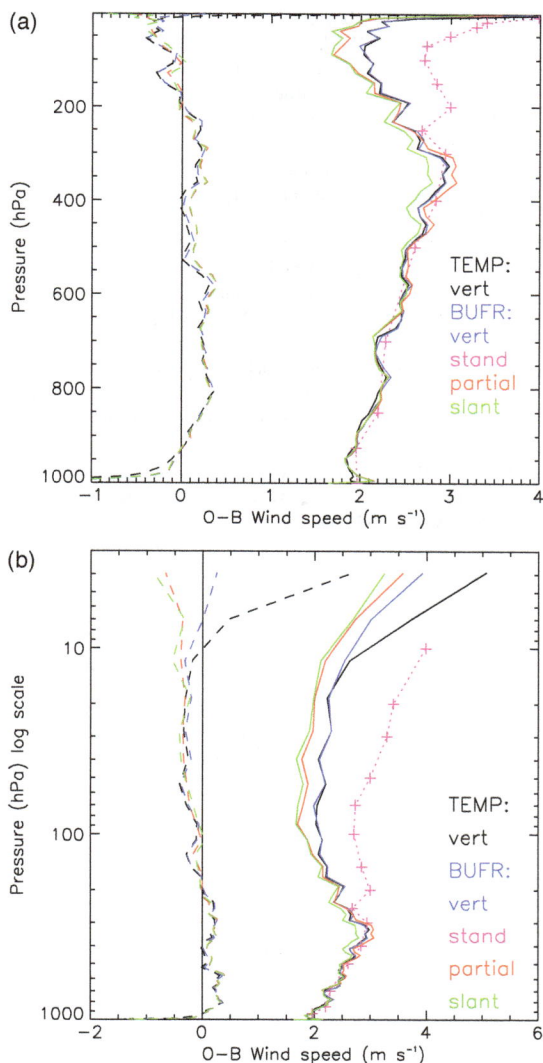

Figure 5. Mean (dashed) and SD (solid) of wind speed (m s^{-1}) $O - B$ for December 2012–February 2013, 00 and 12 UTC ascents; lower plot with logarithmic vertical axis. Except for standard levels (+ symbols, SD only) the observed values are those averaged over model layers and they are plotted at the mean pressure of the layer mid-points. The lines are for vertical processing (TEMP black and BUFR blue); partial (red) and full slant (green) processing of BUFR data.

levels as point values and should see more benefit from use of high-resolution profiles – probably sub-sampled.) High radiosonde vertical resolution is useful for research purposes, such as study of gravity waves (Geller and Love, 2013). The lowest 100 or 200 m of the winds has relatively large errors and should not be used for most purposes. Treatment of radiosonde drift in the interpolation of model fields to the radiosonde profile adds complexity but gives some improvement to $O - B$ statistics in the upper troposphere and lower stratosphere. Partial treatment of the drift (using the correct horizontal positions but treating all levels as valid at the launch time) gives more mixed results – supporting the suggestion by Macpherson (1995) that either both time and spatial displacement should be used or neither. Partly for this reason we question the usefulness of using spatial drift in verification (which generally does not include time interpolation) as suggested by Laroche and Sarrazin (2013). Also the impact of radiosonde drift will be largest on $T + 0$ h (i.e. analysis) statistics: $T + 0$ h fit does not represent verification against independent data if the radiosondes have been assimilated, there would be some impact at $T + 24$ h but beyond that the forecast errors would dominate.

UK high resolution BUFR data (available on the GTS from 30 May 2014) will be assimilated operationally in the UKV from July 2014 – including treatment of balloon drift. Short trials have given neutral impact, perhaps because of the importance of boundary conditions during strong wind/large drift situations (Dow and Macpherson, 2013, pers. comm.). Assimilation in the global system should follow later in 2014 for stations where BUFR data is available and the quality is reasonable. Calculation of balloon drift (where not reported) from the winds is under consideration. Experience suggests that an overlap of more than 6 months and diligent cooperation between observation producers and users is necessary to resolve quality issues with a new radiosonde data stream. In 2016/7 the Met Office will cease using the RS92 and will move to a different radiosonde type (subject to open tender).

stratosphere (but raises some consistency issues, see Appendix S1). Throughout much of the troposphere the SD $O - B$ statistics are dominated by background error (according to the RS92-SGP datasheet, available from http://www.vaisala.com the uncertainties of the sounding are 0.5°C, 5% RH and 0.15 m s^{-1}; although in the lower troposphere the temperature and RH errors should be smaller than this). Because of the vertical averaging the extra vertical resolution of the BUFR reports makes little difference to the $O - B$ statistics (except for some improvement above 10 hPa for winds) although with higher model vertical resolution there might be more difference. (In contrast ECMWF and some other centres treat both standard and significant

References

Durre I, Vose RS, Wuertz DB. 2006. Overview of the integrated global radiosonde archive. *Journal of Climate* **19**: 53–68, DOI: 10.1175/JCLI3594.1.

Geller MA, Love PT. 2013. Research using high vertical-resolution radiosonde data, *SPARC Newsletter* **40**: 29–32. http://www.sparc-climate.org/publications/newsletter/.

Hilton FI, Newman SM, Collard AD. 2012. Identification of NWP humidity biases using high-peaking water vapour channels from IASI. *Atmospheric Science Letters* **13**: 73–78.

Ingleby B. 2014. Global assimilation of air temperature, humidity, wind and pressure from surface stations. *Quarterly Journal of the Royal Meteorological Society*, DOI: 10.1002/qj.2372.

Ingleby NB, Lorenc AC, Ngan K, Rawlins F, Jackson DR. 2013. Improved variational analyses using a nonlinear humidity control variable. *Quarterly Journal of the Royal Meteorological Society* **139**: 1875–1887, DOI: 10.1002/qj.2073.

Jones J. 2010. An assessment of the quality of GPS water vapour estimates and their use in operational meteorology and climate monitoring. PhD thesis, University of Nottingham, UK. http://etheses.nottingham.ac.uk/1287/.

Laroche S, Sarrazin R. 2013. Impact of radiosonde balloon drift on numerical weather prediction and verification. *Weather and Forecasting* **28**: 772–782, DOI: 10.1175/WAF-D-12-00114.1.

Lorenc AC, Ballard SB, Bell RS, Ingleby NB, Andrews PLF, Barker DM, Bray JR, Clayton AM, Dalby T, Li D, Payne TJ, Saunders F. 2000. The Met Office 3-dimensional data assimilation scheme. *Quarterly Journal of the Royal Meteorological Society* **126**: 2991–3012.

Macpherson B. 1995. Radiosonde balloon drift – does it matter for data assimilation? *Meteorological Applications* **2**: 301–305.

McGrath R, Semmler T, Sweeney C, Wang S. 2006. Impact of balloon drift errors in radiosonde data on climate statistics. *Journal of Climate* **19**: 3430–3442, DOI: 10.1175/JCLI3804.1.

Nash J, Smout R, Oakley T, Pathack B, Kurnosenko S. 2006. WMO intercomparison of radiosonde systems, Vacoas, Mauritius, 2–25 February 2005, WMO Instruments and Observing Methods. Report No. 83, Retrieved May 30, 2014. http://www.wmo.int/pages/prog/www/IMOP/publications-IOM-series.html.

Nash J, Oakley T, Vömel H, Wei LI. 2011. WMO intercomparison of high quality radiosonde systems, Yangjiang, China, 12 July–3 August 2010, WMO Instruments and Observing Methods. Report No. 107, Retrieved May 30, 2014. http://www.wmo.int/pages/prog/www/IMOP/publications-IOM-series.html.

Piccolo C, Cullen M. 2012. A new implementation of the adaptive mesh transform in the Met Office 3D-Var system. *Quarterly Journal of the Royal Meteorological Society* **138**: 1560–1570, DOI: 10.1002/qj.1880.

Seidel DJ, Sun B, Pettey M, Reale A. 2011. Global radiosonde balloon drift statistics. *Journal of Geophysical Research* **116**: D07102, DOI: 10.1029/2010JD014891.

Steinbrecht W, Claude H, Schönenborn F, Leiterer U, Dier H, Lanzinger E. 2008. Pressure and temperature differences between Vaisala RS80 and RS92 radiosonde systems. *Journal of Atmospheric and Oceanic Technology* **25**: 909–927, DOI: 10.1175/2007JTECHA 999.1.

Thorne P, Parker D, Tett S, Jones P, McCarthy M, Coleman H, Brohan P, Knight J. 2005. Revisiting radiosonde upper air temperatures from 1958 to 2002. *Journal of Geophysical Research* **110**: D18105, DOI: 10.1029/2004JD005753.

WMO. 2011. WMO Publication No. 306 – Manual on codes. Volume I.1 and Volume I.2. Retrieved May 30, 2014. http://www.wmo.int/pages/prog/www/WMOCodes.html.

26

Measurement of boundary layer ozone concentrations on-board a Skywalker unmanned aerial vehicle

Sam Illingworth,[1] Grant Allen,[1] Carl Percival,[1] Peter Hollingsworth,[2] Martin Gallagher,[1] Hugo Ricketts,[1] Harry Hayes,[1] Paweł Ładosz,[2] David Crawley[3] and Gareth Roberts[2]

[1] The Centre for Atmospheric Science, The School of Earth, Atmospheric and Environmental Science, The University of Manchester, M13 9PL, UK
[2] The School of Mechanical, Aerospace and Civil Engineering, The University of Manchester, M13 9PL, UK
[3] The School of Electrical and Electronic Engineering, The University of Manchester, M13 9PL, UK

*Correspondence to:
S. Illingworth, The Centre for Atmospheric Science, The School of Earth, Atmospheric and Environmental Science, The University of Manchester, Oxford Road, Manchester M13 9PL, UK.
E-mail:
samuel.illingworth@manchester.ac.uk

Abstract

This study demonstrates novel measurements of *in situ* ozone (O_3) concentrations and thermodynamics sampled on-board an instrumented Skywalker Unmanned Aerial Vehicle (UAV). Small spatial and temporal gradients were observed over a localized region, which nearby ground-based *in situ* measurements lack the ability to resolve. It was found that the UAV-measured O_3 concentrations provided a useful additional indicator of O_3 variability at the sub-urban scale. The ability to sample subtle variability over a localized area highlights the important and novel capabilities of UAVs to rapidly characterize local area micrometeorology and chemistry.

Keywords: UAV; ozone; urban scale; atmospheric chemistry; micrometerology

1. Introduction

Unmanned Aerial Vehicles (UAVs) are remotely or autonomously piloted aircraft. While UAVs have recently been most associated with military applications, their use in the field of atmospheric science and environmental monitoring is rapidly growing, from the monitoring of carbon dioxide (CO_2) concentrations (Watai *et al.*, 2006) to observing the spatial distribution of evapotranspiration (Rauneker and Lischeid, 2012). With the continued miniaturization of highly accurate and precise sensors, their potential effectiveness to make low-cost measurements at high spatial resolution is the subject of much scientific and technological interest.

For smaller UAVs the limiting factor in their utility is typically the availability of high quality, miniaturized sensors necessary for their reduced payload capacity, as well as relatively short-duration flight times (typically of the order 1–2 h); while one of the biggest current challenges for larger platforms concerns permission to fly by the appropriate regulatory bodies, such as the Civil Aviation Authority (CAA) for UK airspace.

Low Altitude, Short Endurance (LASE) UAVs are relatively simple to operate, with simple ground-control stations and control mechanisms, requiring only a small crew (Watts *et al.*, 2012). Their small size means that they can be hand-launched from a variety of terrains, and in the UK UAVs with an operating mass of 7 kg or less are exempt from the majority of the regulations that are normally applicable to large and manned aircraft (CAA, 2010). In the atmospheric sciences, UAVs have now been used for a variety of purposes, from making in situ measurements of thermodynamic properties in the planetary boundary layer (Houston *et al.*, 2012) to studying emissions at active volcano sites (Diaz *et al.*, 2010). Measurements of reactive gases such as ozone (O_3) from UAVs offer a novel opportunity to sample the three-dimensional (3D) spatial and temporal variability of such gases to enable process analysis.

Although only about 10% of all atmospheric O_3 is located in the troposphere, it is a principal driver of the photochemical processes regulating many of the gases that are emitted into the atmosphere by either natural or anthropogenic processes. In addition to this, tropospheric O_3 is itself a pollutant with impacts on human health and the environment. In Europe alone it is estimated that O_3 contributes to over 20 000 premature deaths per annum (EEA, 2007). Tropospheric O_3 is also a greenhouse gas and has been reported to contribute a net warming effect to the climate system, with a radiative forcing estimated at 0.35 W m^{-2} (Forster *et al.*, 2007).

Tropospheric O_3 is not emitted; rather it is produced in the atmosphere from reactions involving precursor pollutants such as volatile organic compounds (VOCs) and NO_X (nitrogen oxide and nitrogen dioxide). These rapid chemical interactions can result in large spatial and temporal gradients of O_3 in urban environments, which usually peak in sub-urban areas downwind, with the O_3 production rate generally increasing with NO_X concentrations (Fowler, 2008). However, in urbanized centres, characterized by even higher concentrations of atmospheric NO_X, O_3 production can be inhibited, as a result of the reaction of O_3 with NO and the formation of NO_y (i.e. the sum of NO_X plus the reservoirs of NO_X, such as nitric acid and peroxyacetyl nitrate). Thus, in an urban environment, a reduction in NO_X concentrations can often lead to an increase in tropospheric O_3. With future estimates of NO_X emissions in the UK predicting reductions of 45% over the next decade (Hall *et al.*, 2006), an even greater emphasis might be placed on monitoring urban hazardous O_3 concentrations on the local scale.

Ozone concentrations depend strongly on the spatial distributions of NO_X and VOCs (and their associated emission sources), which vary both from one city to another as well as within the urban environment itself. Prevailing meteorological conditions can also vary rapidly, and mixing by micrometeorological processes such as the urban heat island effect and street-canyon-scale dynamics all act to modulate the chemistry and transport of ozone and its tracers across a range of temporal and spatial scales. Ground-based *in situ* measurements such as the Automatic Urban and Rural Network (AURN) in the UK operated by the Department for Environment, Food and Rural Affairs (Defra) can provide surface-level O_3 concentrations, although such fixed *in situ* measurements are limited in their spatial coverage. AURN currently has 103 active sites across the UK (Figure 1), which while providing high-resolution hourly information, is not able to provide information about surface-level O_3 concentrations on the sub-urban scale. Measurements from ground-based monitoring sites such as AURN are often used as validation datasets for regional air quality models such as the Met Office Air Quality Unified Model. Clearly, the inability of such sites to inform on the sub-urban scale can lead to a poor interpretation (and validation) of urban environments (Savage *et al.*, 2012).

An alternative to ground-based measurements of tropospheric O_3 can be to use aircraft. Typically, airborne measurements around urban environments are made using large aircraft. However, such flight campaigns are not only expensive, but still also lack the required spatial resolution for many applications, e.g. Manchester city centre has a diameter of approximately 2 km, which combined with the science speed of the UK's atmospheric research aircraft ($\sim$100 m s^{-1}), means that a typical 1 Hz instrument would only be able to make approximately 20 measurements during an overpass of the city. Large research aircraft are also restricted by the CAA, which often means that they are unable to fly around urban centres or within the lower boundary layer.

However, UAVs offer an ideal alternative at such scales, bridging the gap between ground-based and traditional airborne methods, with the potential to deliver detailed, high-resolution, and precise measurements of tropospheric and near-surface O_3 concentrations at the local scale. Potentially synergistic instruments such as the CityScan ground-based NO_2 remote sensing system (Roland Leigh, pers. comm.) in development at the University of Leicester could be used in conjunction with UAV sampling to fully characterize the sub-urban scale.

In this paper, we present measurements of planetary boundary layer O_3 concentrations measured *in situ*, thereby demonstrating the development of a system with the capabilities of providing high spatial and temporal resolution sampling, which can be deployed in urban environments. This work describes this development and the first field measurements at a site near to Manchester city centre.

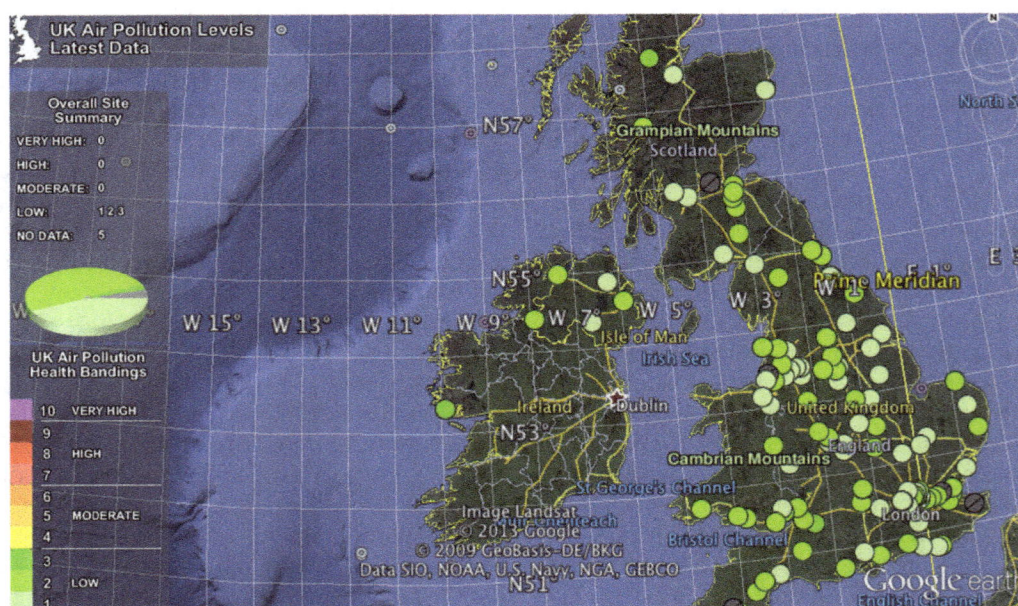

Figure 1. Locations of AURN sites in the UK plotted on Google Earth (Source: 'UK' 53°16′44.20″N and 2°43′36.48″E. Google Earth. April 10, 2013. July 7, 2013). Pollution levels correspond to measurements taken on Friday 7 July 2013; data courtesy of Defra (http://uk-air.defra.gov.uk/)

Table I. Accuracy and precision of the Vaisala RS92-KE radiosonde, when operating from 1080 to 100 hPa.

Quantity	Accuracy	Precision
Pressure (hPa)	1.5	0.1
Temperature (°C)	0.2	0.1
RH (%)	5	1

Table II. Specifications of Skywalker UAV.

Specifications	Skywalker
Length	1100 mm
Wingspan	1880 mm
Payload Bay	3450 cm³
Maximum take-off weight	3.0 kg
Typical cruising s	45 km h⁻¹
Endurance (10 000 mAh battery)	1 h minimum

2. System design

In this study, we have adapted an Electrochemical Concentration Cell (ECC) ozonesonde to fly on a fixed-wing LASE UAV. We now describe the components and integration of this system.

The ECC ozonesonde (manufactured by Science Pump Ltd) comprises a motor, a Teflon pump, and the ECC module. The cell is made up of two chambers (anode and cathode) containing electrodes made of bright platinum and a potassium iodide solution with differing concentration in each cell. Both half-cells also contain potassium bromide in equal concentrations as a buffer. The electrodes quickly polarize, but when O_3-rich air is bubbled through the cathode solution, an electromotive force and current are induced which is proportional to the O_3 partial pressure. Data from the ozonesonde were relayed to a Vaisala RS92-KE radiosonde that simultaneously measured pressure, temperature, and Relative Humidity (RH). These data were transmitted continuously during the flight via

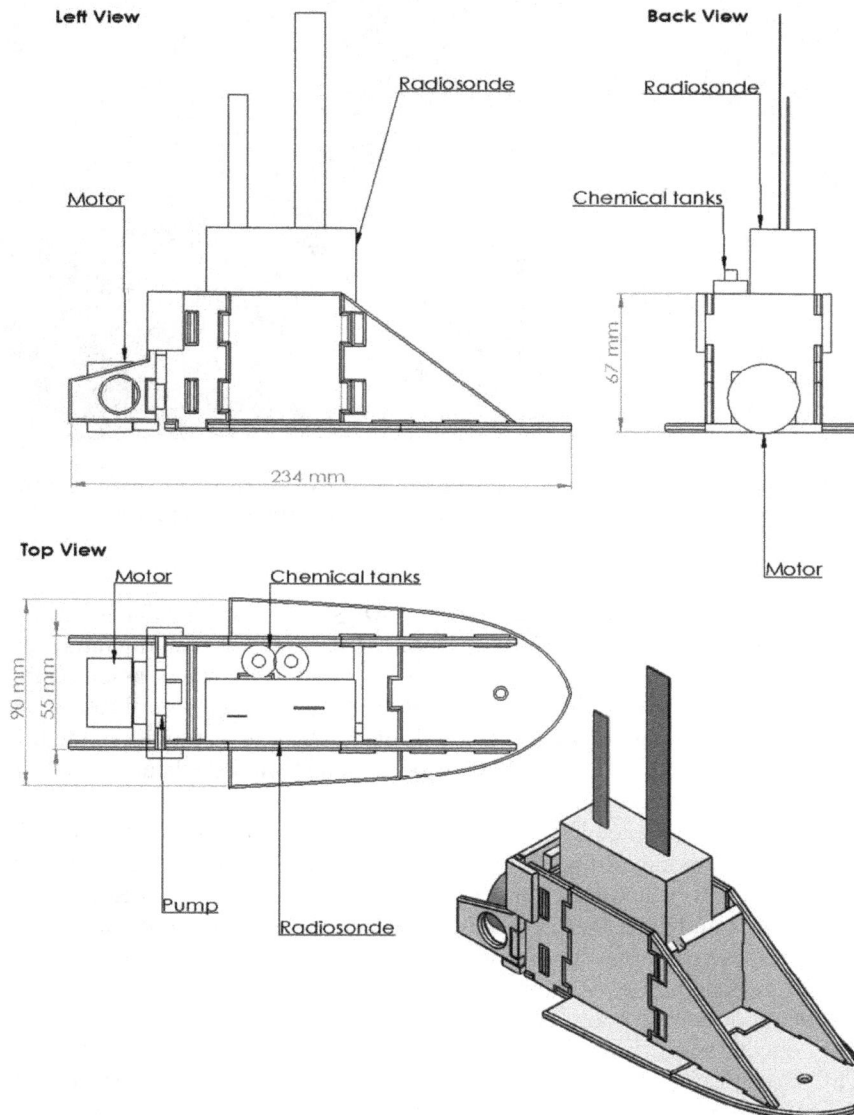

Figure 2. Drawing of final ozonesonde integration configuration.

Figure 3. A picture of the integrated ozonesonde and Skywalker airframe at the Hough End Fields site.

Figure 4. Google Earth (Source: 'UK' 53°25′59.48″N and 2°14′59.19″W. Google Earth. February 6, 2009. October 29, 2013.) image showing location of Hough End Fields site in relation to Manchester city centre (red ellipse). Inset: GPS locations of measured O_3.

radio to a dedicated ground station. The accuracy and precision for these measurements are given in Table I.

Tropospheric O_3 concentrations measured using this system have a typical relative precision of $\pm 3-6\%$ (Smit and Sträter, 2004), corresponding to ~1–2 ppb at concentrations measured here, with a sampling frequency of 0.5 Hz. This was also confirmed in the laboratory before flight. Anode and cathode solutions for the ECC were prepared in the laboratory at known concentrations in triple-distilled water, and the cell current was calibrated and validated for these solutions under ambient and high (saturated) O_3 concentrations. A calibrated Thermo Scientific TE49C photometric ozone analyser measured the ambient concentration in the lab, and a Science Pump Ltd TSC-1 ozoniser unit generated saturated flows. For a further description of the ozonesonde and radiosonde system, see Skrivankova (2004) and references therein.

The Skywalker airframe (the fuselage and wings) was designed and manufactured by Skywalker Technologies, before being assembled at the University of Manchester; its specifications are summarized in Table II. As well as providing the frame onto which the ozonesonde was attached, the UAV also provided GPS information.

Combining an ozonesonde with a UAV controlled via radio therefore allows the collection of real-time O_3, GPS, pressure, temperature, and humidity data.

When integrating the ozonesonde into the UAV, the electro-mechanical integration of the payload equipment needed to be taken into consideration. In addition to incorporating the ozonesonde, the final payload had to house the inlet pump motor to draw air into the ozonesonde for measurement, as well as discrete voltage sources for both the pump motor and ozonesonde. The final integration of the Vaisala probe and Skywalker airframe is shown in Figures 2 and 3.

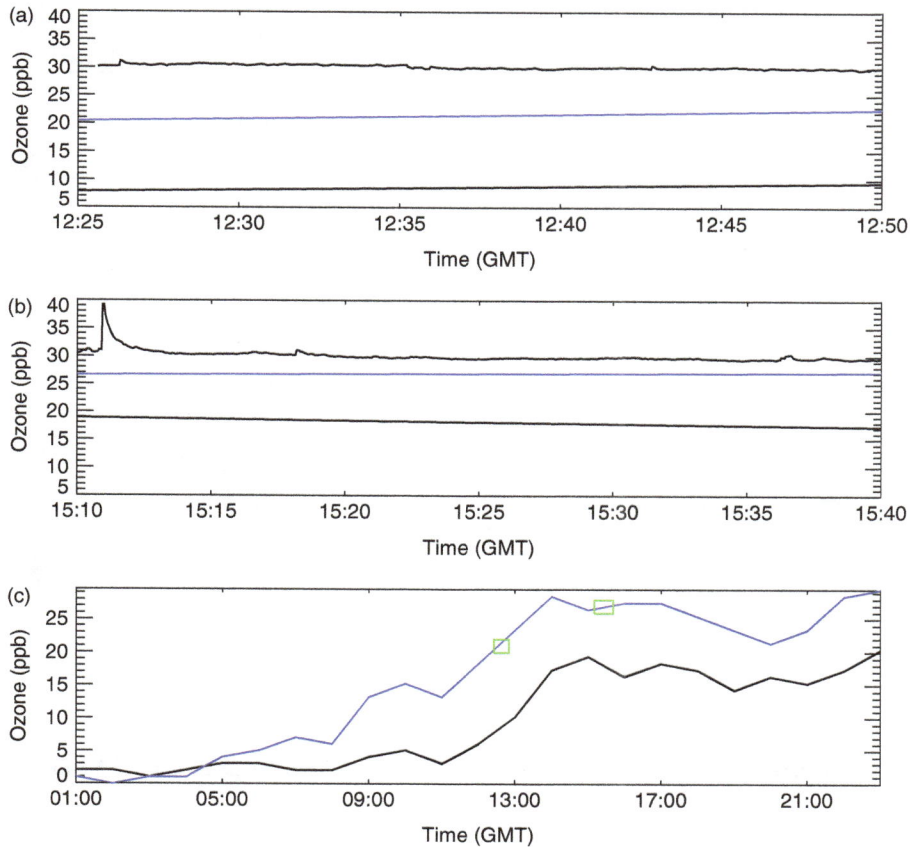

Figure 5. (a) Measured O_3 concentrations from 12:25 until 12:50. UAV measurements are given in black, Piccadilly Gardens AURN site measurements in red, and Manchester South AURN site measurements in blue. (b) Same as (a) but for 15:10 until 15:40. (c) AURN site measurements for the 28 June 2012 between 01:00 and 23:00. The Piccadilly Gardens site is shown in red, and the Manchester South site in blue. The green boxes indicate the data shown in (a) and (b).

3. Results

Test flights were carried out at Hough End Fields, Manchester on 28 June 2012 at midday, with partially cloudy conditions and gusting winds of up to 10 mph as recorded by a handheld anemometer. While the Skywalker UAV (including ozonesonde payload) weighs less than the 20 kg for which a certificate of airworthiness becomes a prerequisite, there are still a number of restrictions that are enforced by the CAA for those UAVs without an approved detect and avoid capability (as was the case here), namely that: the UAV is not allowed to fly in controlled airspace; the UAV can not be flown at an altitude exceeding 400 ft; the UAV cannot be flown within 150 m of any congested area of a city or town, or within 30 m of any person; and that a direct line of sight (500 m maximum range) must be maintained (CAA, 2010). Working under these restrictions, the Hough End Fields site, which is an existing model flying location (Figure 4), was chosen.

The ozonesonde sampled for two flying periods: from 12:25:36 to 12:50:26 and 15:09:50 to 15:41:30. These data are shown in Figure 5, with the corresponding meteorological data (pressure, temperature, and RH) shown in Figure 6. As seen from these plots, the O_3

appears to be reasonably well mixed in the lower boundary layer with a mean and standard deviation across this data set being 30.11 and 0.70 ppb, respectively. From Figure 6, the periods where the pressure was constant at roughly 1005 hPa (e.g. at approximately 12:40) indicate the periods of time when the UAV was grounded.

Peak concentrations of approximately 39 ppb were observed at 15:10. This peak was associated with a short-term shift in the prevailing wind direction to a more north-easterly direction, bringing air from the nearby main road (A5103, Princess Road) and Manchester 'city centre'. Such changes in wind direction were measured on this day using a handheld anemometer and were noted to be coincident with turbulent downdrafts associated with passing non-precipitating cloud. Our ability to sample this variability illustrates the potential for future process studies that relate to air quality, micrometeorology, and local dynamics in regions of strong local sources.

For comparative purposes, the data from two Manchester-based AURN sites is also shown in Figure 5. These sites are situated in Manchester city centre (Manchester Piccadilly: 53.481520°N −2.237881°E), and near to Manchester airport (Manchester South: 53.369026°N −2.243280°E).

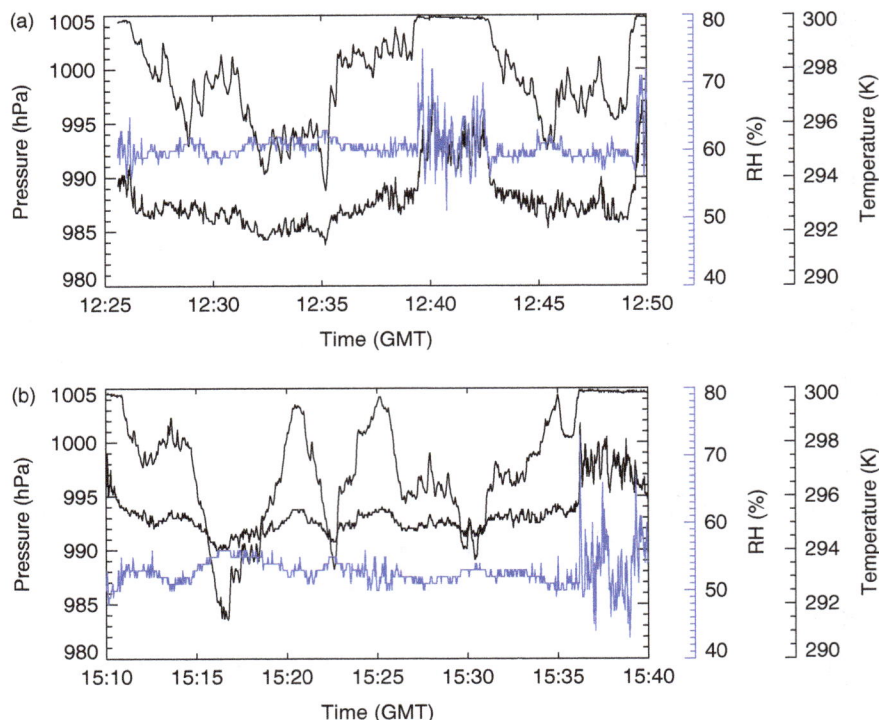

Figure 6. Meteorological data (pressure, temperature, and relative humidity) from (a) 12:25 until 12:50 and (b) 15:10 until 15:40.

Figure 5 illustrates the importance of being able to make numerous measurements of O_3 at a high spatial and temporal resolution for process studies and air quality model validation, because while the AURN sites provide a well-calibrated and consistent dataset they are only able to inform on concentrations of O_3 in their immediate vicinity, at a relatively poor temporal resolution (1 h). Furthermore, as shown in Figure 5(c), the AURN stations show rising ozone concentration throughout this day, especially between the two periods of measurement used in this study (green boxes in Figure 5(c)). Unlike at the AURN sites, the Skywalker measurements do not show a consistent increase in O_3 concentration between the midday and mid-afternoon measurement periods. These large local differences and gradients in species such as O_3 illustrate the need for dense sampling in order to inform and validate air quality models on the sub-urban scale.

Back trajectories can serve to provide information about the general airmass history of air sampled in a particular location. To probe the history of the air masses encountered by the Skywalker system, we used multiple single-particle 3D (vertical motion enabled) back trajectories from the offline Hybrid Single-Particle Lagrangian Integrated Trajectory (HYSPLIT) model (Draxler and Rolph, 2003), initialized using the National Centre for Environmental Prediction (NCEP) reanalysis meteorological wind fields at 1° spatial resolution, at 200 m vertical intervals between the ground and 2 km. 48-h back trajectories (with half-hourly outputs) were calculated with endpoints corresponding to the Hough End Fields site (Figure 7).

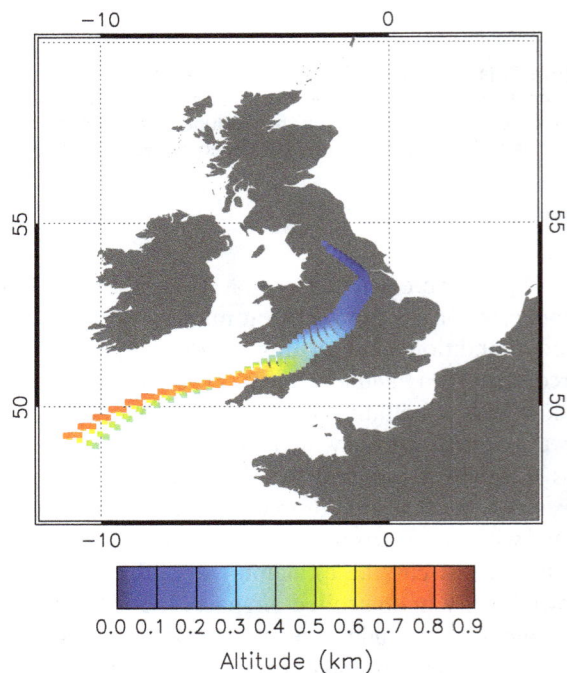

Altitude (km)

Figure 7. Forty eight-hour ensemble 3D back trajectories from the HYSPLIT Lagrangian model, ending at Hough End Fields at midday on 28 June 2012, initialized at 200 m vertical intervals between the ground and 2 km.

We would expect that the observed variability in O_3 over the relatively short timeframe of the measurements in this study would be driven more by local emissions, chemistry, and dynamics than regional scale transport, with the trajectories shown in Figure 7 demonstrating

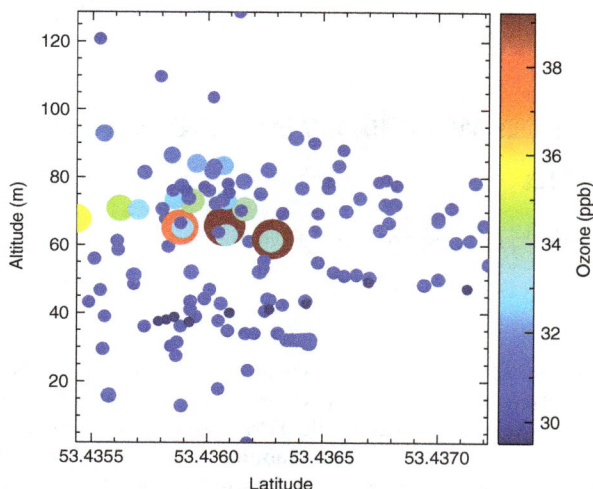

Figure 8. Plot of O_3 along the latitudinal flight track of the UAV. For ease of reference, the sizes of the plotted circles are proportional to the O_3 concentration.

that the prevailing wind direction at the time of measurement was approximately south-easterly (consistent with observations noted at Hough End Fields at the time of flight). This suggests that air sampled at Hough End would have passed near to the Manchester South AURN site; however, the poor sampling resolution (1 h) of the AURN site (and the lack of information between the two sites) means that we are unable to comment on any localized chemistry, although this might be possible in the future using simultaneous measurements of other active chemical species with UAVs.

The use of the UAV also enabled a spatially dense map of surface-level and lower boundary layer O_3 concentrations to be generated, as shown in Figure 8. It should be noted that not all of the data shown in Figure 5 is plotted in Figure 8, as there was a period of time for which the GPS data was not recorded due to a logging fault. The red colours (larger symbols) in Figure 8 show the enhancement in O_3 associated with the change of wind direction noted earlier. This highlights the potential for sampling 3D variability that may enable micrometeorological process analysis in the future using simultaneous measurements of thermodynamics and other trace gases from UAVs.

4. Conclusions

This work has presented the design, data, and analysis of a Skywalker UAV system incorporating an ozonesonde and thermodynamic measurement capability. These measurements demonstrate the capabilities of UAVs to make high-density, 3D measurements of tropospheric O_3, and (more widely) other trace gases on a sub-urban scale. The measurements recorded in this study demonstrate the localized spatial and temporal variability that exists in urban environments, and the difficulty that traditional ground-based instrumentation has in capturing these gradients at sufficient resolution.

Simultaneous measurements of pollutant trace gases are necessary to inform and validate air quality models to provide real-time data for monitoring, and to alert the public to dangerous levels. UAVs equipped with such a suite of instruments could conceivably help to provide such a network, giving vertical as well as ground-based information, and enabling a much clearer 3D picture to be developed.

Acknowledgements

We would like to thank NERC for funding the summer placement of Harry Hayes. We would also like to thank the Royal Meteorological Society for equipment and travel support of the project through their Legacies Fund. We also thank Defra for access to the AURN database.

References

CAA. 2010. Civil Aviation Authority CAP 722 Unmanned Aircraft System Operations in UK Airspace – Guidance, Directorate of Airspace Policy (http://www.caa.co.uk/).

Diaz JA, Pieri D, Arkin CR, Gore E, Griffin TP, Fladeland M, Bland G, Soto C, Madrigal Y, Castillo D. 2010. Utilization of in situ airborne MS-based instrumentation for the study of gaseous emissions at active volcanoes. *International Journal of Mass Spectrometry* **295**(3): 105–112.

Draxler R, Rolph G. 2003. HYSPLIT (HYbrid Single-Particle Lagrangian Integrated Trajectory) model access via NOAA ARL READY website, NOAA Air Resources Laboratory, Silver Spring, MD (http://www.arl.noaa.gov/ready/hysplit4.html).

EEA. 2007. Air pollution in Europe 1990–2004, *European Environmental Agency Report No 2/2007*. Copenhagen (http://www.eea.europa.eu).

Forster P, Ramaswamy V, Artaxo P, Berntsen T, Betts R, Fahey DW, Haywood J, Lean J, Lowe DC, Myhre G. 2007. Changes in atmospheric constituents and in radiative forcing. *Climate Change* **20**: 129–234.

Fowler D. 2008. *Ground-Level Ozone in the 21st Century: Future Trends, Impacts and Policy Implications.* The Royal Society: London.

Hall J, Dore A, Heywood E, Broughton R, Stedman J, Smith R, O'Hanlon S. 2006. Assessment of the environmental impacts associated with the UK Air Quality Strategy.

Houston AL, Argrow B, Elston J, Lahowetz J, Frew EW, Kennedy PC. 2012. The collaborative Colorado-Nebraska unmanned aircraft system experiment. *Bulletin of the American Meteorological Society* **93**(1): 39–54.

Rauneker P, Lischeid G. 2012. Spatial distribution of water stress and evapotranspiration estimates using an unmanned aerial vehicle (UAV), Paper presented at EGU General Assembly Conference Abstracts.

Savage N, Agnew P, Davis L, Ordóñez C, Thorpe R, Johnson C, O'Connor F, Dalvi M. 2012. Air quality modelling using the Met Office Unified Model: model description and initial evaluation. *Geoscientific Model Development Discussions* **5**(4): 3131–3182.

Skrivankova P. 2004. Vaisala radiosonde RS92 validation trial at Prague–Libus. *Vaisala News* **164**: 4–8.

Smit HG, Sträter W. 2004. JOSIE-2000: Jülich ozone sonde intercomparison experiment 2000: the 2000 WMO international intercomparison of operating procedures for ECC-ozone sondes at the environmental simulation facility at Jülich, WMO.

Watai T, Machida T, Ishizaki N, Inoue G. 2006. A lightweight observation system for atmospheric carbon dioxide concentration using a small unmanned aerial vehicle. *Journal of Atmospheric and Oceanic Technology* **23**(5): 700–710.

Watts AC, Ambrosia VG, Hinkley EA. 2012. Unmanned aircraft systems in remote sensing and scientific research: classification and considerations of use. *Remote Sensing* **4**(6): 1671–1692.

27

Orographic disturbances of surface winds over the shelf waters adjacent to South Georgia

J. Scott Hosking,* Daniel Bannister, Andrew Orr, John King, Emma Young and Tony Phillips

British Antarctic Survey, NERC, Cambridge, UK

*Correspondence to:
J. Scott Hosking,
British Antarctic Survey,
Madingley Road, High Cross,
Cambridge, Cambridgeshire CB3
0ET, UK.
E-mail: jask@bas.ac.uk

Abstract

This study seeks to quantify the influence of South Georgia's orography on regional surface winds. A typical case study characterized by large-scale westerly winds is analysed using a high-resolution setup (3.3 km) of the Weather Research and Forecasting (WRF) regional model. The simulation produces significant fine-scale spatial variability which is in agreement with satellite-derived winds. The model simulation indicates that these orography-driven wind disturbances are responsible for strong wind stress curl and enhanced heat flux over the shelf waters surrounding South Georgia. Such surface forcing is entirely absent from the reanalysis, highlighting the need to use high-resolution forcing in regional ocean model simulations.

Keywords: orographic disturbances; regional modelling; South Georgia; surface heat fluxes; wind stress curl; WRF

1. Introduction

South Georgia is an extremely mountainous and narrow island located in the southern Atlantic Ocean (Figure 1(a)). With an approximate southeast–northwest orientation, it is roughly 170 km long and 2–40 km wide, contains 11 peaks exceeding 2000 m, and is separated from the surrounding deep ocean by shallow shelf waters of about 50–150 km width and mostly less than 300 m deep (Figure 1(b)). It lies within a region of strong mean westerly winds, which are variable on short time-scales because of storms caused by intense cyclonic activity. The focus of this article is the influence of South Georgia's orography on these winds and in the associated air–sea interaction.

Orographic-induced disturbance of large-scale winds by islands can result in significant regional near-surface wind variations (Etling, 1989, 1990; Smith *et al.*, 1997), which in turn can generate variations in the dynamical and thermal coupling between the atmosphere and the ocean that are important for the mean state and variability of the oceanic circulation (e.g. Xie *et al.*, 2001; Chelton *et al.*, 2004; Dong and McWilliams, 2007; Pullen *et al.*, 2008; Couvelard *et al.*, 2012). Around South Georgia the shelf waters often show properties that are markedly different from the open ocean waters beyond (Brandon *et al.*, 2000; Meredith *et al.*, 2005), indicating that local processes are also important in dictating shelf water characteristics at this location. Despite the importance of these waters as one of the most biodiverse marine ecosystems on Earth (Hogg *et al.*, 2011), their response to atmospheric forcing at fine spatial scales has not been investigated.

Young *et al.* (2011, 2012, 2014) have studied the circulation of the South Georgia shelf and surrounding deeper ocean using a high-resolution ocean model driven by atmospheric forcing derived from reanalysis data. However, the coarse spatial resolution of reanalyses is not able to adequately represent the orography of South Georgia, nor is it able to explicitly resolve fine-scale wind features (e.g. Yuan, 2004; Risien and Chelton, 2008). To overcome this problem in other study areas, mesoscale atmospheric models with a horizontal resolution of a few kilometres have been used to dynamically downscale the reanalysis data in order to resolve the missing small-scale features generated by the orography. The more realistic atmospheric forcing thus generated can be used to force an ocean model (e.g. Dong and McWilliams, 2007; Couvelard *et al.*, 2012).

The objective of this study is to quantify the impact of orographic forcing on the wind field over the shelf waters adjacent to South Georgia. We address this question by using the Weather Research and Forecasting (WRF) atmospheric model at high spatial resolution to dynamically downscale reanalysis data for the South Georgia region. After characterizing the improvement by comparison with scatterometer measurements of the surface wind field, output from the WRF model is used to determine the associated forcing of the underlying ocean.

2. Methods, model and data

A single case study on 13 September 1999 of westerly flow with a 10 m wind speed of approximately 13–14 m s^{-1} is selected. The pertinence of this case study can be gauged by comparison with the climatological wind rose for 10 m winds at King Edward Point as shown in Figure 2, produced from ERA Interim reanalysis data which has a N128 (0.7° × 0.7°,

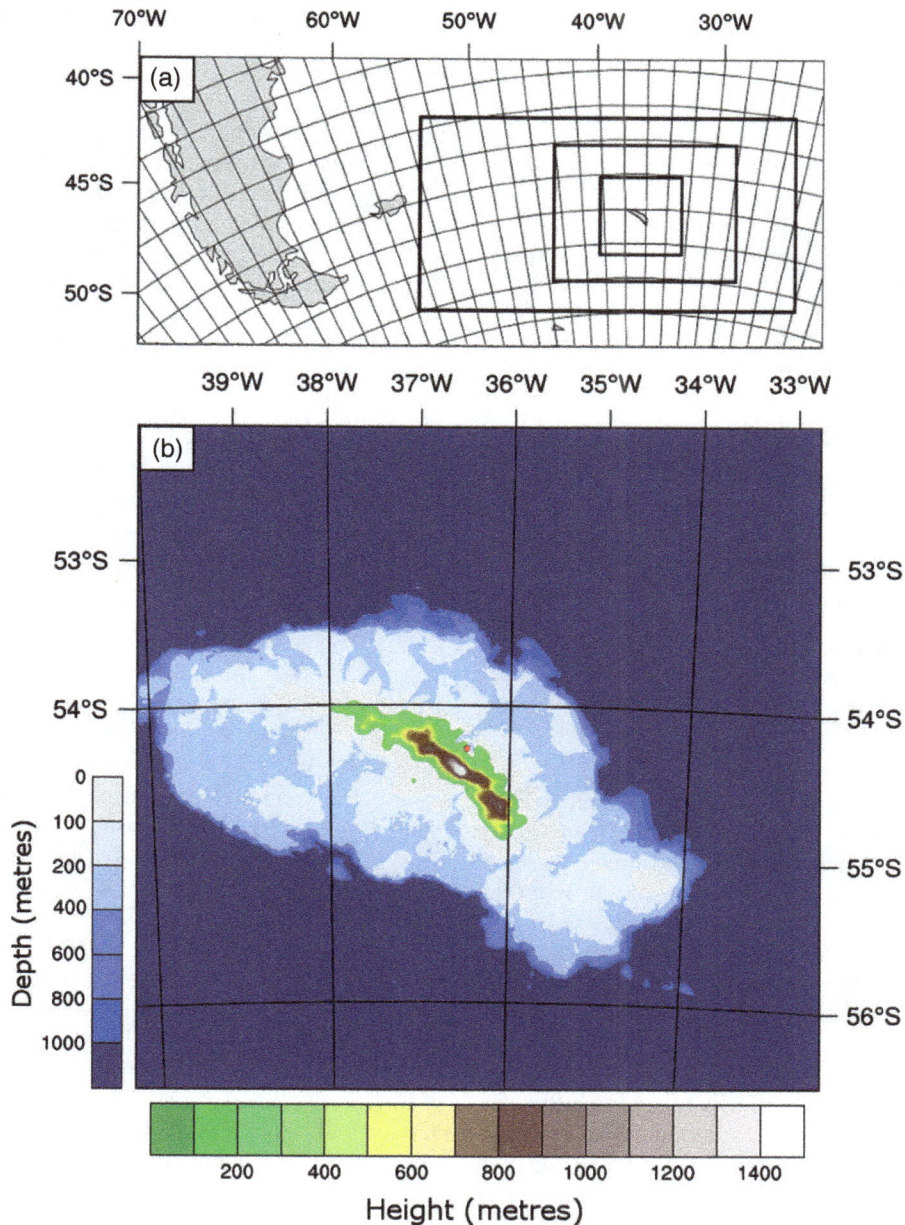

Figure 1. Regional maps to show the nested domain setup used for the South Georgia WRF model simulation. Panel (a) illustrates the lateral boundaries for the 30 km outer domain, 10 km intermediate domain, and the 3.3 km inner domain (black line boxes). A zoomed in map of the innermost domain is illustrated in panel (b), along with the bathymetry of the ocean shelf (blue shading), the island's orographic elevation at 3.3 km resolution (green–brown shading), and the location of King Edward Point (filled red circle).

~45 km × ~80 km near South Georgia) spatial resolution (Dee *et al.*, 2011). This shows that it is representative of atmospheric conditions around South Georgia, which are typified by primarily westerly winds with speeds in the range of 7.5–15 m s^{-1}.

The WRF model is an atmosphere-only, limited-area, nonhydrostatic, mesoscale modelling system (http://www.wrf-model.org). Here we utilize WRF version 3.4.1 (Skamarock *et al.*, 2008). The configuration of the three nested domains (one-way interaction) used in our study is illustrated in Figure 1(a). The outer domain has 75 × 45 grid points at a horizontal grid-spacing of 30 km, covering a relatively large

ocean area in order to better resolve the prevailing westerly winds of the Southern Ocean, and particularly the representation of small mesocyclones (scales of a few hundred kilometres) which are also inadequately represented by reanalysis (Condron *et al.*, 2006). The intermediate domain has 115 × 91 grid points and a horizontal grid-spacing of 10 km. The innermost domain has 151 × 151 grid points at a horizontal grid-spacing of 3.3 km, covering South Georgia and the surrounding shelf sea (illustrated by the zoomed in map in Figure 1(b)). All domains have a total of 70 vertical levels and a model top at 10 hPa. ERA-Interim was used to initialize the surface and lateral boundary conditions

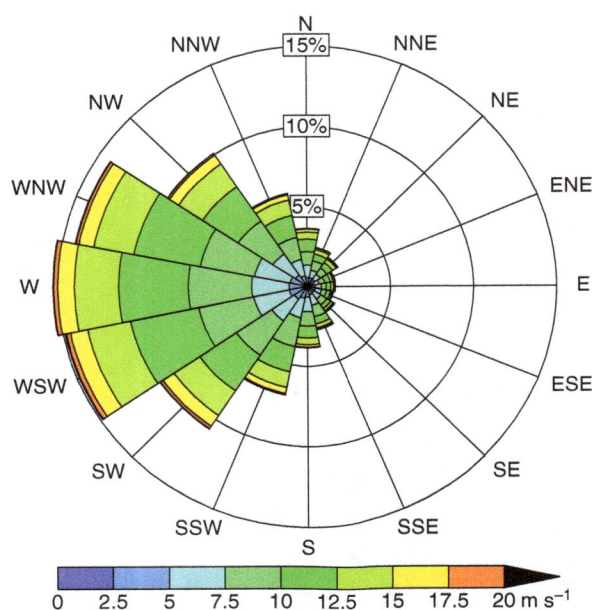

Figure 2. Climatological wind rose to illustrate the direction (°), frequency (%), and speed (m s⁻¹) of large-scale near-surface winds over the period 1979–2012. Data are interpolated from 6-hourly ERA-Interim reanalysis 10 m wind fields at King Edward Point, South Georgia (location indicated in Figure 1(b)).

for the outer domain and to update them every 6 h as the model is integrated forward. The WRF run started at 0000 UTC 12 September 1999 and ended at 1800 UTC 13 September 1999. Only model output from the innermost 3.3 km resolution domain is discussed.

To improve the representation of resolved orographic effects, the model topography was generated using the 90 m resolution Shuttle Radar Topography Mission (Jarvis *et al.*, 2008) dataset (the standard topography in WRF is derived from a dataset which has a resolution of approximately 1 km). Similarly, to improve the representation of the land surface, the model land use classification was generated from 90 m resolution Landsat satellite data which correctly classified South Georgia's higher elevations as 'permanent snow cover' (the standard land-use field in WRF erroneously classified South Georgia as entirely 'bare rock').

Preliminary sensitivity experiments testing various physics options for the WRF model were conducted for a range of contrasting meteorological conditions, with the model output being compared against measurements of pressure, temperature, wind speed, and direction at King Edward Point, South Georgia (see Figure 1(b) for location). On the basis of this the physics choices selected include the New Goddard scheme for longwave and shortwave radiation (Chou and Suarez, 1999), the Mellor–Yamada–Nakanishi–Niino Level 2.5 planetary boundary layer scheme (Nakanishi and Niino, 2004), the unified Noah land-surface model (Chen and Dudhia, 2001), and the Kain–Fritsch cumulus scheme (Kain and Fritsch, 1993). Owing to the steeply sloping terrain and high horizontal and vertical resolution, an adaptive time step (rather than fixed) was required in order to rectify computational instability.

The WRF model results are compared to near-surface (10 m) wind fields obtained from the Quick Scatterometer (QuikSCAT), which were obtained from the Jet Propulsion Laboratory website (ftp://podaac-ftp.jpl.nasa.gov/allData/quikscat/L3/jpl/v2/hdf/). This dataset has a spatial resolution of 0.25° (~25 km), but is unable to provide observations over land and so consequently winds over South Georgia and its nearshore regions are excluded. QuikSCAT wind data are accurate to within $1–2$ m s⁻¹ in speed and $20–30°$ in direction (e.g. Ebuchi *et al.*, 2002; Sousa *et al.*, 2013).

WRF is subsequently used to determine the associated patterns of surface wind stress curl and surface heat flux, which both exert strong controls on ocean circulation and water column structure. To this end the surface wind stress is computed from WRF 10 m winds using the formulation of Large and Pond (1982) to calculate the drag coefficient.

3. Results

Figure 3 shows the observed (a) and simulated 10 m (b) wind field over the shelf waters adjacent to South Georgia. It should be noted that orography-driven disturbances are entirely absent from ERA-Interim wind field (c). This is unsurprising given that South Georgia is not resolved by ERA-Interim, i.e. ERA-Interim treats all the grid cells containing South Georgia as ocean and therefore completely fails to represent its orography and land–sea distribution.

Assuming wind speed $U \sim 14$ m s⁻¹, mountain height $h \sim 2000$ m, buoyancy frequency $N \sim 0.01$ s⁻¹, mountain half-width $D \sim 20$ km, and Coriolis parameter $f \sim -1.2 \times 10^{-4}$ s⁻¹ then the nondimensional mountain height ($\hat{h} = Nh/U$) and the Rossby number (Ro $= U/|f|D$) are $\hat{h} \sim 1.4$ and Ro ~ 5.8, respectively. This suggests that the flow response generated by the orography will be characterized by nonlinear behaviour (i.e. $\hat{h} > 1$) where the effects of rotation are important (i.e. Ro < 10), as reviewed by Orr *et al.* (2005, 2008). Figure 3(a) shows a number of features in the observed wind field which are consistent with this, most notably deceleration of the incident flow, flow splitting around the island (diverting more to the right when looking downwind), a wake of relatively weak winds of approximately $10–11$ m s⁻¹ extending many hundreds of kilometres downstream of the island, and the formation of wind jets of up to 18 m s⁻¹ either side of the wake (with wind speeds greater on the right-hand side than the left-hand side). Comparison with Figure 3(b) shows that the WRF simulation agrees qualitatively well with the QuikSCAT winds, although in the model the wind jets are $1–2$ m s⁻¹ too weak and the modelled wake region does not extend far enough downstream or close enough to South Georgia.

The features described above are readily identifiable in fields of wind stress curl and sensible and latent heat fluxes from WRF, as shown in Figure 4 (including over the near-shore waters where QuikSCAT data

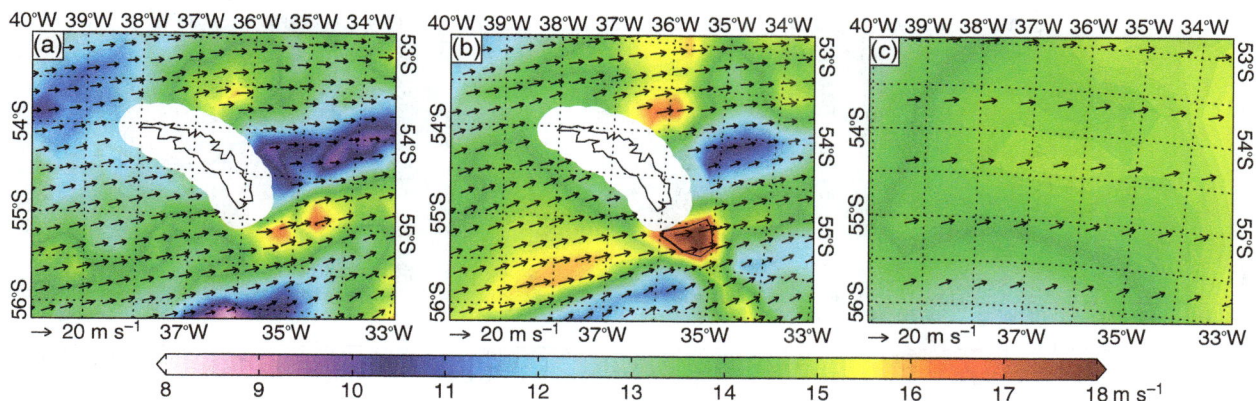

Figure 3. Comparison between observed, modelled, and reanalysis 10 m winds (m s^{-1}) over the shelf waters adjacent to South Georgia for a case study on 13 September 1999 typical of the large-scale flow for the region. The panels show (a) observations from QuikSCAT satellite mission at ~0900 UTC, (b) WRF modelled 3.3 km resolution (re-sampled to the 'QuikSCAT' grid for fair comparison) at 0900 UTC, and (c) ERA-Interim reanalysis at 1200 UTC. In panel (a) the 'blank' area indicates where the QuikSCAT wind data are unavailable, with the South Georgia coastline included for illustrative purposes. This is repeated in panel (b) for comparative purposes. As shown in panel (c), there is no representation of South Georgia within ERA-Interim reanalysis data.

are absent). Here we primarily focus on wind-induced forcing over the shelf sea waters, i.e. the area confined within the 300 m bathymetry depth contour. The high-speed (17–18 ms^{-1}) wind jet off the southern tip of the island, as seen in Figure 3(b) around 35.5°W and 55.2°S, is associated with a large upward (positive) sensible heat flux exceeding 50 W m^{-2}, while surface forcing from wind stress curl and latent heat fluxes are less marked at around 1.6×10^{-5} Nm^{-3} and 70 W m^{-2}, respectively. In the wake of weak winds there are regions of both negative and positive wind stress curl with magnitudes in excess of 4×10^{-5} Nm^{-3}. This coincides with relatively weak downward sensible heat fluxes of between −30 and 0 W m^{-2}, and upward latent heat flux exceeding 100 W m^{-2} (i.e. the net heat flux is upward). Comparison with ERA-Interim (not shown) indicates that magnitudes of wind stress curl are typically two orders of magnitude greater than the corresponding variables in the reanalysis. Furthermore, the WRF sensible heat fluxes are around one order of magnitude larger, while latent heat fluxes are around twice those derived by ERA-Interim.

4. Summary and discussion

This study shows that fine-scale structure in winds, forced by South Georgia's steep orography, contribute to the generation of important surface forcing of the surrounding shelf seas, e.g. wind stress curl of the order 10^{-5} N m^{-3} is the same order of magnitude as that observed over the Irminger Sea forced by the Greenland tip jet (Pickart *et al.*, 2003). Wind effects such as these are consistent with nonlinear (i.e. $\widehat{h} > 1$) flow regimes where the effects of rotation are important (i.e. Ro < 10). For the range of 10 m wind speeds suggested to typify South Georgia (7.5–15 m s^{-1}), the associated values of $\widehat{h}$ and Ro would lie in the range 1.3–2.7 and 3–6, respectively, indicating that South Georgia is

frequently associated with these effects. Such fine-scale wind effects and their subsequent surface forcing were entirely missing from the reanalysis, i.e. the corresponding values of wind stress curl in the reanalysis were two orders of magnitude less than that simulated by WRF (not shown).

Together the high values of wind stress curl and air–sea heat flux would force complex localized ocean circulation features, such as wind-driven gyres or downwelling, which would then drive changes in sea-surface temperature. Additionally, localized sea-surface temperatures would also be affected by, e.g. changes in precipitation and incident solar radiation due to the influence of South Georgia on clouds. However, as we are using an atmosphere-only (uncoupled) model, we are unable to account for how these changes to sea-surface temperature would feed back on the atmosphere (Businger and Shaw, 1984; Caldeira and Tomé, 2013). Addressing such feedbacks would require a coupled atmosphere-ocean regional model, which is an area for further work. Nevertheless, we assert that any such feedback effects would be of secondary significance relative to the gross modification of the regional wind field by the orography, and the impact this would have on the ocean current system.

Understanding of the influence of complex wind-driven oceanic circulation features is vital for insight into local marine ecosystem dynamics. The waters surrounding South Georgia are amongst the most biologically productive in the Southern Ocean, supporting vast local colonies of higher predators, including penguins and fur seals, as well as significant international fisheries (Atkinson *et al.*, 2001; Kock *et al.*, 2007). Previous ocean modelling work using atmospheric forcing derived from reanalysis showed that South Georgia's marine ecosystem dynamics are strongly influenced by variability in the underlying oceanography (Young *et al.*, 2012, 2014). However, our study suggests that using reanalysis data for the

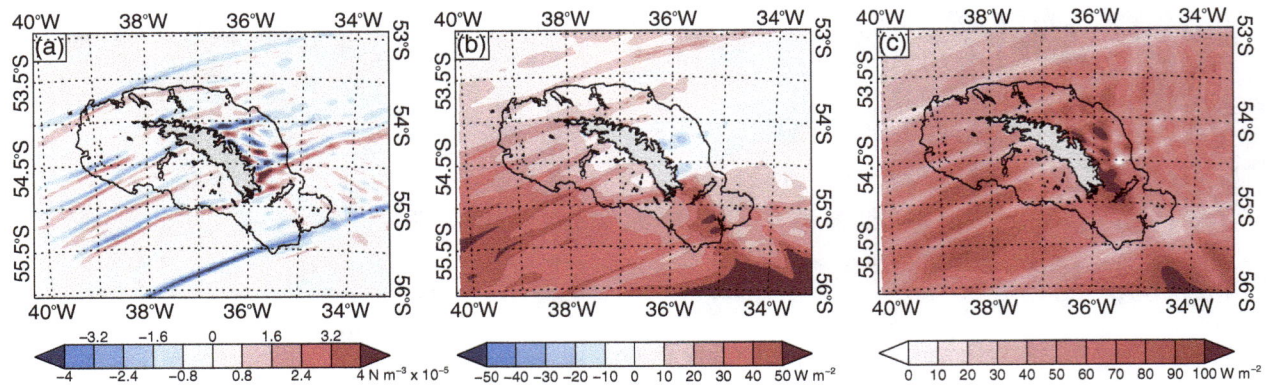

Figure 4. WRF modelled 3.3 km resolution surface forcing at 0900 UTC 13 September 1999 for (a) wind stress curl ($N\,m^{-3} \times 10^{-5}$), (b) sensible heat flux ($W\,m^{-2}$), and (c) latent heat flux ($W\,m^{-2}$). Positive sensible and latent heat fluxes are upward. To highlight the sea shelf region, the 300 m bathymetry depth contour is included (black line).

atmospheric forcing would have resulted in unrealistically weak spatial and temporal variability in the local oceanography due to the failure to capture important wind-driven complex circulation features. This study thus highlights the need to use high spatial resolution wind forcing in a regional ocean model.

Acknowledgements

The case study examined in this article was also a focus of the Jet Propulsion Laboratory 'Photojournal' (http://photojournal.jpl.nasa.gov/catalog/PIA02457). Daniel Bannister was supported by a PhD studentship funded by the UK Natural Environment Research Council.

References

Atkinson A, Whitehouse MJ, Priddle J, Cripps GC, Ward P, Brandon MA. 2001. South Georgia, Antarctica: a productive, cold water pelagic ecosystem. *Marine Ecology Progress Series* **216**: 279–308.

Brandon MA, Murphy EJ, Trathan PN, Bone DG. 2000. Physical oceanographic conditions to the northwest of the sub-Antarctic Island of South Georgia. *Journal of Geophysical Research* **105**: 23983–23996.

Businger JA, Shaw WJ. 1984. The response of the marine boundary layer to mesoscale variations in sea-surface temperature. *Dynamics of Atmospheres and Oceans* **8**: 267–281.

Caldeira RM, Tomé R. 2013. Wake response to an ocean-feedback mechanism: Madeira island case study. *Boundary-Layer Meteorology* **148**: 419–436.

Chelton DB, Schlax MG, Freilich MH, Milliff RF. 2004. Satellite measurements reveal persistent small-scale features in ocean winds. *Science* **303**: 978–983.

Chen F, Dudhia J. 2001. Coupling an advanced land surface-hydrology model with the Penn State-NCAR MM5 modelling system. Part I: model implementation and sensitivity. *Monthly Weather Review* **129**: 569–585.

Chou M-D, Suarez MJ. 1999. A solar radiation parameterization for atmospheric studies. Technical Report Series on Global Modeling and Data Assimilation, MJ Suarez, NASA/TM-1999-104606, Vol. 15, Goddard Space Flight Center, Greenbelt, MD, 42 pp.

Condron A, Bigg GR, Renfrew IA. 2006. Polar mesoscale cyclones in the northeast Atlantic: comparing climatologies from ERA-40 and satellite imagery. *Monthly Weather Review* **134**: 1518–1533.

Couvelard X, Caldeira RMA, Araújo IB, Tomé R. 2012. Wind mediated vorticity-generated and eddy-confinement, leeward of the Madeira Island: 2008 numerical case study. *Dynamics of Atmospheres and Oceans* **58**: 128–149.

Dee DP, Uppala SM, Simmons AJ, Berrisford P, Poli P, Kobayashi S, Andrae U, Balmaseda MA, Balsamo G, Bauer P, Bechtold P, Beljaars ACM, van de Berg L, Bidlot J, Bormann N, Delsol C, Dragani R, Fuentes M, Geer AJ, Haimberger L, Healy SB, Hersbach H, Hólm EV, Isaksen L, Kållberg P, Köhler M, Matricard M, McNally AP, Monge-Sanz BM, Morcrette JJ, Park BK, Peubey C, de Rosnay P, Tavolato C, Thépaut JN, Vitart F. 2011. The ERA-interim reanalysis: configuration and performance of the data assimilation system. *Quarterly Journal of the Royal Meteorological Society* **137**: 553–597.

Dong C, McWilliams JC. 2007. A numerical study of island wakes in the Southern California Bight. *Continental Shelf Research* **27**: 1233–1248.

Ebuchi N, Graber HC, Caruso MJ. 2002. Evaluation of wind vectors observed by QuikSCAT/SeaWinds using ocean buoy data. *Journal of Atmospheric and Oceanic Technology* **19**: 2049–2062.

Etling D. 1989. On atmospheric vortex streets in the wake of large islands. *Meteorology and Atmospheric Physics* **41**(3): 157–164.

Etling D. 1990. Mesoscale vortex shedding from large islands: a comparison with laboratory experiments of rotating stratified flows. *Meteorology and Atmospheric Physics* **43**(1–4): 145–151.

Hogg OT, Barnes DKA, Griffiths HJ. 2011. Highly diverse, poorly studied and uniquely threatened by climate change: an assessment of marine biodiversity on South Georgia's continental shelf. *PLoS ONE* **6**: e19795, DOI: 10.1371/journal.pone.0019795.

Jarvis A, Reuter HI, Nelson A, Guevara E. 2008. Hole-filled SRTM for the globe Version 4. Available from the CGIAR-CSI SRTM 90 m database, Retrieved January 17, 2012. http://srtm.csi.cgiar.org.

Kain JS, Fritsch JM. 1993. Convective parameterization for mesoscale models: the Kain-Fritsch scheme. The representation of Cumulus convection in numerical models. *Meteorological Monographs* **46**: 165–170.

Kock K-H, Reid K, Croxall J, Nicol S. 2007. Fisheries in the Southern Ocean: an ecosystem approach. Philosophical Transactions of the Royal Society of London. *Series B* **362**(1488): 2333–2349, DOI: 10.1098/rstb.2006.1954.

Large WG, Pond S. 1982. Sensible and latent heat flux measurements over the ocean. *Journal of Physical Oceanography* **12**: 464–482.

Meredith MP, Brandon MA, Murphy EJ, Trathan PN, Thorpe SE, Bone DG, Chernyshkov PP, Sushin VA. 2005. Variability in hydrographic conditions to the east and northwest of South Georgia, 1996–2001. *Journal of Marine Systems* **53**: 143–167.

Nakanishi M, Niino H. 2004. An improved Mellor–Yamada level-3 model with condensation physics: its design and verification. *Boundary-Layer Meteorology* **112**(1): 1–31.

Orr A, Hanna E, Hunt JCR, Cappelen J, Steffen K, Stephens AG. 2005. Characteristics of stable flows over Greenland. *Pure and Applied Geophysics* **162**: 1747–1778.

Orr A, Marshall GJ, Hunt JCR, Sommeria J, Wang C-G, van Lipzig NPM, Cresswell D, King JC. 2008. Characteristics of summer airflow over the Antarctic Peninsula in response to recent strengthening of westerly circumpolar winds. *Journal of the Atmospheric Sciences* **65**: 1396–1413.

Pickart RS, Spall MA, Ribergaard MH, Moore GWK, Milliff RF. 2003. Deep convection in the Irminger Sea forced by the Greenland tip jet. *Nature* **424**: 152–156.

Pullen J, Doyle JD, May P, Chavanne C, Flament P, Arnone RA. 2008. Monsoon surges trigger oceanic eddy formation and propagation in the lee of the Philippine Islands. *Geophysical Research Letters* **35**: L07604, DOI: 10.1029/2007GL033109.

Risien CM, Chelton DB. 2008. A global climatology of surface wind and wind stress fields from eight years of QuikSCAT data. *Journal of Physical Oceanography* **38**: 2379–2413.

Skamarock WC, Klemp JB, Dudhia J, Gill DO, Barker M, Duda KG, Powers JG. 2008. A description of the Advanced Research WRF Version 3. NCAR Technical Report TN-475+STR, National Center for Atmospheric Research, Boulder, CO, 1–113.

Smith RB, Gleason AC, Gluhosky PA, Grubišić V. 1997. The Wake of St. Vincent. *Journal of the Atmospheric Sciences* **54**(5): 606–623.

Sousa MC, Alvarez I, Vaz N, Gomez-Gesteira M, Dias JM. 2013. Assessment of wind pattern accuracy from the QuikSCAT satellite and the WRF model along the Galician coast (northwest Iberian Peninsula). *Monthly Weather Review* **141**: 742–753.

Xie S-P, Liu WT, Liu Q, Nonaka M. 2001. Far-reaching effects of the Hawaiian Islands on the Pacific Ocean atmosphere. *Science* **292**: 2057–2060.

Young EF, Meredith MP, Murphy EJ, Carvalho GR. 2011. High resolution modelling of the shelf and open ocean adjacent to South Georgia, Southern Ocean. *Deep Sea Research II* **58**: 1540–1552.

Young EF, Rock J, Meredith MP, Belchier M, Murphy EJ, Carvalho GR. 2012. Physical and behavioural influences on larval fish retention: contrasting patterns in two Antarctic fishes. *Marine Ecology: Progress Series* **465**: 201–204.

Young EF, Thorpe SE, Banglawala N, Murphy EJ. 2014. Variability in transport pathways on and around the South Georgia shelf, Southern Ocean: implications for recruitment and retention. *Journal of Geophysical Research: Oceans* **119**(1): 241–252, DOI: 10.1002/2013JC009348.

Yuan X. 2004. High-wind-speed evaluation in the Southern Ocean. *Journal of Geophysical Research* **109**: D13101, DOI: 10.1029/2003JD004179.

PERMISSIONS

LIST OF CONTRIBUTORS

Theodore L. Allen and Brian E. Mapes
Department of Meteorology and Physical Oceanography, Rosenstiel School of Marine and Atmospheric Science, University of Miami, FL, USA

Nicholas Cavanaugh
Climate and Ecosystem Sciences, Lawrence Berkeley National Laboratory, Berkeley, CA, USA

H. Athar
Department of Meteorology, COMSATS Institute of Information Technology, Islamabad, Pakistan

Dan Brawn
Department of Mathematical Sciences, University of Essex, Colchester, UK

Unnikrishnan C.K., Saji Mohandas, Ashu Mamgain, E. N. Rajagopal and Gopal R. Iyengar
ESSO, MoES, National Centre for Medium Range Weather Forecasting, Noida, India

Biswadip Gharai and P. V. N. Rao
Atmospheric and Climate Sciences Group, Earth & Climate Science Area, National Remote Sensing Centre, ISRO, Hyderabad, India

Elisabeth Callen
Department of Geological and Atmospheric Sciences, Iowa State University, Ames, IA, USA

Donna F. Tucker
Department of Geography and Atmospheric Science, University of Kansas, Lawrence, KS, USA

Ch. Purna Chand, M. V. Rao, I. V. Ramana and M. M. Ali
National Remote Sensing Centre, ISRO, Hyderabad, India

J. Patoux
Department of Atmospheric Sciences, University of Washington, Seattle, WA, USA

M. A. Bourassa
Center for Ocean-Atmospheric Prediction Studies, The Florida State University, Tallahassee, FL, USA

Federico Cossu, Klemens Hocke and Christian Mätzler
Institute of Applied Physics, University of Bern, Switzerland

Oeschger Centre for Climate Change Research, University of Bern, Switzerland

Andrey Martynov and Olivia Martius
Oeschger Centre for Climate Change Research, University of Bern, Switzerland
Institute of Geography, University of Bern, Switzerland

E. Dutra, M. Diamantakis, I. Tsonevsky, E. Zsoter, F.Wetterhall, T. Stockdale, D. Richardson and F. Pappenberger
European Centre for Medium Range Weather Forecasts, Reading, UK

Xia Feng and Paul Houser
Geography and Geoinformation Science, George Mason University, Fairfax, VA, 22030, USA

João V. C. Garcia and Stephan Stephany
National Institute for Space Research (INPE), Sao Jose dos Campos, Brazil

Augusto B. d'Oliveira
Center for Monitoring and Warnings of Natural Disasters (CEMADEN-MCTI), Cachoeira Paulista, Brazil

Gibies George, D. Nagarjuna Rao, Ankur Srivastava and Suryachandra A. Rao
Program for Seasonal and Extended Range Prediction of Monsoon, Indian Institute of Tropical Meteorology, Pune, India

C. T. Sabeerali
Program for Seasonal and Extended Range Prediction of Monsoon, Indian Institute of Tropical Meteorology, Pune, India
The Center for Prototype Climate Modeling, New York University, Abu Dhabi, UAE

Xiaofan Li and Guoqing Zhai
Department of Earth Science, Zhejiang University, Hangzhou, Zhejiang 310027, China

Shouting Gao
Laboratory of Cloud-Precipitation Physics and Severe Storms (LACS), Institute of Atmospheric Physics, Chinese Academy of Sciences, Beijing 100029, China

Xinyong Shen
Key Laboratory of Meteorological Disaster of Ministry of Education, Nanjing University of Information Science and Technology, Jiangsu 210044, China

P. A. Mooney, C. L. Bruyère and D. O. Gill
Mesoscale and Microscale Meteorology Laboratory, National Center for Atmospheric Research (NCAR), Boulder, CO, USA

F. J. Mulligan
Department of Experimental Physics, Maynooth University, Kildare, Ireland

J. A. Martinez
Facultad de Ciencias Marinas, UABC, Ensenada, Mexico

G. A. Passalacqua
Facultad de Ciencias Marinas, UABC, Ensenada, Mexico
Departamento de Oceanografía Física, CICESE, Ensenada, Mexico

J. Sheinbaum
Departamento de Oceanografía Física, CICESE, Ensenada, Mexico

D. R. Pattanaik
India Meteorological Department, Pune, India

Arun Kumar
NOAA/NWS/NCEP, Climate Prediction Centre, College Park, MD, USA

N. K. Sakellariou and H. D. Kambezidis
Atmospheric Research Team, Institute for Environmental Research & Sustainable Development, National Observatory of Athens, Greece

A. A. Scaife, A. Brookshaw, R. Eade, M. Gordon, C. MacLachlan, N. Martin, N. Dunstone and D. Smith
Met Office Hadley Centre, Exeter, UK

A. Yu. Karpechko
Arctic research, Finnish Meteorological Institute, Helsinki, Finland

M. P. Baldwin
Department of Mathematics and Computer Science, University of Exeter, UK

A. H. Butler
Cooperative Institute for Research in Environmental Sciences (CIRES), Boulder, CO, USA
Earth System Research Laboratory, NOAA, Boulder, CO, USA

Andrew Smith, Nigel Atkinson, William Bell and Amy Doherty
Met Office, Exeter, UK

Prashant K. Srivastava, Dawei Han, Miguel A. Rico Ramirez and Tanvir Islam
Water and Environment Management Research Centre, Department of Civil Engineering, University of Bristol, UK

Esa-Matti Tastula and Boris Galperin
College of Marine Science, University of South Florida, St. Petersburg, FL, USA

Semion Sukoriansky
Department of Mechanical Engineering, Ben-Gurion University of the Negev, Beer-Sheva, Israel

Ashok Luhar
CSIRO Marine and Atmospheric Research, Aspendale, Australia

Phil Anderson
Scottish Association for Marine Science, Oban, UK

Warren E. Heilman and Xindi Bian
USDA Forest Service, Northern Research Station, Lansing, MI, USA

Craig B. Clements and Daisuke Seto
San Jose State University, San Jose, CA, USA

Kenneth L. Clark
USDA Forest Service, Northern Research Station, New Lisbon, NJ, USA

Nicholas S. Skowronski
USDA Forest Service, Northern Research Station, Morgantown, WV, USA

John L. Hom
USDA Forest Service, Northern Research Station, Newtown Square, PA, USA

M. Zeri and G. Cunha-Zeri
Brazilian Center for Monitoring and Early Warnings of Natural Disasters (CEMADEN), São José dos Campos, Brazil

V. S. B. Carvalho
Instituto de Recursos Naturais, Universidade Federal de Itajubá, Brazil

J. F. Oliveira-Júnior and G. B. Lyra
Departamento de Ciências Ambientais, Instituto de Florestas, Universidade Federal Rural do Rio de Janeiro, Seropédica, Brazil

E. D. Freitas
Instituto de Astronomia, Geofísica e Ciências Atmosféricas, Universidade de São Paulo, Brazil

M. Zeri and G. Cunha-Zeri
Brazilian Center for Monitoring and Early Warnings of Natural Disasters (CEMADEN), São José dos Campos, Brazil

V. S. B. Carvalho
Instituto de Recursos Naturais, Universidade Federal de Itajubá, Brazil

J. F. Oliveira-Júnior and G. B. Lyra
Departamento de Ciências Ambientais, Instituto de Florestas, Universidade Federal Rural do Rio de Janeiro, Seropédica, Brazil

Yimin Liu
State Key Laboratory of Numerical Modeling for Atmospheric Sciences and Geophysical Fluid Dynamics (LASG), Institute of Atmospheric Physics, Chinese Academy of Sciences, Beijing, China

Haifeng Zhuo
State Key Laboratory of Numerical Modeling for Atmospheric Sciences and Geophysical Fluid Dynamics (LASG), Institute of Atmospheric Physics, Chinese Academy of Sciences, Beijing, China
University of Chinese Academy of Sciences, Beijing, China

Jiming Jin
College of Water Resources and Architectural Engineering, Northwest A&F University, Yangling, China
Department of Watershed Sciences and Plants, Soils and Climate, Utah State University, Logan, UT, USA

Seyda Tilev-Tanriover
Department of Meteorological Engineering, Istanbul Technical University, Istanbul, Turkey

Abdullah Kahraman
Turkish State Meteorological Service, ITU Met-Office, Istanbul, Turkey
Graduate School of Science, Engineering and Technology, Istanbul Technical University, Istanbul, Turkey

David Edwards
Met Office, Exeter EX1 3PB, UK

Bruce Ingleby
Met Office, Exeter EX1 3PB, UK
ECMWF, Reading RG2 9AX, UK

Sam Illingworth, Grant Allen, Carl Percival, Martin Gallagher, Hugo Ricketts and Harry Hayes
The Centre for Atmospheric Science, The School of Earth, Atmospheric and Environmental Science, The University of Manchester, M13 9PL, UK

Peter Hollingsworth, Paweł Ładosz and Gareth Roberts
The School of Mechanical, Aerospace and Civil Engineering, The University of Manchester, M13 9PL, UK

David Crawley
The School of Electrical and Electronic Engineering, The University of Manchester, M13 9PL, UK

J. Scott Hosking, Daniel Bannister, Andrew Orr, John King, Emma Young and Tony Phillips
British Antarctic Survey, NERC, Cambridge, UK

Index